AF581663

Smart Wireless Acoustic Sensor Network Design for Noise Monitoring in Smart Cities

Smart Wireless Acoustic Sensor Network Design for Noise Monitoring in Smart Cities

Editors

Rosa Ma Alsina-Pagès
Patrizia Bellucci
Giovanni Zambon

MDPI • Basel • Beijing • Wuhan • Barcelona • Belgrade • Manchester • Tokyo • Cluj • Tianjin

Editors

Rosa Ma Alsina-Pagès
Universitat Ramon Llull
Spain

Patrizia Bellucci
DIV Research and Development
Italy

Giovanni Zambon
University of Milano-Bicocca
Italy

Editorial Office
MDPI
St. Alban-Anlage 66
4052 Basel, Switzerland

This is a reprint of articles from the Special Issue published online in the open access journal *Sensors* (ISSN 1424-8220) (available at: https://www.mdpi.com/journal/sensors/special_issues/smart_wireless_acoustic_sensor_network).

For citation purposes, cite each article independently as indicated on the article page online and as indicated below:

LastName, A.A.; LastName, B.B.; LastName, C.C. Article Title. *Journal Name* **Year**, *Article Number*, Page Range.

ISBN 978-3-03943-280-6 (Hbk)
ISBN 978-3-03943-281-3 (PDF)

Contents

About the Editors

Rosa Ma Alsina-Pagès has served as Associate Professor in the Research Group of Media Technologies since September 2015, where she coordinates the Signal Processing Research Line. Since 2019 she is the Director of Research of La Salle – URL. She received her MSc degree in Electronics and in Telecommunications in 2002 and 2004, respectively, from La Salle, Universitat Ramon Llull, Barcelona; she later completed her Humanities Degree in 2011 at Universitat Oberta de Catalunya. In 2003, she completed a Project Management MSc Degree at La Salle – URL. She received her PhD in Telecommunications Engineering in July 2012 with a PhD thesis about signal processing in HF long haul link, completed at La Salle – URL (with honors). Her research interests are in digital signal processing, especially in acoustic and adaptive signal processing. From 2007 to 2015, she participated in several research projects on ionospheric communications with Antarctica. From 2016 to 2019 she was the team leader at La Salle in DYNAMAP, a LIFE+ project whose goal is dynamic noise mapping in urban environments. Nowadays, she is the principal investigator of CowTalk, a Llavor project funded by the Catalan Government, and SuaraMap, a CDTI project funded by the Spanish Government.

Patrizia Bellucci has served as Coordinator of Research and Development and Head of the Acoustics and Lightning Lab at ANAS S.p.A., the Italian Road Administration, since 2017 and 2000, respectively. She received her MSc degree in Electronics in 1992 and her PhD in Applied Physics from the University of Rome "La Sapienza" with a thesis on a novel sound absorption measurement system for road pavements based on a digital signal processing technique. In 2007, she also completed a European Project Management degree within the framework of the European project CERTAIN in Brussels. As national and European project coordinator, she leads and participates in research projects, laboratory activities, studies, and providing insights on specific issues. In this role, she has prepared numerous project proposals in collaboration with other national and foreign partners to access European funds within the LIFE program, VII FRAMEWORK PROGRAM, HORIZON 2020, TEN-T, PON RI and CEDR, both as coordinator and member of the consortium. In this context, she was awarded as coordinator of the LIFE13 ENV/IT/001254 DYNAMAP Project and took part in the EU-funded projects INFRAVATION, DYSTANCE, RAFAEL, INFRA4DFUTURE, and COST 354 "Performance Indicators".

Giovanni Zambon graduated in Physics in 1999 at the Physics Department of the University of Milan with a thesis entitled "The Acoustic Quality of the Strehler Theater in Milan: Analysis and Interventions". Since his appointment on 23 December 2004, he has served as researcher in Applied Physics at the Department of Environmental and Territorial Sciences of the University of Milan-Bicocca (DISAT). In this role, he has concentrated his research activity in architectural acoustics and environmental acoustics in relation to the evolution of the DISAT Physics Area programs. He has been the scientific director of numerous projects for direct assignment and public tender in the acoustic sector since 2003, financed by local, national, and international territorial bodies, in particular with the Lombardy Region, the Province of Milan, the Municipality of Milan, the Municipality of Pavia, ARPA Lombardia, CNR, Anas, and the European Commission. Since 2010, he has been in charge of the research group in Environmental Physics of DISAT. In 2018, he obtained the national scientific qualification (ASN) for the role of Associate Professor in Applied Physics (SSD-FIS/07).

In August 2020, the competition for the call-up as Associate Professor at DISAT ended. In the 2004–2005 academic year, he held courses in Environmental Acoustics for the master's degree in Science and Technology for the Environment and the Territory and of the Applied Physics course for the three-year degree course in Environmental Sciences and Technologies. He has supervisd numerous doctoral theses and numerous degree theses. He has been Lecturer in various Master and workshop courses, including internationally. He has served as Director and teacher of the course for Competent Technicians in Acoustics at the University of Milan Bicocca. Since 2001, he has regularly participates, with presentations of memoirs (also by invitation), in the most important international acoustic congresses (Forum Acusticum, ICA, ICSV, Internoise). In some of these congresses, he was chairman of the section. Since 2001, he has regularly participated in the National Acoustics Congress organized by the Italian Acoustics Association (AIA). In some of these congresses he was chairman of the section.In 2008 he organized at Bicocca University, as Scientific Secretary, the Annual Conference of the Italian Acoustics Association.Since 2005 member of numerous commissions of competitions for the awarding of grants for collaboration in research activitiesFrom 2006 to 2016: Member of the teaching staff of the PhD in Environmental Sciences of the University of Milan BicoccaSince 2008 he has been Director of the Acoustics Service Center of Bicocca University (CESAB)Since 2012 member of the "Orientation", "Internship and Work" and "Didactic Planning" commissions of the three-year and master's degree courses of the DISAT.Since 2001 he has been a member of the Italian Acoustics Association (AIA).He is reviewer of a dozen scientific papersAuthor of over 100 national and international scientific works on the themes of environmental acoustics, architectural acoustics and soundscape.

Editorial

Smart Wireless Acoustic Sensor Network Design for Noise Monitoring in Smart Cities

Rosa Ma Alsina -Pagès [1,*], Patrizia Bellucci [2] and Giovanni Zambon [3]

1 GTM–Grup de recerca en Tecnologies Mèdia, La Salle–Universitat Ramon Llull. c/Quatre Camins, 30, 08022 Barcelona, Spain
2 ANAS S.p.A., DIV Research and Development. Via Monzambano, 10-00185 Rome, Italy; p.bellucci@stradeanas.it
3 Department of Earth and Environmental Sciences (DISAT), Universitá degli Studi di Milano Bicocca, Piazza della Scienza 1, 20126 Milano, Italy; giovanni.zambon@unimib.it
* Correspondence: rosamaria.alsina@salle.url.edu

Received: 18 August 2020; Accepted: 21 August 2020; Published: 23 August 2020

Abstract: This Special Issue is focused on all the technologies necessary for the development of an efficient wireless acoustic sensor network, from the first stages of its design to the tests conducted during deployment; its final performance; and possible subsequent implications for authorities in terms of the definition of policies. This Special Issue collects the contributions of several LIFE and H2020 projects aimed at the design and implementation of intelligent acoustic sensor networks, with a focus on the publication of good practices for the design and deployment of intelligent networks in any locations.

Keywords: smart cities; noise monitoring; acoustic sensor design; noise mapping; acoustic event detection; map generation; public information; END; CNOSSOS-EU

---

1. Introduction

The Environmental Noise Directive (END) requires that a five-year updating of noise maps be carried out, and requires checking and reporting on the changes that occurred during the reference period. The updating process is usually achieved using a standardized approach, consisting of collecting and processing information through acoustic models to produce the updated noise maps. This procedure is time consuming and costly, and has a significant impact on the financial statement of the authorities responsible for providing the maps. Furthermore, the END requires that easy-to-read noise maps are made available to the public, to provide information on noise levels, and that subsequent actions be undertaken by local and central authorities to reduce noise impacts.

In order to update the noise maps more easily and in a more effective way, it is convenient to design an integrated system incorporating real-time noise measurement and signal processing to identify and analyze the noise sources present in the mapping area (road traffic noise, leisure noise, etc.), and to automatically generate and present the corresponding noise maps. This wireless acoustic sensor network design requires transversal knowledge, from accurate hardware design for the acoustic sensors, to network structure design and management of the information, signal processing to identify the origin of the measured noise and graphical user interface application design to present the results to end users.

2. Contributions

This Special Issue presents eleven outstanding papers covering all the technologies necessary for the development of an efficient wireless acoustic sensor network, from the first stages of its design to the tests conducted during deployment; its final performance; and possible subsequent implications for authorities in terms of the definition of policies.

Picaut et al. [1] proposed an extensive review of the literature around low cost sensors for urban noise monitoring. Furthermore, they also identified the expected technical characteristics of the sensors to address the problem of noise pollution assessment. Finally, the paper also presents the challenges required to respond to a massive deployment of low-cost noise sensors.

Other authors have centered their work on the sensor network and node design. In Prieto et al. [2], the authors describe the design and implementation of a complete system for a WASN deployed in the city of Linares (Jaén-Spain). The complete system covers the network topology design, and the hardware and software of the sensor nodes elaboration, along with protocols and the web server platform. They provide several metrics about the noise measured in the nodes: L_{Aeq} for a given period of time; L_{den}, L_{day}, $L_{evening}$ and L_{night}; the percentile noise levels (LA01T, LA10T, LA50T, LA90T and LA99T); the temporal evolution representation of noise levels; and the predominant frequency noise. López et al. [3] present a design for a versatile electronic device to measure outdoor noise, designed according to the technical standards of this type of instrument, and it has been tested following the regulations of the calibration laboratories for sound level meters (SLM). The evaluation of the quality of the electronics and the algorithm fully fit the requirements of a type 1 noise measurement instrument. Nevertheless, the use of an electret microphone reduces the technical features of the designed instrument into a type 2 noise measurement instrument. The authors deployed a low-cost sub-network in the city of Málaga (Spain) to analyze the leisure noise. The designed equipment is a two channel instrument, measuring simultaneously 86 sound parameters for each channel, such as L_{eq} (with Z, C and A frequency weighting); the peak level (with Z, C and A frequency weighting); the maximum and minimum levels (with Z, C and A frequency weighting); the impulse, fast and slow time weightings; seven percentiles (1%, 5%, 10%, 50%, 90%, 95% and 99%); and continuous equivalent sound pressure levels in the one-third octave and octave frequency bands.

In terms of instrumentation, a contribution by Bianco et al. [4] studied the development of a measurement instrument that can be fastened by means of holding elements to a moving laboratory, a vehicle, for example. This device overcomes the fact that the measurements are usually on-site, and allows one to perform a continuous spatial characterization of a given pavement in order to yield a direct evaluation of the surface's equality. The system was uncoupled from the vehicle by means of PID controller, evaluated to maintain the system at a fixed distance from the ground and reduced damping. Related to vehicles also, Flor et al. [5] focused on the evaluation of the noise level inside a vehicle by means of statistical tools. They made an experimental setup of microphones and a microcomputer strategically located on the car's panel, and they conducted measurements under different conditions, including car window positions, rain, traffic and various speeds. They discussed the relevance of the variables that contribute to the noise level inside a car.

There are several papers about the analysis of the noise data gathered by means of wireless acoustic sensor networks. In Alías et al. [6], the authors took as a starting point the LIFE DYNAMAP project that developed a WASN-based dynamic noise mapping system based solely on road traffic noise (RTN) levels. After analyzing the bias caused by individual anomalous noise events (ANEs) on the computation of the A-weighted L_{eq}, they evaluated the aggregate impact of the ANEs on the RTN measurements in a real-operation environment. They evaluated this metric over more than 300 h of labeled acoustic data collected by means of two WASNs deployed in the project in the suburban area of Rome and in the urban area of Milan. Benocci et al. [7] also took as a starting point the LIFE DYNAMAP project, and their contribution was the final assessment of the system installed in the area of Milan. The traffic noise data gathered by the nodes, each one of them representative of a number of roads sharing the same characteristics, were used to build-up a real-time noise map. In particular, the analysis focused on the 21 sites belonging to the testing campaign. That allowed them to evaluate the accuracy and reliability of the system by comparing the predicted noise level of the DYNAMAP sensors with field measurements in several randomly selected sites.

Brambilla et al. [8] dealt in their study with the application of the intermittency ratio (IR) to urban road traffic noise data, collected in terms of 1-sec A-weighted sound pressure level (SPL), without any technician attending the measurements, and with continuous monitoring for 24 h in 90 different sites in the city of Milan. They show the IR computed for each hourly dataset of the 251 time series available—24 h each—including different types of roads, from motorways to local roads with low traffic flow. The authors processed the IR using clustering methods to extract the most significant temporal pattern features of IR to propose a criterion to classify the urban sites while taking into account road traffic noise events, which potentially increase the noise annoyance. Guarnaccia, in [9], presented EAgLE, which stands for equivalent acoustic level estimator. It is a new methodology, based on video processing and object detection tools. When the vehicles, their typology and their speed, are recorded, the sound power level of each vehicle is computed according to the EU recommended standard model CNOSSOS-EU [10] and the sound exposure level (SEL) of transit is estimated at the receiver. The L_{eq} is evaluated in the end, summing up the contributions of all the vehicles.

Finally, in the Special Issue there are two contributions based on the prediction and the use of artificial intelligence on acoustic data. Navarro et al. [11] proposed the forecasting of temporal short-term sound levels as a useful tool for urban planners and managers. A long short-term memory (LSTM) deep neural network technique models the temporal behavior of sound levels at a certain location—both SPL and loudness level—in order to predict near-time future values. It can be trained and integrated for every node of a network to provide functionalities as a method of early warning against noise pollution, but also as a backup in case of a single node malfunction. The last contribution, by Glowacz [12], presents a fault detection technique of an electric impact drill (EID), coffee grinder A (CG-A) and coffee grinder B (CG-B) by means of acoustic signals. The three of them use commutator motors; the measurements of acoustic signals were carried out with a microphone. Seven signals from EID were measured, four of CG-A and three of CG-B, and an acoustic analysis was carried out, with good results in efficiency of recognition of all classes.

Funding: This special issue edition has been partially supported by the LIFE DYNAMAP project (LIFE13 ENV/IT/001254).

Acknowledgments: The authors of the submissions have expressed their appreciation to the work of the anonymous reviewers and the Sensors editorial team for their cooperation, suggestions and advice. Likewise, the editors of this Special Issue thank the staff of Sensors for the trust shown and the good work done.

Conflicts of Interest: The authors declare no conflict of interest.

Abbreviations

The following abbreviations are used in this manuscript:

ANE	Anomalous Noise Events
CNOSSOS-EU	Common Noise Assessment Methods in Europe
CG-A	Coffee Grinder A
CG-B	Coffee Grinder B
EID	Electric Impact Drill
END	Environmental Noise Directive 2002/49/EC
IR	Intermittency Ratio
LSTM	Long Short-Term Memory
RTN	Road Traffic Noise
SEL	Sound Exposure Level
SLM	Sound Level Meters
SPL	Sound Pressure Level
WASN	Wireless Acoustic Sensor Networks

References

1. Picaut, J.; Can, A.; Fortin, N.; Ardouin, J.; Lagrange, M. Low-Cost Sensors for Urban Noise Monitoring Networks—A Literature Review. *Sensors* **2020**, *20*, 2256. [CrossRef] [PubMed]
2. Fernandez-Prieto, J.A.; Gadeo-Martos, M.A. Wireless Acoustic Sensor Nodes for Noise Monitoring in the City of Linares (Jaén). *Sensors* **2020**, *20*, 124. [CrossRef] [PubMed]
3. López, J.M.; Alonso, J.; Asensio, C.; Pavón, I.; Gascó, L.; de Arcas, G. A Digital Signal Processor Based Acoustic Sensor for Outdoor Noise Monitoring in Smart Cities. *Sensors* **2020**, *20*, 605. [CrossRef] [PubMed]
4. Bianco, F.; Fredianelli, L.; Lo Castro, F.; Gagliardi, P.; Fidecaro, F.; Licitra, G. Stabilization of a pu Sensor Mounted on a Vehicle for Measuring the Acoustic Impedance of Road Surfaces. *Sensors* **2020**, *20*, 1239. [CrossRef] [PubMed]
5. Flor, D.; Pena, D.; Pena, L.; A de Sousa, V.; Martins, A. Characterization of Noise Level Inside a Vehicle under Different Conditions. *Sensors* **2020**, *20*, 2471. [CrossRef] [PubMed]
6. Alías, F.; Orga, F.; Alsina-Pagès, R.M.; Socoró, J.C. Aggregate Impact of Anomalous Noise Events on the WASN-Based Computation of Road Traffic Noise Levels in Urban and Suburban Environments. *Sensors* **2020**, *20*, 609. [CrossRef] [PubMed]
7. Benocci, R.; Confalonieri, C.; Roman, H.E.; Angelini, F.; Zambon, G. Accuracy of the dynamic acoustic map in a large city generated by fixed monitoring units. *Sensors* **2020**, *20*, 412. [CrossRef] [PubMed]
8. Brambilla, G.; Confalonieri, C.; Benocci, R. Application of the intermittency ratio metric for the classification of urban sites based on road traffic noise events. *Sensors* **2019**, *19*, 5136. [CrossRef] [PubMed]
9. Guarnaccia, C. EAgLE: Equivalent Acoustic Level Estimator Proposal. *Sensors* **2020**, *20*, 701. [CrossRef] [PubMed]
10. Kephalopoulos, S.; Paviotti, M.; Anfosso-Lédée, F. Common Noise Assessment Methods in Europe (CNOSSOS-EU). *Publ. Off. Eur. Union* **2002**, *49*, 1–180.
11. Navarro, J.M.; Martínez-España, R.; Bueno-Crespo, A.; Martínez, R.; Cecilia, J.M. Sound Levels Forecasting in an Acoustic Sensor Network Using a Deep Neural Network. *Sensors* **2020**, *20*, 903. [CrossRef] [PubMed]
12. Glowacz, A. Fault detection of electric impact drills and coffee grinders using acoustic signals. *Sensors* **2019**, *19*, 269. [CrossRef] [PubMed]

Review

Low-Cost Sensors for Urban Noise Monitoring Networks—A Literature Review

Judicaël Picaut [1,*], Arnaud Can [1], Nicolas Fortin [1], Jeremy Ardouin [2] and Mathieu Lagrange [3]

1 UMRAE, Univ Gustave Eiffel, IFSTTAR, CEREMA, F-44344 Bouguenais, France; Arnaud.Can@univ-eiffel.fr (A.C.); Nicolas.Fortin@univ-eiffel.fr (N.F.)
2 Wi6Labs, F-35510 Cesson-Sévigné, France; jeremy.ardouin@wi6labs.com
3 LS2N, UMR CNRS 6004, Ecole Centrale de Nantes, F-44321 Nantes, France; mathieu.lagrange@cnrs.fr
* Correspondence: Judicael.Picaut@univ-eiffel.fr

Received: 20 March 2020; Accepted: 12 April 2020; Published: 16 April 2020

Abstract: Noise pollution reduction in the environment is a major challenge from a societal and health point of view. To implement strategies to improve sound environments, experts need information on existing noise. The first source of information is based on the elaboration of noise maps using software, but with limitations on the realism of the maps obtained, due to numerous calculation assumptions. The second is based on the use of measured data, in particular through professional measurement observatories, but in limited numbers for practical and financial reasons. More recently, numerous technical developments, such as the miniaturization of electronic components, the accessibility of low-cost computing processors and the improved performance of electric batteries, have opened up new prospects for the deployment of low-cost sensor networks for the assessment of sound environments. Over the past fifteen years, the literature has presented numerous experiments in this field, ranging from proof of concept to operational implementation. The purpose of this article is firstly to review the literature, and secondly, to identify the expected technical characteristics of the sensors to address the problem of noise pollution assessment. Lastly, the article will also put forward the challenges that are needed to respond to a massive deployment of low-cost noise sensors.

Keywords: noise; low-cost sensors; networks

---

1. Introduction

Noise pollution is a major environmental pollution whose impact on health is now widely recognized [1]. As a result, many countries have implemented policies and strategies, for many years, to reduce noise pollution and to preserve quiet areas. Moreover, these policies are increasingly introducing citizens into the actuation process, by giving them the opportunity to be informed about their exposure to noise but also to be active in the management of their noise environment. In Europe in particular, Directive 2002/49/EC [2] introduces many rules on the assessment and management of noise environments, including the production of strategic noise maps, which are the starting point for the implementation of action plans to reduce noise pollution, but also as a tool for communicating between the different stakeholders. While this directive applies to large European cities, their general application to any community is obviously permitted.

The production of realistic noise maps is therefore a major challenge to ensure the proper implementation of the European directive, as well as, the relevance of the proposed actions to control noise pollution. Currently, noise maps are obtained by using sound mapping software tools, based on standards for calculating acoustic emission and propagation in the environment. As soon as the input data that are necessary for the calculation are available, the undeniable advantage of this method is its ease of implementation, making it possible to produce noise maps over a very large area. Auditing action plans using such noise maps can only be relevant if the limits of the methodology considered to produce the maps is taken into account during the process. In particular, only transportation noise sources and industrial noise are taken into account, on the basis of noise emission data that are very largely simplified and averaged over a large time period. It can also be pointed out that the propagation models used in software are based on many approximations (for example, lack of consideration of diffusion through facades and fitting objects; limited consideration of urban micro-meteorological conditions and vegetation). In addition, such noise maps cannot account for temporal dynamics, which nevertheless plays a role in the way inhabitants perceive the quality of sound environments [3,4].

In situ measurements would therefore be an immediate solution to make these maps more realistic. Nevertheless, given the urban scale considered and the spatial variability of the sound environment, the number of measurement points to be considered would be very large to model the relevant variability of the sound environment [5]. While noise observatories currently exist in many cities around the world, with networks of up to 150 measurement points [6], their use is mostly intended to provide objective and enforceable data on sound environments in certain strategic locations in order to better understand why inhabitants of this area are concerned with the quality of their sound environment. In view of the cost of a professional measurement point using Class-1 devices [7], enlarging the number of sensors of this type of measurement networks would be very costly. As a consequence, it is not realistic to aim at producing strategic noise maps on the technical basis of such observatories.

Many major technical developments have emerged in the last decade, making it possible to develop capturing devices integrating transducers of different kinds, embedded processing systems and wired/wireless communication systems, while optimizing power consumption and reducing their size. At the same time, this technological development has been accompanied by a significant reduction in the costs of electronic components and products, making these capture devices more affordable. Consequently, environmental monitoring tools based on Internet of things (IoT) have emerged in many related areas: atmospheric pollution, agriculture, transportation, smart cities [8–10]. Application of IoT to the monitoring of the sound environment has also a great potential. However, it should be noted that the development of low-cost sensors for acoustic measurement and the deployment of an ad-hoc sensor network can be complex to implement. Indeed, the high spatial and temporal variability of sound environments requires a high density of sensors, and advanced processing capabilities within the sensors [11].

Thus, the use of low-cost Sensor Networks (SN) can be a solution to the current limits of the noise observatories, mentioned above, by making it possible to reach a density of measurement points in a territory that is capable of providing a very rich acoustic information. The use of such a noise SN also opens up many additional interesting opportunities, such as to assess and manage road traffic noise [12–14], to enhance the traditional strategic noise mapping process [15–19], to produce dynamics noise maps [20] or to capture the sound sources of interest or acoustic events within the signal [21–29]. One can mention also that low-cost SN may be used in other fields in acoustics, such as sound source localisation [30] and biodiversity monitoring [31–33].

The relevance of such a system relies on many elements, among which, the acoustic measurement quality, the resilience of the sensors, the implementation of a communicating and smart sensor network, the deployment of a compatible data infrastructure able to manage the considerable amount of obtained information, the maintenance of the devices and infrastructures, the development of powerful algorithms to manipulate data and many others. Each of these topics are the subject of a considerable number of articles in the literature in the field of sensor networks for environmental monitoring. Being a more recent subject of the application of sensor networks, the literature specific to the sound environment subject is currently focused on the development of sensors themselves, but it is clear that the other subjects will see a growing of interest in the future.

The purpose of this article is thus to focus on the main element of such data gathering systems, namely the sensor. Solutions for the implementation of a complete information technology (IT) infrastructure for sensor, network and data management is not developed in this article, as this topic would in itself merit its own literature review in order to address the important issues that are associated (data encryption, privacy, security, scalability, management of a large number of sensors, interfacing, database management, storage, data visualization, etc.). Nevertheless, the reader can already refer to recent publications showing examples of full noise SN deployment [34,35].

For the past fifteen years, many researchers have proposed, evaluated, and in the best cases, deployed technical solutions for different noise applications. It therefore seems interesting to make a detailed point of current research on this subject, and in particular to highlight the essential characteristics that must be considered for this new generation of acoustic sensors in order to respond to the identified issues and the future ones. Compared to recent articles that already present a review of the existing system [35–39], the present contribution focuses on a detailed analysis of the technical solutions developed in the literature, highlighting their strengths and weaknesses, and showing how the rapid evolution of technologies can now fully meet the requirements for a successful deployment of modern noise SN.

The paper is organized as follows. Section 2 presents a review of the literature concerning the development of low-cost sensors for the deployment of noise monitoring networks. After a brief description of the SN terminology, the main technical solutions are detailed and synthesized. This synthesis then allows to identify the characteristics and performances that are expected for low cost sensors, in order to meet current and future noise monitoring challenges (Section 3). A conclusion then closes this article.

2. Literature Review

2.1. Sensor Networks: Definitions

A SN structure can be abstracted into a composition of a set of *nodes* and a *sink* node (more simply the *sink*, also called *gateway*. A node collects information, performs pre-processing and transmits the produced data to the sink. The sink collects all the data from the nodes and transmits them to servers for storage and further processing. On the basis of this simple architecture, many variants can be designed, depending on the needs and targeted applications, each variant giving rise to specific technical constraints. Among the most common variants are the following [40] (Figure 1):

- The network can be made up of several sinks. In this case, a group of identified nodes transmit the produced data to a specific sink. All sinks then transmit data to the servers. Another possible option is to consider that a node can choose the sink according to particular constraints, such as availability, proximity, sink load...
- The transmission of data from one node to the sink can be relayed using one or more nodes. The node then acts simultaneously as a sensor and a *relay*. This defines a *multi-hop* sensor network, as opposed to *single-hop* sensor network. The management of data transmission from the nodes to a given sink is then governed by relatively complex routing protocols that depend on the selected topology, such as point-to-point, star or mesh topologies.
- The nodes and the sinks may be mobile. The network is then defined *mobile* sensor network, as opposed to a *static* network. The term 'mobile' must be considered in two ways: (1) continuously 'mobile': a node moves continuously over time (like a sensor installed on a vehicle); (2) occasionally: a node is moved from one static position to another static position, for a long measurement time, in which case the network is always considered as a static network.
- Data transmission from a node to a sink can be performed in wired or wireless mode. In the latter case, the network is defined as a wireless acoustic sensor network (WASN). Nowadays, the wireless transmission mode is almost the main part of sensor networks for environmental monitoring. Data transmission from a sink to the server can also be carried out using one of these two transmission modes. Nodes and sinks may also simultaneously include several wireless transmission protocols, where some protocols get involved in the case of failure of the main protocol.

- The type of power supply of the nodes can also give rise to several variants: via a public or private power grid, by exchangeable battery, power supply by rechargeable battery from an external renewable energy source (solar, wind).
- Several families of nodes can also be considered, each with its own technical characteristics (measurement characteristics, processing power, power supply mode...). In this case, we are talking about a *heterogeneous* sensor network, to be opposed to a *homogeneous* network.

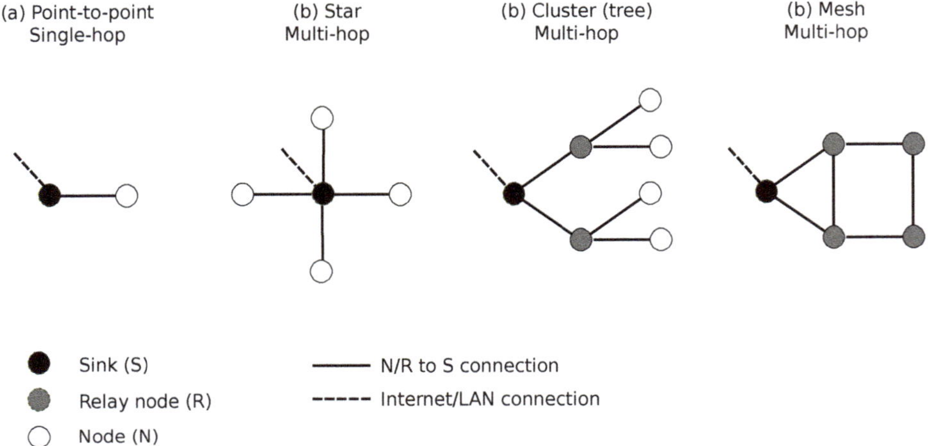

Figure 1. Sensor network definitions and topology examples.

2.2. Low-Cost Noise Sensors: Literature Review

Since the mid-2000s, with the emergence of low-cost electronic components and computing processors, many researchers have been working on the development of low-cost noise SN, mainly for environmental acoustics applications. Three SN main families can be distinguished: (1) fixed sensor networks, (2) mobile participatory measurement networks (mainly with smartphones), (3) mobile sensor networks. In the context of this study, we mainly rely on low-cost sensors specifically developed for environmental acoustic measurements, allowing to set up a controlled measurement network. Participatory measurement with smartphones is an interesting approach, complementary to that of low-cost and professional static networks, but beyond the scope of this review. Nevertheless, it can be noted that the approach shares common technical aspects with static noise SN, such as data pre-processing, the choice of acoustic indicators, the implementation of a data management infrastructure [41] ... Lastly, the third family, concerning mobile sensors, although generating significant additional complexities in relation to data transmission and sensor autonomy, considers specifically designed noise sensors, and thus will also be included in this review.

The goal of the present section is to focus on references that give detailed information of the proposed low-cost noise sensors and additional information concerning their validation and tests. Some technical aspects are deliberately developed in order to highlight the diversity of the proposed solutions and to be able to subsequently identify the technologies, solutions and challenges that have strong potential for future research.

A first study about the use of sensor networks considered as an alternative to traditional "professional" networks for noise monitoring have been presented in [42–44]. Initially, the authors used a Tmote Sky platform (Moteiv), made of a microcontroller unit (MCU) (Texas Instruments MSP430 with a 16-bit RISC central processing unit (CPU)) equipped with a multimodality sensor board (EasySense SBT80) and an omnidirectional condenser microphone (ECM) (EM6050P-423) [43]. In a second step, they turned to a Tmote Invent platform made of a MCU (Texas Instruments MSP430) [42]. Sensors were connected to a sink via the IEEE 802.15.4 radio protocol. However, the authors have quickly highlighted the limitations of the microprocessors used (which limits the sample frequency to 8 kHz for the sound environment study), the rapid battery consumption at the relevant sampling frequencies and in simultaneous radio transmission, the dispersion and poor quality of the microphones used. They decided, in the first, instance to develop their own sound level meter [44], and later to use a Class-2 commercial sound level meter (Extech 407740) [36], both connected to the Tmote Sky platform via an analog input channel, mainly to overcome the problem of limited computational resources. Despite the technical limitations encountered, the authors showed that the use of wireless sensor SN was technically possible and could address the issue of environmental noise assessment in the future.

At about the same time, McDonald et al. [12] developed a noise sensor at a cost of about 130 GBP, based on a Triton XXS platform, using an Intel XScale PXA255 32-bit MCU. The acoustic measurement was performed through an ECM connected to the MCU via a 16-bit analog-to-digital converter (ADC) (sampling at 49 kHz). Equivalent sound levels over a given time period were calculated using a A-weighting digital filter. The sensors have been deployed on site, but unfortunately no information is given on the final acoustic characteristics of the sensors, their calibration and accuracy.

Although the first works were published from 2008, it seems that the origin of the studies conducted by Barham et al. [45–47] dates back to 2004. The work carried out seems to be particularly successful. The authors have very early highlighted the interest of using micro-electrical-mechanical systems (MEMS) microphones for environmental acoustic measurement, including their own design of a MEMS microphone with optimized performance for the environmental issue. In addition, numerous tests have been carried out to evaluate both the intrinsic acoustic performance of the MEMS used and the evolution of such solution under operating conditions (effect of atmospheric conditions, wind, cold, vibrations, etc.). The techniques developed were sufficiently advanced to deploy a hundred of sensors on several urban sites, and to start analyzing the collected data. Only little information is provided by the authors on the nature of the electronic components that where considered, excepted that they used a Floating-Point-Gate-Array (FPGA) system instead of a MCU, for an efficient use of the battery. Developments continued thereafter and allowed the implementation and deployment of nodes consisting of a specifically developed MEMS microphone package, connected to Raspberry Pi (R-Pi) 2 Model B with wired power supply, and using global system for mobile communications (GSM) connection [48].

One of the first works showing the design of a fully operational low-cost noise sensor network was detailed in [49–51]. The sensor was based on a CiNet platform, composed of a 8-bit MCU (Microchip ATmega128) with a 10-bit ADC running at a sample frequency of 33 kHz, using a IEEE 802.15.4 compliant chip (CC4420 Texas Instruments) and a high-quality omnidirectional ECM (MONACOR MCE-400). The authors implement a real-time A-weighting analog filter, as well as two pre-amplification channels (Maxim MAX4524) in order to overcome to the limited theoretical dynamic imposed by the ADC. Measurements were compared to Class-2 and Class-1 sound level meters, in indoor and outdoor environments, showing deviations of less than 2 dB. Beyond the technical aspect of the sensor developed, the work also focused on the development of a fully functional multi-hop large-scale network infrastructure, composed of nodes connected to a sink through relay nodes, all perfectly synchronized [52]. Later, the sensor network was successfully deployed for long-term measurements in a school yard, showing a good agreement by comparison with a Class-2 sound level meter [53].

In 2013, Tan and Jarvis [54] focused their works mainly on the energy harvesting using solar panel. The sensor node was based on an ultra-low power wireless sensor module TelosB [55] made of a 16-bit Texas Instruments MSP430 MCU, running on TinyOS, with the IEEE 802.15.4 protocol for the radio transmission and an electret microphone (Sparkfun BOB-09964). Various technical issues were highlighted, including the poor acoustic performances of the ECM in terms of signal-to-noise ratio (SNR) and dynamic, as well as the limitations of the MCU (memory size) and of the analog-to-digital converter (ADC). To overcome some of the limitations, the authors proposed latter to use a MEMS microphone (ADMP401) [56], offering better acoustic performances. The results mainly showed the need for sufficient energy storage capacity to ensure the proper functioning of the sensor over the duration of a measurement (depending of the kind of acoustic outputs that are required), as well as the need of power consumption management algorithms to perform autonomous measurements.

Mariscal-Ramirez et al. [57,58] presented in 2011 and 2015 the implementation of a fuzzy noise indicator and the development of frequency based algorithm for the calculation of noise indicators. The sensor node was based on a Sun Spot node with an electret microphone, using a 8-bit ADC and the IEEE 802.15.4 data transmission protocol. The node was tested on several signals, showing a deviation lower than 4% in term of sound levels, in comparison with a Class-2 sound level meter.

Segura-Garcia et al. [59] compared two low-cost acoustic measurement platforms, in terms of psycho-acoustic metrics instead of classical equivalent sound level. The first solution was based on the Tmote Invent platform as already used in [42]. The second was an R-Pi platform, composed of a Broadcom BCM2835 chip with an ARM1176JZF-S MCU and a Wi-Fi adaptater (IEEE 802.11 b/g/n). Acoustic Measurements were carried out using a universal serial bus (USB) sound card (Logilink UA0053) with an omnidirectional ECM. As already mentioned in [42], the Tmode Invent platform presents serious limitations for acoustic measurement, which definitively precludes it from any relevant operational solution. On the other hand, the data produced by the R-Pi shows a very good correlation with reference measurements made in the real-world, in terms of sound levels and psychoacoustic indices. This solution was used later to collect data in a city of Spain and to test a geo-statistical methodology for noise level prediction in urban areas [60]. The network was made of 39 nodes, some at fixed locations, others being moved to different locations. In order to collect and manage data, the authors successfully implement a specific network infrastructure, using a routing protocol (Babel/Quagga) for the wireless mesh network, a client–server suite (OwnCloud) for synchronizing data and a reliable and interoperable server (OpenCPU) for data analysis based on the statistical computing software R.

In 2016, Noriega-Linares and Navarro Ruiz [61] have proposed the design of an advanced acoustic sensor on the basis of R-Pi 2 Model B platform (Broadcom BCM2836 chip with ARM Cortex-A7 MCU) in order to compute in real-time several standard noise level indicators, such as instantaneous and equivalent sound levels, percentile levels, and 1/3 octave sound pressure levels. Unlike the solutions proposed so far, the authors have chosen a wired connection to the network, using a Power over Ethernet (POE) injector/splitter, which avoids potential interference between the hardware within the sensor and the radio transmission. Audio acquisition was carried out using a USB omnidirectional ECM (T-Bone GC 100 USB), including an internal ADC. According to the authors, this embedded solution offered better performances in terms of noise than using an external microphone with an USB sound card. Microphone frequency response equalisation, microphone calibration, 1/3 octave band filtering and real time implementation are mentioned by the authors, but are not detailed in their paper. The comparisons of measurements with a reference sound level meter seem to demonstrate a good performance of the sensor.

In a paper dating from 2016, Alsina-Pagès et al. [37] describes the design of a low-cost mobile noise sensors network. The corresponding article only describes the principle of a low-cost sensor and of the ad-hoc network, resulting from a detailed analysis of possible technical solutions. The proposed sensor is composed of a ARM 32-bit MCU on a NXP FRDM-KL25Z chip with a 12-bit ADC, using an omnidirectional electret microphone (CMA-4544PF-W) with a microphone amplifier (Maxim MAX9814 with automatic gain control). In order to optimize the data transmission from the mobile sensors to the data server, depending of the networks availability, the authors recommend to consider two data transmission protocols, a Wi-Fi connection (using the Wi-Fi ESP8266 module) and a GSM network (using the Adafruit FONA 808 cellular module, including a GPS). Since the design is very innovative, this research is worth mentioning. However, the implementation of this sensor in an operational system seems particularly difficult given the technical challenge of such a mobile network, such as ensuring a full and permanent connectivity between mobile nodes.

In 2017, Mydlarz et al. [62] have presented a low-cost noise sensor based on mini personal computer Tronsmart MK908ii with a Rockchip RK3188 quad core Cortex A9 CPU, a Wi-Fi connectivity and a wired connection to the public power supply. Acoustic signal acquisition was performed using an analog MEMS microphone Knowles SPU0410LR5H-QB with a eForCity USB audio interface on a specifically designed printed circuit board (PCB) and installed on a specific mount to limit potential harms due to weather. The set of acoustical tests show that the sensor generally meets the Class-2 criteria of the IEC 61672-1 standard [7]. This sensor was recently updated [35], replacing the processing unit by a R-Pi 2 Model B device in order to increase the computational performances and using a digital MEMS in order to reduce the radio-frequency (RF) interference. This sensor was successfully deployed across New-York City, producing a huge acoustic data collection in terms of sound level indicators and audio signal samples.

Although the sensors were designed for indoor use, the network presented in [38] in 2018 can be detailed here, since such device could be adapted for outdoor environments. In their paper, the authors use an analog MEMS microphone (SiSonic^{TM} SPM0408LESH-TB) with an analog amplification gain of 20 dB, mounted on a MCU (ARM 32-bit Cortex STM32F050K6U6A, STMicroelectronics), while the sink node is carried out using a PC-based system on a R-PI with a ETRX357 ZigBee module. Although the authors pointed out the possible defects of the frequency response curve of the MEMS microphone, they considered that the cost of an equalization filter was too high on a low-cost sensor, preferring an *a posteriori* calibration in frequency and level, on the basis of comparison with a reference Class-1 sound level meter. A digital A-weighting filter was however designed. Indoor test measurements were compared with a Class-1 sound level meter showing a very good correlation on the measured sound levels, with a mean difference of 1.6 dB over a 12 mn testing time-period.

In 2018, Peckens et al. [63] have proposed a sensor based on the Teensy 3.6 platform made of a 32-bit ARM Cortex-M4 MCU, on a MK20DX256VLH7 chip, using a XBee-PRO ZigBee Modules (S2B), and an external board for the acoustic measurement (connected to the 16-bit ADC of the MCU, using a sample frequency of 20 kHz). The measurement system was composed of an omnidrectional ECM (PUI POW-1644P-B-R), an amplifier circuit with variable gain and an A-Weighing analog filter circuit. It is interesting to note that the calibration, the A-Weighing filter and the amplifier gain have been implemented using analog circuits, instead of digital processing, in order to reduce the power consumption. Comparisons with sound level measurements using a Class-1 reference device, show a good agreement, with deviations around ±1.5 dB. This system with four nodes and one sink (based on a R-Pi platform), was tested with success in a real environment. The ability of the proposed approach to monitor noise was validated. Some limitations are however highlighted such as insufficient dynamics (50 dB), high residual noise level, as well as power supply issues generating artifacts in the time signal.

In the context of the European Life MONZA project [64], two different low-cost noise sensors have been presented in [65,66], in 2017 and 2019 respectively. The sensors are not clearly described, but it is likely that one is using a digital MEMS (mounted on 1/2 inch support) on a Mini PC platfom, and the other one with an electret microphone (mounted on 1/4 inch support) on a MCU board. Data are transmitted using the GSM 2G/3G protocol and power is supplied by several means: battery, solar panel, electricity network. A digital filter was carried out in order to proceed to the A-weighting filtering, as well as a 1/3 octave band analysis. Ten prototype sensors have been installed in a pilot area in the City of Monza (Italy) since 2017, and seem to provide consistent data in comparison with reference sound level meters, with however a systematic offset around 3 dB. One interesting choice by the authors is the use of digital MEMS, provided an embedded ADC, which limits power consumption and the noise generated by possible interference. However, they observed a reduction of sensitivity of the MEMS during time, requiring to continuously correct the raw data at least until the end of the running-in period of the first few months of use.

In the context of a French national research project (CENSE), Ardouin et al. [67] have presented in 2018 two complementary sensors: firstly, a node using a STMicroelectronics STM32L4 MCU with a ARM Cortex-M4 powered using a solar panel and, secondly, a gateway, based an R-Pi 3 platform, wired powered, acting simultaneously as a sensor and a sink for the nodes. Data transmission between the node and the gateway is performed using the 802.15.4 standard (6LoWPAN MAC layer). The gateways which are located on streetlights are connected to Internet using a wired connection to a GSM router, through a power-line communication (PLC) built using the public lightening system. Acoustic measurements were performed using an omnidirectionnal digital MEMS (Invensense ICS-43432) integrated on a Mini PC, including additional sensors (temperature and humidity). A third octave band analysis and an A-weighting are performed using digital filtering. In addition, the data are encoded to reduce the memory load [26]. The influence of the air temperature have been evaluated in a climatic chamber, showing a very small deviation. However, no details are given concerning the acoustic performances of the sensors.

Recently, an alternative approach was detailed in [68], based on a digital signal processor (DSP) (Texas Instrument TMS320VC5502), to which is connected an acoustic acquisition chain, composed of a Panasonic WM 63-PR electret microphone, an analog conditioning and an ADC (Cirrus Logic CS5344). The processing capabilities of the DSP thus allow the real time calculation of many acoustic indicators as well as a 1/3 or 1/1 octave spectral analysis, simultaneously on two microphone channels. This technical approach opens up many perspectives, due to its high computing capacity in flash memory, which allows the integration of complex additional acoustic processing, with low power consumption. Another special feature of the system is the use of standard interfaces to control the sensor from an external CPU based system, for example a wireless NRG 2 panStamp (with a 868 MHz radio transmission) in the corresponding article. The performance of the sensor, compared to the expected characteristics of the IEC 61672 standard [7], is Class-1 for the DSP part and Class-2 for the whole system, due to the reduced performance of the microphone that was used. Eight sensors were deployed for 3 months in the city of Málaga (Spain), monitored over time and compared to Class-1 sensors. Although the prospects for the used of such a sensor are very interesting, the authors encountered problems of electromagnetic interference, material damage on some sensors, as well as variability due to meteorological phenomena. The authors also recommend the use of MEMS to improve the overall performance of the sensor.

Some other works related to the development of sensor networks for various applications and purposes can also be mentioned, even-though they are not sufficiently detailed to give a complete description. Tan et al. [69] considers a floating-point DSP to perform frequency and time weighting filtering, instead of using the MCU. Botteldooren et al. [21,22] tested noise sensor nodes, based on an ALIX single board computer and an ECM, using the ZigBee radio protocol, to develop sound monitoring networks and algorithms in order to access sound perception. Bell and Galatioto [13] proposed to adapt sensor nodes initially developed for environmental monitoring within the framework of the MESSAGE project [70]. It features 8-bit MCU, a condenser microphone, and an IEEE 802.15.4 data transmission protocol. Fifty nodes were deployed in the cities of Leicester (UK) in order to study road traffic noise. This experiment demonstrated that such technology can be used for urban noise assessment and management. The use of low-cost sensors was also proposed by the City of Barcelona (Spain) as a complementary network of a main Class-1 monitoring network [71].

2.3. *Synthesis*

2.3.1. General Considerations

The study of the literature highlights some key aspects and respective design choices, which are summarized in Tables 1 and 2. As indicated in the Introduction, our review focuses only on the first element of the chain, namely the sensor, and not on the infrastructure to implement the entire network.

Table 1. Main characteristics of low-cost noise sensors detailed in the literature: node based development platform, quantization level of the MCU, 'N-2-S' node-to-sink data transmission protocol, 'Mic' microphone type (ECM, analog MEMS (a-M), digital MEMS (d-M)), ADC quantization and sample frequency, power supply (battery (B), AA and LR20 cells (AA-B or LR20-B), battery with solar panel (B/S), wired connection (W) and energy autonomy in parenthesis when concerned and available (in days or hours)), pre-processing (Z, A or C weighting (Z-w, A-w or C-w), gain amplification (G), calibration (Cal), frequency equalization (Eq), 1/3 or 1/1 octave band analysis (1/3 or 1/1), encoding (enc)), price (in Euros (EUR), United States dollars (USD), Great Britain pounds (GBP)), goal of the corresponding article (proof-of-concept (POC), sensor design only, operational WSN (Op-WSN)).

Reference	Node Plateform	MCU	N-2-S	Mic.	ADC	Power	Pre-Processing	Cost	Goal
Barham and Goldsmith [45] (2008)	FPGA		GSM	a-M		B (15d)	A-w, C-w, Tc	100 EUR	Op-WSN
Santini et al. [42] (2008)	Tmote	16-bits	802.15.4	ECM	12-bit (8 kHz)	AA-B			POC
McDonald et al. [12] (2008)	Triton	32-bits	802.11b	ECM	16-bit (49 kHz)		A-w	130 GBP	Op-WSN
Hakala et al. [49] (2010)	CiNet	8-bits	802.15.4	ECM	10-bit (33 kHz)	AA-B (ds)	A-w, G, Cal		Op-WSN
Tan and Jarvis [54] (2013)	TelosB	16-bits	802.15.4	ECM	12-bit (33 kHz)	B/S			POC
Tan and Jarvis [56] (2014)	TelosB	16-bits	802.15.4	a-M	12-bit (33 kHz)	B/S			POC
Segura-Garcia et al. [59] (2015)	Tmote	16-bits	802.15.4	ECM	12-bit (8/20 kHz)	B (78d)	Cal	41.45 EUR	POC
Segura-Garcia et al. [59] (2015)	R-Pi	32-bits	802.11	ECM	16-bit (22.05 kHz)	LR20-B (39h)	Cal		Op-WSN
Noriega-Linares and Navarro Ruiz [61] (2016)	R-Pi	32-bits	wired LAN	ECM		W	Cal, Eq, 1/3	121 USD	POC
Alsina-Pagès et al. [37] (2016)	NXP chip	32-bits	Wi-Fi/GSM	ECM	12-bit (nc)	B			Design only
Mydlarz et al. [62] (2017)	mini-PC	32-bits	Wi-Fi	a-M	16-bit (44.1 kHz)	W	Eq	100 USD	POC
Risojević et al. [38] (2018)	STM32F0 series	32-bits	ZigBee	a-M		B (7d)	A-w, G, Cal	41.45 EUR	Op-WSN
Peckens et al. [63] (2018)	Teensy USB	32-bits	XBee	ECM	16-bit (20 kHz)	B (7d)	A-w, G, Cal	135 USD	POC
Ardouin et al. [67] (2018)	STM32L4 series	32-bits	802.15.4	d-M	16-bit (32 kHz)	B/S	A-w, 1/3, enc		POC
Ardouin et al. [67] (2018)	R-Pi	32-bits	802.15.4	d-M	16-bit (32 kHz)	W	A-w, 1/3, enc		POC
Silvaggio et al. [66] (2019)	mini-PC		GSM	d-M		B/S,W	A-w, 1/3		Op-WSN
Silvaggio et al. [66] (2019)	MCU		GSM	ECM		B/S,W	A-w, 1/3		Op-WSN
Mydlarz et al. [62] (2019)	R-Pi	32-bits	Wi-Fi/POE	d-M	16-bit (48 kHz)	W	A-w, C-w, 1/3	80 EUR	Op-WSN
López et al. [68] (2020)	DSP Board	32-bits	radio (868 MHz)	ECM	24-bit (108 kHz)	B	Z-w, A-w, C-w, 1/3, 1/1		Op-WSN

Table 2. Mean features of low-cost noise sensors detailed in the literature: frequency range, sound level dynamic, sound level range, residual noise, output acoustic indicators (equivalent sound level $L_{eq,T}$ or $L_{Aeq,T}$, maximum (Max) and minimum (Min) levels, percentiles L_N, 1/3 or 1/1 octave band spectrum, audio signal, psychoacoustic metrics, Peak detection). Data with the symbol * design measured sensor performances while the other ones are estimated.

Reference	F-Range	Dynamic	L-Range	Residual Noise	Outputs
Barham and Goldsmith [45] (2008)	20–20k Hz	70 dB	30–100 dB	25 dB	$L_{eq,T}$, L_N (10 mn)
Santini et al. [42] (2008)					L_{1s}
McDonald et al. [12] (2008)					$L_{Aeq,T}$
Hakala et al. [49] (2010)	<16.5 kHz		30–90 dB		$L_{eq,125ms}$, $L_{eq,1s}$
Tan and Jarvis [54] (2013)	<5 kHz *		93 dB *	60 dB *	
Tan and Jarvis [56] (2014)			100 dB *	50–60 dB *	Peak
Segura-Garcia et al. [59] (2015)	<20 kHz	96 dB			Psychoacoustic metrics
Noriega-Linares and Navarro Ruiz [61] (2016)	125–8k Hz * (1/3)				$L_{eq,T}$, L_N (N=10,50,90), 1/3
Alsina-Pagès et al. [37] (2016)					
Sevillano et al. [72] (2016)			35–115 dB		$L_{eq,1s}$, audio
Piper et al. [48] (2017)					$L_{Aeq,0.2s}$
Mydlarz et al. [62] (2019, 2017)	20–20k Hz	88.1 dBA		29.9 dBA *	Audio (10 s)
Risojević et al. [38] (2018)	up to 16 kHz *	72 dB	50–100 dB *		$L_{eq,250ms}$
Peckens et al. [63] (2018)	<10 kHz *	50 dB *		50 dB *	$L_{eq,125ms}$ (10 mn each 1 hour)
Ardouin et al. [67] (2018)	20–16k Hz		35–105 dBA		$L_{eq,125ms}$, $L_{eq,1s}$, 1/3
Silvaggio et al. [66] (2019)	20–20k Hz	70 dB	30(40)–100(110) dB	30–35 dBA	$L_{eq,1s}$, 1/3
Mydlarz et al. [62] (2019)			32–100 dBA		$L_{eq,125ms}$, $L_{eq,1s}$, 1/3, audio (10s)
López et al. [68] (2020)	up to 8 kHz		39.1–120.1 dB		$L_{eq,125ms}$, $L_{eq,1s}$, Peak, Max, Min, L_N (N=1,5,10,50,90,95,99), 1/3, 1/1

A full comparison of the proposed solutions is unfortunately not feasible for all points of interest. In particular, there is a consistent lack of description of the acoustic performance of the sensors, in terms of residual noise, sound level dynamic or frequency range. Most of the time, the information mentioned generally corresponds to the expected characteristics extracted from the data sheets of the sensor components, which can be assumed to be a higher bound in terms of performance. Indeed, assembled within the sensor with a wide variety of other hardware components and subjected to environmental stress, those performances are likely to be degraded.

Also, different levels of maturity of the proposed low-cost sensors are found. Some developments were limited to the design of the sensors alone (without prototypes), others proposed proof-of-concept (POC) (prototypes and tests), and some others have proposed to deploy several sensors in real urban areas in a quasi-operational framework.

Cost estimates of the sensors given by the authors show that the objective of obtaining a low-cost acoustic sensor (less than 150 EUR) is clearly achieved, with relatively high signal processing capabilities when considering most recent studies [35,66].

The question of the autonomy of wireless noise sensors, while being a fundamental aspect of the design, is in the end not extensively studied in the literature. This is linked in particular to the operating mode of the sensor, such as activity time/sleep time, duration of measurements, number of calculated indicators, and so forth. All those aspects are generally not detailed. However, it should be noted that most of the latest achievements [35,66,67] mention noise sensors directly powered by an electrical network, which seems to illustrate the difficulty of developing wireless sensors with acoustic performances that are relevant to the task at hand.

2.3.2. Sensor Platform

The choice of the sensor platform determines the main functionalities and characteristics of the sensors. Three main families can be distinguished: (1) MCU based existing platforms; (2) specifically developed electronic boards; (3) Mini PC.

The use of existing platforms (1) simplifies the sensor development by using components that have already been optimized in terms of energy consumption (TelosB, CiNet, Teensy USB, Tmote) [33,42,54,56,59,63]. Those platforms generally include all the components needed to develop an environmental sensor (radio communication module, ADC, storage memory, connectors for other sensors, etc.). Such platforms have played a key role in demonstrating the feasibility of developing and deploying noise sensors. In addition, programming the MCU is facilitated by the use of dedicated libraries, but requires a physical connection to a computer.

However, the lack of autonomy and the reduced computing capabilities for calculating relevant acoustic indicators have motivated researchers to develop their own electronic boards [37,38,67]. A specific design makes it possible to select the processor and components best suited to the expected requirements in terms of autonomy and computing power. However, the measured autonomy of the sensors remains to be few days in the best cases, and is very dependent on the periodicity and the duration of the measurements as well as on the data processing that is performed.

The main interest of using a Mini PC, lies mainly in increased computing power [35,59,62,66,67], allowing more advanced digital audio processing, easy integration of external components, of remote updating (without physical connection to a computer), but at the expense of higher energy consumption. In most cases, these sensors are directly connected to the power grid as using a battery limits operation to only a few hours.

2.3.3. Data Transmission Protocol

Choosing a relevant communication protocol for a sensor network normally depends on several parameters, such as the distance between nodes and sinks, data rate, quantity and type of measurements, network topology, global architecture, latency and reliability, cost... Although this problem of choosing a communication protocol and optimizing it has been widely discussed in other areas of sensor network applications, very little has been said about noise sensor networks. The experiments detailed in Table 1 show that different protocols have been used, most often imposed by the choice of the sensor development platform. More information about data transmission protocol are given in Section 3.2.6.

2.3.4. Microphones

As mentioned by several authors [68,73], the microphone is a critical element of the noise sensor. A wrong choice of microphone will impact the quality of the acoustic indicators produced which cannot be overcome. The literature review shows that three types of microphones are considered: (1) electret condenser microphones, and the more recent MEMS microphones that can be either (2) analog or (3) digital. The replacement of ECM microphones using MEMS ones was justified by the authors on the basis of their acoustic performances that were *a priori* more interesting for their use in acoustic measurements [56,62]. In addition, MEMS microphones have reduced dimensions, are relatively reliable and durable, and above all are produced at a lower cost. The use of MEMS therefore was considered particularly relevant in the context of the implementation of low-cost sensor networks, specifically for urban noise monitoring. To our knowledge, the first work mentioning the potential interest of MEMS for applications in acoustic measurements in urban environments, can be attributed to the National Physical Laboratory (NPL) in the UK [45] in 2004, which will later lead to the DREAMSys sensor prototype. While the overall results are rather positive, some authors have highlighted some limitations and problems that are mainly related to the technology and layout of MEMS used in sensors. In order to fully understand these limitations, the following paragraph provides a brief description of MEMS microphone technology.

A MEMS microphone is composed of a sensor (MEMS sensor) and integrated circuits. The MEMS sensor is a silicon capacitor made of two electrically isolated surfaces. One surface, called the backplate is fixed, and is covered by an electrode and made of many holes, that is, acoustic holes. The other is movable and is called the membrane or the diaphragm. Sound wave passing through the acoustic holes of the backplate will set the diaphragm in motion, creating a change of the capacitance between the two corresponding surfaces, which is converted in an electrical signal by the application specific integrated circuit (ASIC). The ASIC delivers an analog output or a digital output, depending of the microphone type (analog or digital). The MEMS microphone and the ASIC are packaged together, surrounded by a substrate and a lid, forming a cavity. A sound inlet (acoustic port) is present either in the substrate (bottom port configuration) or in the lid (top port configuration), and, most of time, positioned directly in the MEMS cavity. For analog MEMS microphones, the electrical output signal from the ASIC is sent to an external pre-amplifier, also in charge of converting the output to a signal that can be used as input of an acoustic chain. For digital MEMS microphones, the ASIC output is sent to an internal analog-to-digital converter (ADC) to provide a digital signal, either as a pulse density modulated PDM format (1-bit high sample rate data stream) or I2S format (same as PDM microphone but including a decimation filter and a serial port in order to produce a standard audio sample rate).

Acoustic performance of MEMS depends on mainly technical aspects, such as the location of the sound inlet [74]. Considering the bottom configuration where the sound inlet is above the cavity (front chamber), the interaction between the air in the sound inlet (back chamber) and the air in the MEMS front chamber creates an Helmholtz resonance, whose frequency increases as the air volume decreases, positioning the resonant frequency at the upper part of the frequency band of interest, and thus, leading to a flatter frequency response. Additionally, locating the sound inlet above the MEMS sensor allows easier interaction between the diaphragm and the sound wave, thus increasing the sensitivity and the SNR. Conversely, considering the sound inlet at the top will create a larger air volume in the cavity between the lid and the backplate (front chamber) in comparison with the air volume in the cavity between the diaphragm and the substrate (back chamber); it generates a lower resonant frequency, thus possible resonant frequency in the frequency range of interest. Such unwanted effects were observed, for example, in [45,62]. In addition, due to a smaller volume in the back chamber for the top configuration in comparison with the bottom configuration, it will be more difficult for the diaphragm to move, leading to worst sensitivity and SNR. Thus, it appears that the bottom configuration provides better acoustic performance.

The thickness of the PCB on which the microphone is soldered can also modify the volume of the front chamber and the inlet length altering the upper frequency response of the microphone [38,45]. This effect can also be more pronounced if a cover is added on the device, as a protection [75]. A possible solution is to mount the MEMS and the ASIC in the inner side of the lid, for the top configuration, leading to the same expected performance than for the bottom configuration. Barham and Goldsmith [45] also mentioned that the background noise of MEMS microphones was significantly higher than classical microphones. In practice, experiments described in the literature mention noise levels between 20 and 30 dB, depending on the frequency [76]. However, recent works on the development of new generation MEMS microphones suggest the possibility of developing noise sensors with high acoustic performance [77], which makes it the ideal component in the future for the development of noise high-performance sensors.

2.3.5. Frequency Weighting

Most acoustic indicators for the assessment of environmental noise require frequency weighting (generally A-weighting) to take into account the sensitivity of the human ear to certain frequencies. Since the calculation of acoustic indicators, such as equivalent sound levels, is integrated within the sensors, this weighting should be done as a pre-processing, using analog filtering [45,49,63] or digital filtering [38,61,62,66,67,69]. Analog filtering makes it possible to overcome the computing limitations of the microprocessor used, reduces energy consumption, can be processed in real time and avoids any bias, such as rounding errors during digital computation [78]. However, analog filtering of the audio signal constrains the nature of the acoustic indicators at the output of the sensors, unless the filter is replaced. Conversely, digital filtering offers more flexibility, but at the cost of reducing the possibility of real time computation if the microprocessor is not powerful enough, and potentially reducing autonomy [63]. Risojevic et al. [38] recently compared several techniques for digital A-weighting filtering and showed that using a matched z-transformation filter with a cascade form implementation was relevant for a small processor core, and slightly better than a bilinear transformation [67], when comparing with an analog filter.

2.3.6. Frequency Equalization

The acoustic acquisition chain generally has a frequency response curve that is not as "flat" as expected, often due to the frequency characteristics of the microphone. This creates a bias on the measured audio signal, and therefore on the calculated acoustic indicators. Some authors therefore propose to compensate the frequency response of the acquisition chain by implementing an equalization filter [61,62]. However, this filtering requires a significant workload on the microprocessor, with an impact on the sensor autonomy. Another solution proposed by [48], implemented in their own MEMS design in stainless steel tube, was to use a patented acoustic filter made of a resistive element and a closed volume, the whole system acting as a low-pass filter.

Low-cost microphones can exhibit significant variability in frequency response. An equalization filter should therefore be determined for each microphone, which would be ideal, but is not possible when deploying a large number of sensors. Mydlarz et al. [62] thus propose to generate the equalization filter on the basis of an average of a limited number of microphone impulse responses. The authors observe a good behaviour of the implemented real-time time-domain filter (based on an inverse linear-phase finite impulse response filter), however with an increasing of residual noise between 20 Hz and 400 Hz.

2.3.7. Calibration

At the output of sensors, sound level indicators must be adjusted to give measures that are as close as possible to sound levels measured by a reference device, such as an acoustic calibrator. In the best case, this adjustment takes into account the variability of the microphone sensitivity, but this adjustment can also correct linearity defects in the acquisition system or in the digital processing chain. Two calibration methods are considered in the literature, either by using an acoustic calibrator (mainly 94 dB at 1000 Hz) [59,66], when the microphone mounting device allows it (matching the diameter of the microphone mount with the acoustic calibrator), or by comparing with a reference sound level meter under the same measuring conditions [49,57,62,63]. In the simplest case, the same correction is applied to the entire temporal signal, without distinction of frequency or amplitude. In a more advanced way, this correction can also correct linearity defects in level and frequency [38] and temperature [73]. The correction can be taken into account either within the sensor or in post-processing once the data has been collected on a server. In most cases, this correction is applied to the digital signal, but it can also be integrated into an analog circuit [63].

2.3.8. Noise Indicators

The choice of output acoustic indicators is very important for designing the sensors in terms of expected computational and power resources.

A temporal integration in order to obtain an equivalent sound level over a given integration time (1 s for example), will require far fewer resources (energy and calculation) than the calculation of frequency band spectra. Such time integration can easily be processed by a system based on an MCU with battery, while a frequency analysis will require more resources, as proposed today by a mini PC, with a wired power supply [35,62]. Most authors have therefore limited themselves to produce equivalent sound levels, generally with an integration time of 125 ms, that is, fast time weighting, and 1 s, that is, slow time weighting. However, with the help of technological developments, it can be seen that some authors put forward solutions that *a priori* make it possible to carry out frequency band analysis with an MCU [66,67].

While the measurement duration is often indicated in the literature, the temporal periodicity is rarely specified. It is also quite difficult to determine whether the calculations are carried out continuously and in real time, or over periods interrupted to calculate noise indicators, to transmit data and to save power resources during sleep periods.

The transmission of audio signals is clearly not a priority, for obvious technical reasons: network bandwidth, storage on sensors and servers, and energy consumption but more importantly for privacy concerns. The calculation of acoustic indicators directly within the sensors, and then their transmission, is the only relevant solution. However, one can cite a few exceptions—Mydlarz et al. [35] propose to store and transmit audio samples, but only for testing purposes to validate machine learning algorithms for identifying sound sources. Similarly, Sevillano et al. [72] recorded audio data by connecting the sensor to a digital recorder, in order to test an acoustic event detection algorithm. In both cases, the final objective is to integrate, after optimization, these algorithms directly into the sensors. On the other hand, biodiversity monitoring rather emphasizes the need to preserve raw audio signals, which creates constraints for continuously capturing and transmitting data in a fully autonomous wireless SN [31,33,79].

2.3.9. Meteorological and Outdoor Conditions Effects

Renterghem et al. [73] have studied the effect of temperature, humidity and wind, on the sound levels measured by electret and MEMS microphones. It was shown that applying an air temperature correction may have a positive effect on the long-term measurements. For example, by comparison with a reference microphone, such temperature correction (acting as a gain on the signal) applied on an ECM was able to reduce the deviation of the global error from 1.6 dBA to 0.8 dBA, over the full period of observation (several months). In order to proceed to such correction, the authors proposed to consider an additional sensor for measuring the temperature. Because relative humidity and temperature are highly correlated, the temperature correction may also include the humidity correction. The authors have also noted inconsistent behaviour of the MEMS microphone at temperatures below 20 °C, high relative humidity and high wind speed, but no explanation has been given.

The effect of ambient temperature on the sensitivity of a MEMS sensor was also investigated by Barham and Goldsmith [45]. From their results, it seems that the variation in sensitivity would increase with frequency and temperature (tested between −5°C–40°C), in the order of ±1 dB over the frequency range between 100 Hz and 8 kHz. Conversely, [67] have not seen significant variation of MEMS microphone sensibility with temperature.

In another study, Bartalucci et al. [80] report that they observed a reduction of acoustic sensitivity of MEMS microphones during the initial running-in phase, on the order of 1–2 dB on a time period of 4 months.

A procedure was proposed in [76] in order to evaluate the evolution of acoustic performances of MEMS and electret microphones, when exposing to stressing conditions. Microphones were installed inside a salt spray chamber in order to exaggerate the possible damage in outdoor environments (such extreme conditions can however not be reproduced in real environments). Authors observe a slightly better stability for MEMS microphone in term of frequency responses and noise floor. However, anomalies like spike occurrences have been observed in the audio signal measured by some MEMS microphones, whose impact is low in terms of equivalent sound level, but which was remained totally unexplained.

Li et al. [81] have studied the reliability of MEMS microphones, through accelerated life tests in a corrosion test chamber and in a pneumatic shock testing. The second testing is more related to the use of a MEMS microphone in smartphones, and has showed that failures of the diaphragm and backplate of the MEMS can be observed, but only when an impact is generated in the direction normal to the diaphragm and for very high acceleration level (greater than those observed in real life). Considering the corrosion test, wire bond corrosion and membrane embrittlement were observed after 90 days in the test chamber, but with a very slight impact on the frequency response of the microphone.

All the results mentioned above should be considered as points of attention, and not as "absolute" results, since the observed effects may vary due to many design factors. The generation, the type of microphone, but also the presence or not of wind and rain protection devices, all will have an impact on the experimental results. In our opinion, the conclusion to be drawn is that it seems mandatory to carry out specific tests, before the sensors deployment, in a climatic chamber (as also suggested in [62]) or a corrosion chamber for example, in order to evaluate the temperature effects on the sensor, to assess whether a correction should be applied. In addition, protection of the overall sensor (not only the microphone) against wind, chemical agents, dusts, pollutants, shock... seems essential, and their impact must also be evaluated [76,81].

3. Noise Sensor Design for Low-Cost Networks

The literature review highlights the rapid evolution of low-cost sensors, mainly driven by the more global need for electronic and computer systems dedicated to the development of smart sensors. The technical solutions available today seem both sufficiently stabilized and adapted to offer low-cost acoustic sensors that can meet current and future applications for the assessment and management of environmental noise.

In this section, on the basis of the literature review detailed previously, we propose to highlight the technical characteristics that we believe are essential for the correct design of a low-cost acoustic sensors, offering a level of performance in line with the needs in noise monitoring. Depending on the objectives and applications considered, two levels of expected sensors characteristics can be distinguished, ranging from minimal to optimal (Table 3). It is these characteristics that condition the technical elements to be considered for the realisation of the noise sensors.

Table 3. Minimal and optimal expected characteristics for the noise sensors.

Property	Minimal Target	Optimal Target
Measurement range	30–105 dB(A)	30–105 dB(A)
Frequency range	100–12k Hz	100–16k Hz
Integrated sound level	$L_{A,eq,1s}$	$L_{A,eq,125ms}$ $L_{A,eq,1s}$
Spectrum	None	1/3 octave bands
Measurement frequency		Continuous
Pre-processing	A-weighting Calibration	(A, Z)-weighting Calibration 1/3 octave bands analysis Frequency equalization
Other indicators		Source recognition Noise event detection
Additional sensors	Temperature Hygrometry	Temperature Hygrometry
Price	50 EUR	150 EUR

3.1. Expected Characteristics of Noise Sensors

3.1.1. Acoustic Measurement Accuracy

Referring to conventional measuring systems and historical practices in environmental acoustics, one could try to compare the acoustic performance of low-cost sensors with Class-1 or Class-2 devices for 'expertise' or 'control' environmental noise measurements [7]. However, as pointed by [46], there are situations where Class-1 or even Class-2 systems give measurement results whose very high accuracy is not necessarily in line with the practical use made of these data. This is particularly the case in the strictest application of the European Directive 2002/49/EC on the assessment of environmental noise. Informing the public through "coarse" noise maps, such as the establishment of action plans to reduce noise pollution, does not require a very high precision in the performance of acoustic measurements.

This is also highlighted in [22], mentioning that low-cost acoustic sensor characteristics are already quite consistent, in term of metrological capabilities, with what is expected from a strategic noise map: it is not necessary to seek a measurement accuracy of less than 1 dBA; the sound spectrum is rather centred around 1000 Hz, that is, over a frequency range for which a low-cost sensor is not lacking; a rather high minimum noise level, therefore beyond the residual noise of the sensor (except, maybe, for quiet areas).

For more specific needs, the most important guideline is to be aware with the technical limitations of the systems developed, and to ensure that the exploitation of the collected data is consistent with these limitations.

3.1.2. Acoustic Indicators

Indicators of interest to describe urban noise environments are calculated on the basis of 1 s or 125 ms data. They cover at least equivalent sound levels as well as statistical indicators, and, sometimes, emergence indicators such as the number of exceedances at given thresholds. Finally, some authors have introduced more demanding indicators for specific uses, such as the time and frequency second derivative (TFSD) [82] to describe voice and bird sounds, which requires a good temporal and spectral response of the sensors. Consequently, the expected characteristics of noise sensors are guided by the amplitude of the sound levels encountered and the spectral information of the sources to be characterized. The characterization of urban noise environments, from quiet areas to the noisiest events, assumes a linear sensor response in a range from 30 to 105 dB(A). The interest in bio-phonic sources, especially bird sounds, suggests that we should also be able to accurately measure high frequencies, up to 16 kHz.

3.2. Sensor Platform and Components

The choice of the platforms and components for the sensor development is mainly determined by the questions of how the sensor is connected to the network, how the sensor is powered, and what are the expected sensor output indicators. This last question also conditions the first two questions.

A sensor that transmits data by radio will be limited by the maximum data rate of the transmission protocol, as well as the distance and visibility to the nearest gateways or relays. The power supply mode will determine the computational and storage features of the sensor, as well as the operating conditions depending on the energy recovery mode (battery change or power supply using renewable energy).

Several technical solutions can be considered, each with different components/functionalities—a sensor connected with a wired connection to the electrical network and to the data network; a sensor powered to the electrical network through a wired connection, but transmitting data by radio wave; an autonomous energy sensor (possibly also acting as a relay) transmitting data by radio wave.

3.2.1. Wired Sensor Platform

With regard to the experiments presented in the previous literature review, the choice of a Mini PC constitutes an optimal choice with regard to the low-cost, the computing and storage capacities, the connectivity with other modules (radio, other sensors...), the remote maintenance and update of the system, the change of some modules (since not all modules are integrated into the motherboard of the Mini PC but just connected). There are several Mini PC solutions available at a very affordable price, with fairly similar features, and with different operating systems. Among these, it is clear that the R-Pi family seems an excellent choice given the many accessories and modules available, but also given the presence of a very active community. The latest R-Pi models use very powerful 64-bit microprocessors, but at the expense of higher power consumption. The choice can be made for an older model (model A+ or Zero), very cheaper, but with a better power/energy consumption ratio if it were to be run on battery power [83].

3.2.2. Wireless Sensor Platform

The development of a stand-alone sensor is more complex as it must meet many requirements. The nature of the acoustic indicators to be produced (continuous sound levels and spectra) requires a powerful microprocessor; the dynamics of the sound levels to be measured requires quantification by the ADC of at least 16 bits (96 dB of dynamics), ideally 24 bits to take advantage of a wider dynamic range (144 dB), which needs the use of a 24 or 32 bits MCU, as in the STM32 series already used by several authors [38,67].

3.2.3. Microphone and ADC

The acquisition chain (microphone, gain amplifier, ADC) is the other essential element to consider. Most of the achievements have focused either on ECMs or on MEMS, combined with an external ADC. Feedback from the literature review has shown the sensitivity of the acoustic signal to electrical and radio frequency interference, causing an increase in the residual noise. This is why some authors have recently turned to digital MEMS (the analog-to-digital conversion is performed inside the microphone) [35,66,67], which seems today an optimal choice. In addition, using a digital MEMS microphone with an I2S interface, it is unnecessary to use an external codec [84].

The choice of sampling frequency depends mainly on the spectral band of analysis, which depends on the expected sensor application. The optimum corresponds to the audible frequency band 20–20k Hz, covered by most MEMS microphones, which implies a standard sampling frequency of 44.1 or 48 kHz. While such a sampling frequency is not a problem for a wired sensor (such as a R-Pi), it is more problematic with MCU, and even more if a real-time processing is required. Currently, the only reference for a stand-alone node with a digital MEMS node mentions a sampling frequency of 32 kHz [67], but the paper does not described the final sensor performances.

3.2.4. Noise Floor Enhancement

As mentioned by several authors, residual noise is one of the elements of the measurement chain that can limit a sensor ability to perform measurements at low levels, as in quiet spaces. The observed residual background noise levels are generally much higher than the value indicated by the manufacturer for the microphone and are caused, for example, by interference on analog electronic circuits or by the limited performance of some ADCs. The solution to reduce the residual noise is to optimize the electronic components of the sensors. Another way, proposed for example in [73], is to combine several microphones on a same sensor and to reduce the residual noise by applying a noise reduction method based on cross-correlation techniques. Such procedures may be however power and computational consuming. It must be noted that this use of several microphones simultaneously to reduce background noise would also make it possible to envisage other applications of this type of acoustic sensor, such as locating sound sources.

3.2.5. Mass Storage

Sensors can also have mass storage capacities to store various information, for sensor maintenance, but also to temporarily store the collected data when the connection to the gateway or data server is interrupted. The sizing of this memory must take into account the duration of temporary data backup, and potentially the ability of the sensor to transmit a large amount of data once the connection is established, while simultaneously collecting and processing new data.

The type of storage device is the second element to be considered. The more relevant choice is to use a flash memory (memory card, USB flash device, SSD), offering lower power consumption, easier maintenance, higher transfer speed, no operating noise, but at the expense of less mass storage than a traditional hard disk (HD), but also shorter lifetime due to a limited number of write cycles and higher cost.

3.2.6. Data Transmission Protocol

Depending on the type of output noise indicators (such as 1/3 octave band analysis [35,66,67] or audio capture [35]), measurements may require a very high frequency, inducing a huge quantity of data to be transmitted. Conversely, a temporal integration of the audio signal does not generate an enormous amount of data to be transmitted.

With regard to the applications that are currently envisaged for these noise sensors (see Introduction) and the expected output indicators (Section 3.1.2), the needs in terms of data rate are rather increasing. The targeted useful data rate should be up to 10 kb/s with a maximum range around 100 m to be able to transmit compressed data from one node to a sink, in a urban configuration, taking into account trees, buildings, cars and trucks impacting the radio signal propagation.

The use of a wired network is obviously the simplest and most effective solution to ensure data transfer under ideal conditions. This solution will be probably feasible in the future, given the growing number of cities developing smart and connected systems. Nevertheless, the expected spatial density of noise sensors requires dating the use of radio transmission, as most past experiments have envisaged.

Considering the amount of data to be transmitted from a noise sensor, the implementation of a very simple topology, that is, a direct transmission from a node to a sink, seems the most obvious solution. Topologies involving one or more relays do not seem to be possible today, given the technical solutions that are available. In the literature, one recent reference [67] mentions the use of sensors that also act as relays for other sensors, but this has not been developed further.

As a first solution, the 3G/4G and other GSM protocols could be considered. However, such transmission protocols are not cost effective for very dense sensor networks because with one subscriber identity module (SIM) card is required per sensor (or two SIM cards in order to ensure the relay of data transmission in the event of a GSM network failure). Regarding other data transmission protocols without subscription, several solutions are possible, depending of the range/data rate compromise that is expected for the sensors networks (Table 4).

Table 4. Radio transmission protocols specifications.

Protocol	**Bluetooth** [85]	**Bluetooth LE** [85]	**Wi-Fi** [86]	**Wi-Fi** [86]	**Zigbee and 6LoWPAN** [87]	LoRaWAN [88]	Sigfox [89]
Specification	802.15.1	802.15.1	802.11g	802.11n	802.15.4	LoRa Alliance	Sigfox
Frequency	2.4 GHz	2.4 GHz	2.4 GHz	2.4 GHz 5 GHz	868 MHz (EU) 915 MHz (US) 2.4 GHz	Sub-GHz ISM band 868 MHz in EU	Sub-GHz ISM band 868 MHz in EU
Range indoor (m)	30	10	25	50	30	>100	>100
Range max (m)	100	50	75	125	1500	>10,000	>10,000
Data speed max	3 Mbit/s	1 Mbit/s	54 Mbit/s	540 Mbit/s	250 kbit/s	11 kbit/s	100 bit/s
Data speed typ.	2.1 Mbit/s	270 kbit/s	25 Mbit/s	200 Mbit/s	150 kbit/s	300–11k bit/s	100 bit/s
Peak current	150 mA	20 mA	150 mA	150 mA	50 mA	25 mA	25 mA
Sleep current	5 mA	1 µA	100 µA	100 µA	5 µA	4 µA	4 µA
Battery life	Month	Year	Day	Day	Month/Year	Years	Years
Network topologies	Star	Star	Star		Star, Tree, Mesh	Star	Star
Applications	Headsets Computer peripherals	Mobile phones Sport trackers eHealth devices Wireless sensors	PC (networking) WLAN	Same as 802.11g with improved performances Outdoor LAN	Smart home Wireless sensor networks Smart metering	Smart building Smart city	Smart building Smart city

Looking at the available data rates, the LPWAN technologies like LoRaWAN and Sigfox would not allow data transmissions with a sufficient efficiency. Even for the LoRaWAN, the maximum data rate is higher than the useful data rate due to the overhead of the protocol. In terms of data rate, the Wi-Fi and the Bluetooth protocols would be more efficient, but the battery life will be too limited for an application without constant energy and for an efficient coverage of an urban area. Zigbee and 6LoWPAN, based on 802.15.4 specification, present both maximum range and data rates that are compatible with the noise SN.

3.2.7. Additional Sensors

As mentioned in Section 2.3.9, the knowledge of the air temperature can be sometimes useful to calibrate noise sensors. More generally, knowledge of atmospheric and meteorological conditions can be interesting for a better use of data. For example, the presence of rain or a strong wind can cause disturbances to the measured acoustic signal, which, if not identified, can lead to misinterpretation of the collected data. The measurement of these atmospheric conditions (temperature, humidity, wind speed and direction) at the same time as the acoustic signal seems relevant, particularly because of the low cost of the components and of the limited additional data it can generate. More generally, the possibility of connecting other types of sensors (traffic, ambient light, air pollution, video, accelerometer, etc.) [38,49,60] would make it possible to develop a global and multi-disciplinary environmental approach [9,70] by pooling technical resources. The integration of multiple sensors adds additional constraints in terms of maintenance, data storage and transmission, as well as energy consumption, which must be anticipated.

3.3. *Sensor Life*

One important issue is to determine the expected lifetime of a low-cost sensor. Knowing that the lifetime of a Class-1 sound level meter can extend to more than 10 years under normal conditions of use [90], a lifetime of a few years (typically 5 years) already seems an ambitious goal considering the overall cost of a low-cost noise sensor and the quality of its internal components.

There are several components that can affect the lifetime of a sensor, mainly the measuring microphone, the data storage elements, and, if applicable, the battery. All other electronic components embedded in a sensors are designed most of time to operate for more than 10 years without problems in outdoor conditions, except when experiencing unplanned event, such as mechanical damage, high level of humidity, or really extreme temperature conditions out of the expected −20/+55 °C range.

As detailed above, recent microphones and especially MEMS have a fairly good resistance to atmospheric conditions and have a limited drift over time. Newer mass storage devices also have a longer life expectancy given today's permitted read/write cycles. Finally, as far as autonomous sensors are concerned, the most sensitive component is undoubtedly the battery. Either the battery is removable, in which case an on-site intervention is required, or the battery is rechargeable and in this case the life cycle is defined by its ability to recharge, generally by using solar panels, while maintaining optimal properties.

Most solar panel lifespan is around 20 years, with a power output decrease of less than 1% per year [91]. The sensor will then be given around 80% of the initial energy after this time. Environmental conditions, will have also an impact on the longevity of lithium batteries (i.e., the type of battery most commonly used in electronic devices): the worst case is for high temperature (above 40 °C). Most of time, battery packs do not die suddenly, but the runtime gradually shortens as the capacity fades. The capacity of the battery will also decrease during its life, starting from 95% of its nominal capacity it quickly decreases to around 80% in less than a year (with 250 charge/discharge cycles). In addition, the depth of discharge will have also an impact of the battery durability: considering smaller discharges will prolongs the battery life. The lifetime of such batteries can typically range from a few years (typically 5 years) for consumer products to more than 10 years for industrial products.

3.4. Power Resources

As pointed out in [54,56], autonomous nodes, made with a battery and a solar panel, should present enough storage capacity to store the energy that is required over the duration of the measurement. For low cost sensors, the dimensions of the solar panel as well as the dimensions of the battery are a tradeoff between the energy consumption of the sensor and the overall price of the sensor.

Considering an average power consumption of the sensor around 75 mWh, which seems sufficient for a MCU based noise sensor already offering significant computing power, the solar panel should be able to provide enough energy to power 24 hours of energy request, even during the worst month of the year (not in extreme conditions). If during this period, only 3 hours of sun are available, a solar panel should provide $24 \times 0.075 = 1.8$ W, which seems very reasonable in terms of cost and space requirements. To be able to power the sensor during a few days without sun, the battery should have the biggest possible capacity. A 2600 mAh battery will be able to provide enough energy to a sensor during a few days, even if there is no sun available to recharge the battery.

It is also essential to consider the progressive degradation of the properties of the solar panels and batteries in order to ensure the correct operation of the sensor over the envisaged lifetime. From the initial design stage, this means overestimating the capacities of the panels and batteries to ensure trouble-free operation of the sensors over the expected service life. Lastly, one can also mention that the technical improvement of the energy system must also be accompanied by the development of algorithms and procedures to optimize or reduce energy consumption [33].

3.5. Acoustic Calibration

Regardless of the intrinsic performance of the sensors, calibration is an essential operation for any "controlled" acoustic measurement. At a minimum, the sensor calibration should be performed, using an acoustic calibrator, for example, 94 dB at 1000 Hz, to determine the sensitivity correction at the selected frequency, under controlled conditions. The use of a multi-frequency calibrator can be useful in determining a frequency correction, unless the frequency response has been corrected using an equalization filter within the sensor. Considering the expected accuracy of a low-cost sensor, the use of a Class-2 calibrator seems sufficient. The use of a sound calibrator requires that the microphone be mounted on a cylindrical support with a compatible diameter.

As pointed out in [66], it is important to regularly check this sensitivity correction throughout the period of use in order, if necessary, to take into account the variations in sensitivity of the measuring microphone. Taking into account the sensitivity correction directly within the sensor, in pre-processing, (rather than retrospectively on the post-processing data server) seems relevant to ensure that the acoustic indicators at the sensor output are consistent with reality. However, even in this case, the verification procedure with a sound calibrator is required. Such correction was for example proposed in [63], using an electronic circuit, but the it can also be carried out by applying a correction on the raw data.

Because of the variability between low-cost microphones, particularly in terms of frequency response, it may be tempting to determine, in laboratory, a sensitivity correction for each sensor. However, this can quickly become tedious to do for a large number of sensors. It seems more appropriate to estimate a correction value on a sample of sensors, then calculate an average correction that will be applied to all sensors, as proposed in [62].

Moreover, the question of the calibration of a very large number of sensors, particularly in situ, is a very hot subject of research [92] that could be considered to noise sensor networks, for example by developing automatic calibration of sensors without human intervention.

3.6. Additional Challenges

The development of a low-cost acoustic sensor for long-term acoustic measurement is only the first step in a comprehensive approach to the development of a sensor network for noise monitoring. Many other aspects, such as the development of an optimal technical and IT infrastructure for network and data management, anomaly detection, optimization of sensor positions, spatial and temporal data sampling, and the management of hybrid networks, are important challenges, which are still relatively open in the field of noise monitoring as of today. The complete review of all those issues that surround the sole design of the sensor is left for future work. However, for the reader to grasp some aspects of those challenges, we present in the following some related advances and discussions.

3.6.1. Detecting Network Defaults

It cannot be expected from a low-cost sensor the same performance as a professional sensor in terms of reliability and durability. Therefore, and as has been mentioned in several studies [50,73,93], the probability of malfunctioning of a low-cost sensor must be carefully considered. If extreme cases (for example, due to hardware malfunctions, a sensor no longer returns data) can be immediately detected, others irregularities (such as abnormal acoustic data behaviour due to certain weather conditions [73] or sensor power loss problem [63] for example) are more difficult to identify. The implementation of advanced algorithms for dysfunction detection is therefore essential [35,73]. The subject of automatic fault and anomalies detection is particularly developed in the literature about wireless SN, but only little for noise monitoring. The description of these methods is outside the scope of this study; the reader may refer to recent references about this subject [94–97].

3.6.2. Temporal Sparse Sampling Strategies

The question of the duration and frequency of acoustic measurement is crucial since it concerns different aspects of the definition of sensor characteristics, such as memory space for data storage, computing capacities for real-time processing an data transmission rate. Reducing the measurement time therefore makes it possible to be less demanding on the characteristics of the sensor, and thus to reduce its cost and to increase its lifetime. Thus, it may be particularly interesting to study the temporal structures of noise levels in the environment as well as their spatial dependencies, in order to potentially reduce the sampling duration.

As an example, Can et al. [98] investigated the variations of hourly noise levels at the week scale, showing that a matrix of relationships between hourly noise levels can be defined for each measurement point, from which noise levels at any period can be deduced from measurements at other periods. The temporal trends are however different depending on the site. A method for stratifying urban space has also been proposed in [99]. Four categories were considered were arterial roads outside the central zone, arterial roads in the central zone, two-way roads connecting different zones, and one-way roads. The interest of such stratification lies in the fact that temporal variations in noise levels are correlated from one point to another within the same category.

However, these studies are still insufficient to optimize the sensors in terms of their temporal measurement dynamics and should be continued.

3.6.3. Optimizing Sensor Locations and Network Deployment

(1) Spatial Representativeness and Interpolation

Even if the very low intrinsic cost of individual sensors may encourage to not limit the number of measurement points to be integrated into the network, in practice, it must be limited to reduce the overall cost of the networks, mainly in terms of maintenance. Thus, dealing with a limited number of sensors, the choice of the 'best' location of each sensor may be of major importance, since many locations are possible in a large urban areas. In addition, authors have also questioned the spatial representativeness of noise, in order to limit the number of sensors, and then to consider ways to interpolate noise indicators between the measurement locations.

Can et al. [100] showed that interpolation methods were defective when the spacing between sensors was too large (about one measurement point every 250 m in the study). The explanation given is that they do not offer a sufficient covering of the network, and assume spatial variations that are not coherent with traffic dynamics or street configurations. Indeed, in urban areas, a distance of 250 m can see a succession of very varied environments.

The study of the spatial characteristics of sound environment variations help defining interpolation functions. Gozalo et al. [101] showed that a stratification of roads based on their functionality was helpful before interpolating sound levels showed similarly, based on a measurement campaign in the city of Plasencia (Spain), that the characteristics of sound level variations follow the categories formed with road functionalities.

Liu et al. [102] analyzed the sound environments of the city of Rostock, Germany, and observed that spatial variation of urban soundscape patterns was explained by underlying landscape characteristics, while temporal variation was mainly driven by urban activities.

Zuo et al. [103], based on measurements in the city of Toronto (Canada), observed that noise variability was predominantly spatial in nature, rather than temporal: spatial variability accounted for 60% of the total observed variations in traffic noise.

Two examples of spatial interpolation of noise levels based on a dense sensor network can be found in the literature. In [60], a fix grid of 78 sensors was deployed in the city of Algemes (Spain). The network covered 1.8 km^2, which is about a square grid of 50 m on each side. For the purposes of the study, 10 sensors were removed in which levels were estimated at five 3-hours periods of the day by an interpolation method, namely an Ordinary Kriging in which noise levels are described by a logarithmic function. The study shows that under this sensors density the kriging method seems an efficient method to interpolate noise levels, within a RMSE of 3.5 dB(A). In addition, the residuals are spatially correlated except for the [19 h–22 h] period, probably because it entails specific noise behaviors (leisure noise activities, etc.).

In [5], the impact of the density of observation points and the performance of four spatial interpolation methods were presented. Mobile measurements have been performed while walking multiple times in every street of the XIIIrd district of Paris (France), to construct a reference map, which is estimated by adaptively constructing a noise map based on these measurements. The four interpolation methods were constructed by combining two algorithms: (i) the Kriging method, either Ordinary Kriging or Universal Kriging (which consists in adding a linear trend, defined from the distance between each location and its closest road in each amongst four categories) and (ii) the definition of the distance between locations, either Euclidian or computed from the road network.

(2) Best Sensor Location

Huang et al. [104] proposed a hybrid model, based on a K-means clustering algorithm and an immune technology particle swarm optimization algorithm, to define the best locations for measurements stations. The methodology was applied to a real urban areas, showing that 28 optimization measurement points could replace the original 100 grid noise survey points.

In the more general context of sensor networks for environmental monitoring, Reis et al. [105] suggest that models can also be used to optimize the deployment of sensors by identifying areas of interest according to specific metrics. Applied to the noise monitoring field, noise modelling could be used, for example, to detect whether locations in a set of potential measurement points are highly correlated, which in this case would reduce the number of measurement points.

Beyond the metric itself associated with the measurement (i.e., the acoustic measurement in the present case), other elements can also be considered in the deployment and of the optimization of the whole network, such as the connectivity between node/relay/sink, the spatial domain coverage or the network life and energy efficiency [106,107].

3.6.4. Considering Hybrid Networks

Accessing to complementary noise data, that is, data produced by other sensors, can be useful to increase the relevancy of end-user applications.

A low-cost SN can used for example to complement an existing professional quality network (i.e., using Class-1 measurement systems). Data can be shared within the same database and used simultaneously to further evaluate sound environments. This is, for example, the choice of the city of Barcelona in Spain [71]. This makes it possible to extend the spatial coverage of observation at a lower cost. Provided that the data from two networks are in a compatible format, the processing and analysis of the data from the low-cost sensor network can thus take advantage of the existing tools for the professional network, which again limits the cost of the investment.

Merging noise data produced by smartphones in the framework of a crowd-sourcing noise approach [108], with data obtained using a low-cost SN, can increase the quality of the soundscape evaluation. In opposition to 'static' sensors, smartphones can be seen as 'mobile' sensors. In a complementary way, noise sensors located on cars, bus or bicycles [109] are also considered as mobile sensors. Such sensors could be used to increase the relevance of the sound environment database and to reduce the cost of the network. However, considering both mobile and static sensors in a noise network can introduce significant challenges in terms of mobile sensors detection/localization and data transfer to a static gateway [110].

This hybrid network approach can also be extended to networks dealing with multiple pollution. Noise sensors, at the lower cost of environmental pollutant sensors, can then be used as a proxy to estimate airborne pollutant concentrations or number of fine or ultra-fine particles with a limited number of sensors [111,112]. One of the ambitions of these multidimensional treatments is to establish possible confounders in the characterization of the health impacts of noise and air pollutants [113], or noise and fine and ultra-fine particles [114], as expected by epidemiologists. The use of acoustic indicators as proxies for estimating air pollution quantities, motivated by the lower cost of acoustic sensors, faces however many obstacles [115], including a different dispersion behavior that leads to very variable correlation coefficients.

4. Conclusions

Given the major problem of evaluating and controlling sound environments, the development of low-cost sensor networks is today an interesting alternative solution, complementary, to more traditional solutions such as modelling and "professional" observation networks. Numerous researchers have thus focused on the development of low-cost sensors over the last fifteen years, ranging from proof-of-concept to the deployment of operational networks [29].

From a technical point of view, Table 1 illustrates fairly well the evolution of low-cost sensors, from the adaptation of existing sensors (but with limited resources) to the use of mini-PCs and MCUs (with more extensive computing and measurement capabilities). If the sensors can be directly powered by an existing electrical network, mini-PCs are the most relevant solution up-today, especially in view of the modularity and real-time processing capabilities they offer. For stand-alone sensors, most recent MCUs offer interesting performances, but their overall capacities remain very dependent on their power supply and recharging mode. From an acoustic measurement point of view, the use of a digital MEMS with a sampling frequency of 44.1 kHz now seems to be a technically affordable solution, not very sensitive to electrical and electromagnetic interference, that meets the challenges of noise monitoring. Among the possible technical evolution, the development of sensors composed of several microphones would offer new perspectives for the localization and the tracking of sound sources, as well as for measuring 3D audio [30,68,116]. Concerning radio data transmission, Zigbee and 6LoWPAN protocols, based on 802.15.4 specification, present both maximum range and data rates that are compatible with noise measurements. Setting aside the problems of sealing against weather and pollutants, as well as the mechanical protection of the sensor, which can be solved by integrating the electronic components in a specially designed container, the service life of the electronic components, including memory and battery, is now potentially fully compatible with long-term acoustic measurement.

The individual cost of a sensor must be put in relation to the overall cost of an infrastructure consisting of a very large number of sensors [93], potentially requiring a high level of maintenance. The question of the best location of sensors is therefore an important issue for the future. In addition, the automatic detection of anomalies in the network, whether to identify a hardware malfunction or an abnormal set of data, are also subjects that will have to be addressed to improve the database quality. The multitude of such data also raises the question of developing appropriate data infrastructures for their representation and processing [35,48].

There are many opportunities that enhance the value of these measurement networks and the collected data. Environmental services of cities can, for example, use data to dynamically adapt their policies, since they are able to measure directly the effects of the policies tested. Another example is that local residents associations, with the help of specialized services, can understand the environmental quality of their neighborhood and use it to alert the authorities or become a source of proposals. To do this, it is also important that current networks be enriched with perceptive data, in order to better describe the impacts of noise on citizens.

Author Contributions: writing–original draft preparation, J.P.; writing–review and editing, J.P., A.C., N.F., J.A., M.L.; supervision, J.P.; project administration, J.P.; funding acquisition, J.P., A.C. All authors have read and agreed to the published version of the manuscript.

Funding: This research was funded by the french National Agency for Research (Agence Nationale de la Recherche) grant number ANR-16-CE22-0012.

Conflicts of Interest: The authors declare no conflict of interest.

Abbreviations

The following abbreviations are used in this manuscript:

ACI	Acoustic complexity index
ADC	Analog-to-digital converter
AOP	Acoustic overload point
ASIC	Application specific integrated circuit
CPU	Central processing unit
ECM	Electret condenser microphone
EIN	Equivalent input noise
GSM	Global system for mobile communications
FFT	Fast Fourier transformation
FGPA	Floating-point-gate-array
HD	Hard disk
I2S	Integrated interchip sound
IT	Information technology
IoT	Internet of things
LAN	Local area network
MCU	Microcontroller unit
MEMS	Micro-electrical-mechanical systems
NPL	National Physical Laboratory
NSN	Noise sensors networks
OS	Operating system
PC	Personal computer
PCB	Printed circuit board
PDM	Pulse density modulated
PLC	Power-line communication
POC	Proof-of-concept
POE	Power over Ethernet
PSR	Power supply rejection
PSRR	Power supply rejection ratio

RF	Radio-frequency
R-Pi	Raspberry Pi
SD	Secure digital
SIM	Subscriber identity module
SN	Sensors networks
SNR	Signal-to-noise ratio
SSD	Solid-state drive
TFSD	Time and frequency second derivative
THD	Total harmonic distortion
USB	Universal serial bus
Wi-Fi	Wireless fidelity
WASN	Wireless acoustic sensor network

References

1. WHO Regional Office for Europe. *Environmental Noise Guidelines for the European Region*; World Health Organization Regional Office for Europe: Copenhagen, Denmark, 2018.
2. Cox, P.; Palou, J. Directive 2002/49/EC of the European Parliament and of the Council of 25 June 2002 relating to the assessment and management of environmental noise—Declaration by the Commission in the Conciliation Committee on the Directive relating to the assessment and management of environmental noise. *Annex I OJ* **2002**, *189*, 2002.
3. Morel, J.; Marquis-Favre, C.; Dubois, D.; Pierrette, M. Road Traffic in Urban Areas: A Perceptual and Cognitive Typology of Pass-By Noises. *Acta Acustica Acustica* **2012**, *98*, 166–178. [CrossRef]
4. Gille, L.A.; Marquis-Favre, C.; Klein, A. Noise Annoyance Due To Urban Road Traffic with Powered-Two-Wheelers: Quiet Periods, Order and Number of Vehicles. *Acta Acustica Acustica* **2016**, *102*, 474–487. [CrossRef]
5. Aumond, P.; Can, A.; Mallet, V.; De Coensel, B.; Ribeiro, C.; Botteldooren, D.; Lavandier, C. Kriging-based spatial interpolation from measurements for sound level mapping in urban areas. *J. Acoust. Soc. Am.* **2018**, *5*, 2847–2857. [CrossRef] [PubMed]
6. Mietlicki, F.; Mietlicki, C.; Sineau, M. An Innovative Approach for long term environmental noise measurement: RUMEUR Network in the Paris Region. In Proceedings of the 10th European Congress and Exposition on Noise Control Engineering, Maastricht, The Netherlands, 31 May–3 June 2015.
7. International Electrotechnical Commission Electroacoustics—Sound Level Meters—Part 1: Specifications (IEC 61672-1). 2013. Available online: https://webstore.iec.ch/publication/5708 (accessed on 14 April 2020).
8. Hart, J.K.; Martinez, K. Environmental Sensor Networks: A revolution in the earth system science? *Earth Sci. Rev.* **2006**, *78*, 177–191. [CrossRef]
9. Barrenetxea, G.; Ingelrest, F.; Schaefer, G.; Vetterli, M. Wireless Sensor Networks for Environmental Monitoring: The SensorScope Experience. In Proceedings of the 2008 IEEE International Zurich Seminar on Communications, Zurich, Switzerland, 12–14 March 2008; pp. 98–101.
10. Rashid, B.; Rehmani, M.H. Applications of wireless sensor networks for urban areas: A survey. *J. Netw. Comput. Appl.* **2016**, *60*, 192–219. [CrossRef]
11. Domínguez, F.; Dauwe, S.; Cuong, N.T.; Cariolaro, D.; Touhafi, A.; Dhoedt, B.; Botteldooren, D.; Steenhaut, K. Towards an Environmental Measurement Cloud: Delivering Pollution Awareness to the Public. *Int. J. Distrib. Sens. Netw.* **2014**, *10*, 541360. [CrossRef]
12. McDonald, P.; Geraghty, D.; Humphreys, I.; Farrell, S. Assessing Environmental Impact of Transport Noise with Wireless Sensor Networks. *Transp. Res. Rec.* **2008**, *2058*, 133–139. [CrossRef]
13. Bell, M.C.; Galatioto, F. Novel wireless pervasive sensor network to improve the understanding of noise in street canyons. *Appl. Acoust.* **2013**, *74*, 169–180. [CrossRef]
14. Wang, C.; Chen, G.; Dong, R.; Wang, H. Traffic noise monitoring and simulation research in Xiamen City based on the Environmental Internet of Things. *Int. J. Sustain. Dev. World Ecol.* **2013**, *20*, 248–253. [CrossRef]
15. Manvell, D.; Ballarin Marcos, L.; Stapelfeldt, H.; Sanz, R. Sadmam – Combining measurements and calculations to map noise in Madrid. In Proceedings of the 33rd International Congress and Exposition on Noise Control Engineering, Prague, Czech Republic, 22–26 August 2004.

16. Murphy, E.; King, E.A. Smartphone-based noise mapping: Integrating sound level meter app data into the strategic noise mapping process. *Sci. Total Environ.* **2016**, *562*, 852–859. [CrossRef] [PubMed]
17. Bellucci, P.; Peruzzi, L.; Zambon, G. LIFE DYNAMAP project: The case study of Rome. *Appl. Acoust.* **2017**, *117 Part B*, 193–206. [CrossRef]
18. Ventura, R.; Mallet, V.; Issarny, V. Assimilation of mobile phone measurements for noise mapping of a neighborhood. *J. Acoust. Soc. Am.* **2018**, *144*, 1279–1292. [CrossRef] [PubMed]
19. Benocci, R.; Confalonieri, C.; Roman, H.E.; Angelini, F.; Zambon, G. Accuracy of the Dynamic Acoustic Map in a Large City Generated by Fixed Monitoring Units. *Sensors* **2020**, *20*, 412. [CrossRef] [PubMed]
20. Wei, W.; Van Renterghem, T.; De Coensel, B.; Botteldooren, D. Dynamic noise mapping: A map-based interpolation between noise measurements with high temporal resolution. *Appl. Acoust.* **2016**, *101*, 127–140. [CrossRef]
21. Botteldooren, D.; De Coensel, B.; Oldoni, D.; Renterghem, T.; Dauwe, S. Sound monitoring networks new style. In Proceedings of the Acoustics, Gold Coast, Australia, 2–4 November 2011.
22. Botteldooren, D.; Van renterghem, T.; Oldoni, D.; Samuel, D.; Dekoninck, L.; Thomas, P.; Wei, W.; Boes, M.; De Coensel, B.; De Baets, B.; et al. The internet of sound observatories. *Proc. Mtgs. Acoust.* **2013**, *19*, 040140.
23. Oldoni, D.; De Coensel, B.; Boes, M.; Rademaker, M.; De Baets, B.; Van Renterghem, T.; Botteldooren, D. A computational model of auditory attention for use in soundscape research. *J. Acoust. Soc. Am.* **2013**, *134*, 852–861. [CrossRef]
24. Salamon, J.; Bello, J.P. Unsupervised feature learning for urban sound classification. In Proceedings of the 2015 IEEE International Conference on Acoustics, Speech and Signal Processing (ICASSP), Brisbane, QLD, Australia, 19–24 April 2015.
25. Socoró, J.C.; Alías, F.; Alsina-Pagès, R.M. An Anomalous Noise Events Detector for Dynamic Road Traffic Noise Mapping in Real-Life Urban and Suburban Environments. *Sensors* **2017**, *17*, 2323. [CrossRef] [PubMed]
26. Gontier, F.; Lagrange, M.; Aumond, P.; Can, A.; Lavandier, C. An Efficient Audio Coding Scheme for Quantitative and Qualitative Large Scale Acoustic Monitoring Using the Sensor Grid Approach. *Sensors* **2017**, *17*, 2758. [CrossRef]
27. Offenhuber, D.; Auinger, S.; Seitinger, S.; Muijs, R. Los Angeles noise array—Planning and design lessons from a noise sensing network. *Environ. Plan. B Urban Anal. City Sci.* **2018**. [CrossRef]
28. Gloaguen, J.R.; Can, A.; Lagrange, M.; Petiot, J.F. Road traffic sound level estimation from realistic urban sound mixtures by Non-negative Matrix Factorization. *Appl. Acoust.* **2019**, *143*, 229–238. [CrossRef]
29. Bello, J.P.; Silva, C.; Nov, O.; Dubois, R.L.; Arora, A.; Salamon, J.; Mydlarz, C.; Doraiswamy, H. SONYC: A System for Monitoring, Analyzing, and Mitigating Urban Noise Pollution. *Commun. ACM* **2019**, *62*, 68–77. [CrossRef]
30. Faraji, M.M.; Shouraki, S.B.; Iranmehr, E.; Linares-Barranco, B. Sound Source Localization in Wide-range Outdoor Environment Using Distributed Sensor Network. *IEEE Sens. J.* **2019**, *1*. [CrossRef]
31. Luo, L.; Cao, Q.; Huang, C.; Wang, L.; Abdelzaher, T.F.; Stankovic, J.A.; Ward, M. Design, Implementation, and Evaluation of EnviroMic: A Storage-centric Audio Sensor Network. *ACM Trans. Sen. Netw.* **2009**, *5*, 22:1–22:35. [CrossRef]
32. Sethi, S.S.; Ewers, R.M.; Jones, N.S.; Orme, C.D.L.; Picinali, L. Robust, real—time and autonomous monitoring of ecosystems with an open, low-cost, networked device. *Methods Ecol. Evol.* **2018**, *9*, 2383–2387. [CrossRef]
33. Sheng, Z.; Pfersich, S.; Eldridge, A.; Zhou, J.; Tian, D.; Leung, V.C.M. Wireless acoustic sensor networks and edge computing for rapid acoustic monitoring. *IEEE J. Autom. Sin.* **2019**, *6*, 64–74. [CrossRef]
34. Dauwe, S.; Van Renterghem, T.; Botteldooren, D.; Dhoedt, B. Multiagent-Based Data Fusion in Environmental Monitoring Networks. *Int. J. Distrib. Sens. Netw.* **2012**, *8*, 324935. [CrossRef]
35. Mydlarz, C.; Sharma, M.; Lockerman, Y.; Steers, B.; Silva, C.; Bello, J.P. The Life of a New York City Noise Sensor Network. *Sensors* **2019**, *19*, 1415. [CrossRef]
36. Santini, S.; Ostermaier, B.; Adelmann, R. On the Use of Sensor Nodes and Mobile Phones for the Assessment of Noise Pollution Levels in Urban Environments. In Proceedings of the 6th International Conference on Networked Sensing Systems, Pittsburgh, PA, USA, 17–19 June 2009.
37. Alsina-Pagès, R.M.; Hernandez-Jayo, U.; Alías, F.; Angulo, I. Design of a Mobile Low-Cost Sensor Network Using Urban Buses for Real-Time Ubiquitous Noise Monitoring. *Sensors* **2016**, *17*, 57. [CrossRef]

38. Risojević, V.; Rozman, R.; Pilipović, R.; Češnovar, R.; Bulić, P. Accurate Indoor Sound Level Measurement on a Low-Power and Low-Cost Wireless Sensor Node. *Sensors* **2018**, *18*, 2351. [CrossRef]
39. Alías, F.; Alsina-Pagès, R.M. Review of Wireless Acoustic Sensor Networks for Environmental Noise Monitoring in Smart Cities. *J. Sens.* **2019**, *13*. [CrossRef]
40. McGrath, M.J.; Scanaill, C.N. Sensor Network Topologies and Design Considerations. In *Sensor Technologies: Healthcare, Wellness, and Environmental Applications*; McGrath, M.J., Scanaill, C.N., Eds.; Apress: Berkeley, CA, USA, 2013; pp. 79–95.
41. Guillaume, G.; Can, A.; Petit, G.; Fortin, N.; Palominos, S.; Gauvreau, B.; Bocher, E.; Picaut, J. Noise mapping based on participative measurements. *Noise Mapp.* **2016**, *3*, 140–156. [CrossRef]
42. Santini, S.; Ostermaier, B.; Vitaletti, A. First Experiences Using Wireless Sensor Networks for Noise Pollution Monitoring. In Proceedings of the Workshop on Real-World Wireless Sensor Networks, Glasgow, Scotland, 1 April 2008.
43. Santini, S.; Vitaletti, A. Wireless sensor networks for environmental noise monitoring. *Fachgespräch Sensornetzwerke* **2007**, *3*, 98.
44. Filipponi, L.; Santini, S.; Vitaletti, A. Data Collection in Wireless Sensor Networks for Noise Pollution Monitoring. In Proceedings of the 4th IEEE International Conference on Distributed Computing in Sensor Systems, Santorini Island, Greece, 11–14 June 2008.
45. Barham, R.; Goldsmith, M. Performance of a new MEMS measurement microphone and its potential application. In Proceedings of the Institute of Acoustics, Spring Conference, Reading, UK, 10–11 April 2008; pp. 370–377.
46. Barham, R.; Goldsmith, M.; Chan, M.; Simmons, D. Development and performance of a multi-point distributed environmental noise measurement system using MEMS microphones. In Proceedings of the 8th European Conference and Exhibition on Noise Control, Euronoise 2009, Edinburgh, Scotland, UK, 26–28 October 2009; pp. 1039–1046.
47. Barham, R.; Chan, M.; Cand, M. Practical experience in noise mapping with a MEMS microphone based distributed noise measurement system. *INTER-NOISE NOISE-CON Congr. Conf. Proc.* **2010**, *2010*, 4725–4733.
48. Piper, B.; Barham, R.; Sheridan, S.; Sotirakopoulos, K. Exploring the 'big acoustic data' generated by an acoustic sensor network deployed at a crossrail construction site. In Proceedings of the 24th International Congress on Sound and Vibration (ICSV24), London, UK, 23–27 July 2017.
49. Hakala, I.; Kivelä, I.; Ihalainen, J.; Luomala, J.; Gao, C. Design of Low-Cost Noise Measurement Sensor Network: Sensor Function Design. In Proceedings of the 2010 First International Conference on Sensor Device Technologies and Applications, Venice/Mestre, Italy, 18–25 July 2010; pp. 172–179.
50. Hakala, I.; Tikkakoski, M.; Kivelä, I. Wireless Sensor Network in Environmental Monitoring—Case Foxhouse. In Proceedings of the 2008 Second International Conference on Sensor Technologies and Applications (sensorcomm 2008), Cap Esterel, France, 25–31 August 2008; pp. 202–208.
51. Kivelä, I.; Gao, C.; Luomala, J.; Ihalainen, J.; Hakala, I. Design of Networked Low-Cost Wireless Noise Measurement Sensors. *Int. J. Sens. Transducers* **2011**, *10*, 171–190.
52. Kivelä, I.; Gao, C.; Luomala, J.; Hakala, I. Design of noise measurement sensor network: Networking and communication part. In Proceedings of the 5th International Conference on Sensor Technologies and Applications, Nice/Saint Laurent du Var, France, 21–27 August 2011.
53. Kivelä, I.; Hakala, I. Area-based environmental noise measurements with a wireless sensor network. In Proceedings of the 10th European Congress and Exposition on Noise Control Engineering, Maastricht, The Netherlands, 31 May–3 June 2015; pp. 218–220.
54. Tan, W.M.; Jarvis, S.A. Energy harvesting noise pollution sensing WSN mote: Survey of capabilities and limitations. In Proceedings of the 2013 IEEE Conference on Wireless Sensor (ICWISE), Kuching, Sarawak, Malaysia, 2–4 December 2013; pp. 53–60.
55. Polastre, J.; Szewczyk, R.; Culler, D. Telos: Enabling ultra-low power wireless research. In Proceedings of the Fourth International Symposium on Information Processing in Sensor Networks, Los Angeles, CA, USA, 25–27 April 2005; pp. 364–369.
56. Tan, W.M.; Jarvis, S.A. On the design of an energy-harvesting noise-sensing WSN mote. *EURASIP J. Wirel. Commun. Netw.* **2014**, *2014*, 167. [CrossRef]

57. Mariscal-Ramirez, J.A.; Fernandez-Prieto, J.A.; Gadeo-Martos, M.A.; Canada-Bago, J. Knowledge-based wireless sensors using sound pressure level for noise pollution monitoring. In Proceedings of the 11th International Conference on Intelligent Systems Design and Applications, Córdoba, Spain, 22–24 November 2011; pp. 1032–1037.
58. Mariscal-Ramirez, J.A.; Fernandez-Prieto, J.A.; Canada-Bago, J.; Gadeo-Martos, M.A. A New Algorithm to Monitor Noise Pollution Adapted to Resource-constrained Devices. *Multimedia Tools Appl.* **2015**, *74*, 9175–9189. [CrossRef]
59. Segura-Garcia, J.; Felici-Castell, S.; Perez-Solano, J.J.; Cobos, M.; Navarro, J.M. Low-Cost Alternatives for Urban Noise Nuisance Monitoring Using Wireless Sensor Networks. *IEEE Sens. J.* **2015**, *15*, 836–844. [CrossRef]
60. Segura-Garcia, J.; Pérez-Solano, J.J.; Cobos-Serrano, M.; Navarro-Camba, E.A.; Felici-Castell, S.; Soriano-Asensi, A.; Montes-Suay, F. Spatial Statistical Analysis of Urban Noise Data from a WASN Gathered by an IoT System: Application to a Small City. *Appl. Sci.* **2016**, *6*, 380. [CrossRef]
61. Noriega-Linares, J.E.; Navarro Ruiz, J.M. On the Application of the Raspberry Pi as an Advanced Acoustic Sensor Network for Noise Monitoring. *Electronics* **2016**, *5*, 74. [CrossRef]
62. Mydlarz, C.; Salamon, J.; Bello, J.P. The implementation of low-cost urban acoustic monitoring devices. *Appl. Acoust.* **2017**, *117*, 207–218. [CrossRef]
63. Peckens, C.; Porter, C.; Rink, T. Wireless Sensor Networks for Long-Term Monitoring of Urban Noise. *Sensors* **2018**, *18*, 3161. [CrossRef]
64. Bartalucci, C.; Borchi, F.; Carfagni, M.; Furferi, R.; Governi, L.; Lapini, A.; Volpe, Y.; Curcuruto, S.; Mazzocchi, E.; Marsico, G.; et al. LIFE MONZA: Project description and actions' updating. *Noise Mapp.* **2018**, *5*, 60–70. [CrossRef]
65. Bartalucci, C.; Borchi, F.; Carfagni, M.; Furferi, R.; Governi, L.; Silvaggio, R.; Curcuruto, S.; Nencini, L. Design of a prototype of a smart noise monitoring system. In Proceedings of the 24th International Congress on Sound and Vibration (ICSV24), London, UK, 23–27 July 2017; pp. 23–27.
66. Silvaggio, R.; Curcuruto, S.; Bellomini, R.; Luzzi, S.; Borchi, F.; Bartalucci, C. Noise Low Emission Zone implementation in urban planning: results of monitoring activities in pilot area of LIFE MONZA project. In Proceedings of the 2019 International Congress on Acoustics (ICA), Aachen, Germany, 8–13 September 2019.
67. Ardouin, J.; Charpentier, L.; Lagrange, M.; Gontier, F.; Fortin, N.; Écotière, D.; Picaut, J.; Mietlicki, F. An innovative low cost sensors for urban sound monitoring. In Proceeding of the Inter-Noise 2018, Chicago, IL, USA, 26–29 August 2018.
68. López, J.M.; Alonso, J.; Asensio, C.; Pavón, I.; Gascó, L.; de Arcas, G. A Digital Signal Processor Based Acoustic Sensor for Outdoor Noise Monitoring in Smart Cities. *Sensors* **2020**, *20*, 605. [CrossRef]
69. Tan, Q.; Liu, X.; Chen, X.; Yu, D. Digital Environmental Noise Monitoring System Based on B/S Architecture and Floating-Point DSP. In Proceedings of the 2nd International Congress on Image and Signal Processing, Tianjin, China, 17–19 October 2009; pp. 1–5.
70. Blythe, P.; Neasham, J.; Sharif, B.; Watson, P.; Bell, M.C.; Edwards, S.; Suresh, V.; Wagner, J.; Bryan, H. An environmental sensor system for pervasively monitoring road networks. In Proceedings of the IET Road Transport Information and Control Conference and the ITS United Kingdom Members' Conference (RTIC 2008), Manchester, UK, 20–22 May 2008; p. 91.
71. Farrés, J.C. Barcelona noise monitoring network. In Proceedings of the 10th European Congress and Exposition on Noise Control Engineering, Maastricht, The Netherlands, 1–3 June 2015; pp. 2315–2320.
72. Sevillano, X.; Socoró, J.C.; Alías, F.; Bellucci, P.; Peruzzi, L.; Radaelli, S.; Coppi, P.; Nencini, L.; Cerniglia, A.; Bisceglie, A.; et al. DYNAMAP – Development of low cost sensors networks for real time noise mapping. *Noise Mapp.* **2016**, *3*. [CrossRef]
73. Renterghem, T.V.; Thomas, P.; Dominguez, F.; Dauwe, S.; Touhafi, A.; Dhoedt, B.; Botteldooren, D. On the ability of consumer electronics microphones for environmental noise monitoring. *J. Environ. Monit.* **2011**, *13*, 544–552. [CrossRef]
74. Widder, J. Basic Principles of MEMS Microphones. 2014. Available online: https://www.edn.com/basic-principles-of-mems-microphones/ (accessed on 14 April 2020).

75. STMicroelectronics. AN4427—Application Note—Gasket Design for Optimal Acoustic Performance in MEMS Microphones. Available online: https://www.st.com/content/ccc/resource/technical/document/application_note/e9/86/75/b2/8e/fd/48/69/DM00103201.pdf/files/DM00103201.pdf/jcr:content/translations/en.DM00103201.pdf (accessed on 13 November 2019).
76. Nencini, L.; Bellucci, P.; Peruzzi, L. Identification of failure markers in noise measurement low cost devices. In Proceedings of the 45th International Congress and Exposition of Noise Control Engineering, Inter-Noise, Hamburg, Germany, 21–24 August 2016; pp. 6362–6369.
77. Shah, M.A.; Shah, I.A.; Lee, D.G.; Hur, S. Design Approaches of MEMS Microphones for Enhanced Performance. *J. Sens.* **2019**, *2019*, 26. [CrossRef]
78. Gubbi, J.; Marusic, S.; Rao, A.S.; Law, Y.W.; Palaniswami, M. A pilot study of urban noise monitoring architecture using wireless sensor networks. In Proceedings of the 2013 International Conference on Advances in Computing, Communications and Informatics (ICACCI), Mysore, India, 22–25 August 2013.
79. Chen, H.; Jin, H.; Guo, L.; Wu, S.; Gu, T. Audio-on-demand over wireless sensor networks. In Proceedings of the 2012 IEEE 20th International Workshop on Quality of Service, Coimbra, Portugal, 4–5 June 2012.
80. Bartalucci, C.; Borchi, F.; Carfagni, M.; Furferi, R.; Governi, L.; Lapini, A.; Bellomini, R.; Luzzi, S.; Nencini, L. The smart noise monitoring system implemented in the frame of the Life MONZA project. In Proceedings of the 11th European Congress and Exposition on Noise Control Engineering, Heraklion, Crete, Greece, 27–31 May 2018; pp. 27–31.
81. Li, J.; Broas, M.; Raami, J.; Mattila, T.T.; Paulasto-Kröckel, M. Reliability assessment of a MEMS microphone under mixed flowing gas environment and shock impact loading. *Microelectron. Reliab.* **2014**, *54*, 1228–1234. [CrossRef]
82. Aumond, P.; Can, A.; De Coensel, B.; Ribeiro, C.; Botteldooren, D.; Lavandier, C. Modeling soundscape pleasantness using perceptual assessments and acoustic measurements along paths in urban context. *Acta Acustica Acustica* **2017**, *103*, 430–443. [CrossRef]
83. Raspberry Pi 4 Specs and Benchmarks. Available online: https://magpi.raspberrypi.org/articles/raspberry-pi-4-specs-benchmarks (accessed on 4 December 2019).
84. Lewis, J. Analog and Digital MEMS Microphone Design Considerations. 2013. Available online: https://www.analog.com/media/en/technical-documentation/technical-articles/Analog-and-Digital-MEMS-Microphone-Design-Considerations-MS-2472.pdf (accessed on 4 December 2019).
85. Bluetooth Technology Website. Available online: https://www.bluetooth.com/ (accessed on 28 February 2020).
86. Wi-Fi Alliance. Available online: https://www.wi-fi.org/ (accessed on 28 February 2020).
87. Zigbee Alliance. Available online: https://zigbeealliance.org/ (accessed on 28 February 2020).
88. LoRa Alliance. Available online: https://lora-alliance.org/ (accessed on 28 February 2020).
89. Sigfox—The Global Communications Service Provider for the Internet of Things (IoT). Available online: https://www.sigfox.com/en (accessed on 28 February 2020).
90. Guide: Lifetime Cost of Ownership for Class 1 Sound Level Meters. Available online: https://blog.bksv.com/guide-lifetime-cost-of-ownership-for-class-1-sound-level-meters (accessed on 28 February 2020).
91. Jordan, D.C.; Kurtz, S.R. Photovoltaic Degradation Rates-an Analytical Review: Photovoltaic degradation rates. *Prog. Photovoltaics Res. Appl.* **2013**, *21*, 12–29. [CrossRef]
92. Delaine, F.; Lebental, B.; Rivano, H. In Situ Calibration Algorithms for Environmental Sensor Networks: A Review. *IEEE Sens. J.* **2019**, *19*, 5968–5978. [CrossRef]
93. Farrés, J.C.; Novas, J.C. Issues and challenges to improve the Barcelona Noise Monitoring Network. In Proceedings of the 11th European Congress and Exposition on Noise Control Engineering, Heraklion, Crete, Greece, 27–31 May 2018.
94. Dauwe, S.; Oldoni, D.; Baets, B.D.; Renterghem, T.V.; Botteldooren, D.; Dhoedt, B. Multi-criteria anomaly detection in urban noise sensor networks. *Environ. Sci. Process. Impacts* **2014**, *16*, 2249–2258. [CrossRef] [PubMed]
95. Laso, P.M.; Brosset, D.; Puentes, J. Analysis of quality measurements to categorize anomalies in sensor systems. In Proceedings of the 2017 Computing Conference, London, UK, 18–20 July 2017.
96. Elsayed, W.; Elhoseny, M.; Sabbeh, S.; Riad, A. Self-maintenance model for Wireless Sensor Networks. *Comput. Electr. Eng.* **2018**, *70*, 799–812. [CrossRef]

97. Cheng, Y.; Liu, Q.; Wang, J.; Wan, S.; Umer, T. Distributed Fault Detection for Wireless Sensor Networks Based on Support Vector Regression. *Wirel. Commun. Mob. Comput.* **2018**, *2018*, 8. [CrossRef]
98. Can, A.; Aumond, P.; De Coensel, B.; Ribeiro, C.; Botteldooren, D.; Lavandier, C. Probabilistic Modelling of the Temporal Variability of Urban Sound Levels. *Acta Acustica Acustica* **2018**, *104*, 94–105. [CrossRef]
99. Barrigón Morillas, J.M.; Gómez Escobar, V.; Méndez Sierra, J.A.; Vílchez Gómez, R.; Trujillo Carmona, J. An environmental noise study in the city of Cáceres, Spain. *Appl. Acoust.* **2002**, *63*, 1061–1070. [CrossRef]
100. Can, A.; Dekoninck, L.; Botteldooren, D. Measurement network for urban noise assessment: Comparison of mobile measurements and spatial interpolation approaches. *Appl. Acoust.* **2014**, *83*, 32–39. [CrossRef]
101. Gozalo, G.R.; Morillas, J.M.B.; Escobar, V.G.; Vilchez-Gómez, R.; Sierra, J.A.M.; Rio, F.J.C.D.; Gajardo, C.P. Study of the Categorisation Method Using Long-term Measurements. *Arch. Acoust.* **2013**, *38*, 397–405. [CrossRef]
102. Liu, J.; Kang, J.; Luo, T.; Behm, H.; Coppack, T. Spatiotemporal variability of soundscapes in a multiple functional urban area. *Landsc. Urban Plan.* **2013**, *115*, 1–9. [CrossRef]
103. Zuo, F.; Li, Y.; Johnson, S.; Johnson, J.; Varughese, S.; Copes, R.; Liu, F.; Wu, H.J.; Hou, R.; Chen, H. Temporal and spatial variability of traffic-related noise in the City of Toronto, Canada. *Sci. Total Environ.* **2014**, *472*, 1100–1107. [CrossRef]
104. Huang, B.; Pan, Z.; Yang, H.; Hou, G.; Wei, W. Optimizing stations location for urban noise continuous intelligent monitoring. *Appl. Acoust.* **2017**, *127*, 250–259. [CrossRef]
105. Reis, S.; Seto, E.; Northcross, A.; Quinn, N.W.T.; Convertino, M.; Jones, R.L.; Maier, H.R.; Schlink, U.; Steinle, S.; Vieno, M.; et al. Integrating modelling and smart sensors for environmental and human health. *Environ. Modell. Softw.* **2015**, *74*, 238–246. [CrossRef] [PubMed]
106. Abdollahzadeh, S.; Navimipour, N.J. Deployment strategies in the wireless sensor network: A comprehensive review. *Comput. Commun.* **2016**, *91–92*, 1–16. [CrossRef]
107. Ketshabetswe, L.K.; Zungeru, A.M.; Mangwala, M.; Chuma, J.M.; Sigweni, B. Communication protocols for wireless sensor networks: A survey and comparison. *Heliyon* **2019**, *5*, e01591. [CrossRef]
108. Picaut, J.; Fortin, N.; Bocher, E.; Petit, G.; Aumond, P.; Guillaume, G. An open-science crowdsourcing approach for producing community noise maps using smartphones. *Build. Environ.* **2019**, *148*, 20–33. [CrossRef]
109. Bennett, G.; King, E.A.; Curn, J.; Cahill, V.; Bustamante, F.; Rice, H.J. Environmental noise mapping using measurements in transit. In Proceedings of the ISMA 2010, Leuven, Belgium, 20–22 September 2010.
110. Di Francesco, M.; Das, S.K.; Anastasi, G. Data Collection in Wireless Sensor Networks with Mobile Elements: A Survey. *ACM Trans. Sen. Netw.* **2011**, *8*, 7:1–7:31. [CrossRef]
111. Can, A.; Dekoninck, L.; Rademaker, M.; Van Renterghem, T.; De Baets, B.; Botteldooren, D. Noise measurements as proxies for traffic parameters in monitoring networks. *Sci. Total Environ.* **2011**, *410–411*, 198–204. [CrossRef]
112. Can, A.; Rademaker, M.; Van Renterghem, T.; Mishra, V.; Van Poppel, M.; Touhafi, A.; Theunis, J.; De Baets, B.; Botteldooren, D. Correlation analysis of noise and ultrafine particle counts in a street canyon. *Sci. Total Environ.* **2011**, *409*, 564–572. [CrossRef]
113. Ross, Z.; Kheirbek, I.; Clougherty, J.E.; Ito, K.; Matte, T.; Markowitz, S.; Eisl, H. Noise, air pollutants and traffic: Continuous measurement and correlation at a high-traffic location in New York City. *Environ. Res.* **2011**, *111*, 1054–1063. [CrossRef]
114. Weber, S. Spatio-temporal covariation of urban particle number concentration and ambient noise. *Atmos. Environ.* **2009**, *43*, 5518–5525. [CrossRef]
115. Khan, J.; Ketzel, M.; Kakosimos, K.; Sørensen, M.; Jensen, S.S. Road traffic air and noise pollution exposure assessment—A review of tools and techniques. *Sci. Total Environ.* **2018**, *634*, 661–676. [CrossRef] [PubMed]
116. Zalles, G.; Kamel, Y.; Anderson, I.; Lee, M.; Neil, C.; Henry, M.; Cappiello, S.; Mydlarz, C.; Baglione, M.; Roginska, A. A Low-Cost High-Quality MEMS Ambisonic Microphone. *Audio Eng. Soc.* **2017**, *143*, 9857.

Article

Wireless Acoustic Sensor Nodes for Noise Monitoring in the City of Linares (Jaén)

Jose-Angel Fernandez-Prieto *, Joaquín Cañada-Bago and Manuel-Angel Gadeo-Martos

Telematic Engineering System Research Group, CEATIC Center of Advanced Studies in Information and Communication Technologies, University of Jaén, Campus Científico-Tecnológico de Linares, C.P. 23700 Linares, Spain; jcbago@ujaen.es (J.C.-B.); gadeo@ujaen.es (M.-A.G.-M.)

* Correspondence: jan@ujaen.es

Received: 25 October 2019; Accepted: 21 December 2019; Published: 24 December 2019

Abstract: Noise pollution is a problem that affects millions of people worldwide. Over the last few years, many researchers have devoted their attention to the design of wireless acoustic sensor networks (WASNs) to monitor the real data of continuous and precise noise levels and to create noise maps in real time and space. Although WASNs are becoming a reality in smart cities, some research studies argue that very few projects have been deployed around the world, with most of them deployed as pilots for only days or weeks, with a small number of nodes. In this paper, we describe the design and implementation of a complete system for a WASN deployed in the city of Linares (Jaén), Spain, which has been running continuously for ten months. The complete system covers the network topology design, hardware and software of the sensor nodes, protocols, and a private cloud web server platform. As a result, the information provided by the system for each location where the sensor nodes are deployed is as follows: L_{Aeq} for a given period of time; noise indicators L_{den}, L_{day}, $L_{evening}$, and L_{night}; percentile noise levels (L_{A01T}, L_{A10T}, L_{A50T}, L_{A90T}, and L_{A99T}); a temporal evolution representation of noise levels; and the predominant frequency of the noise. Some comparisons have been made between the noise indicators calculated by the sensor nodes and those from a commercial sound level meter. The results suggest that the proposed system is perfectly suitable for use as a starting point to obtain accurate maps of the noise levels in smart cities.

Keywords: noise monitoring; real-time noise mapping; wireless sensor networks

---

1. Introduction

Noise pollution is a problem that affects millions of people worldwide. Different studies have shown that it is currently one of the greatest environmental threats to people's health, leading to increased risk of cardiovascular disorders, hypertension, sleep disturbance, stress, etc., and it is negatively influencing productivity and social behavior [1]. According to the World Health Organization (WHO), noise pollution is responsible for 50,000 heart attacks each year in Europe. Moreover, 1.8% of total heart attacks can be attributed to traffic noise levels greater than 60 dBA. In the particular case of Andalusia (Spain), in the studies carried out for the last Ecobarometer of Andalusia [2], citizens considered noise pollution as one of the main environmental problems in cities and towns that has caused a considerable degradation in the quality of life.

Nevertheless, until the 1990s, policies to reduce environmental noise always had a lower priority than policies regarding the pollution of water or air. In 1993, The Fifth European Commission (EC) Environmental Action Program [3] marked the beginning of attention being paid to the problem of noise pollution, and noise reduction programs began to be developed. The first step in the development of this program was in 1996, when the EC published the first policy to reduce environmental noise—"Future noise policy: European commission green paper" [4].

The Environmental Noise Directive 2002/49/EC (END) [5] required European member states to provide and publish accurate mappings of noise levels and action plans every five years throughout large agglomerations, all major roads, railways, and major airports.

Currently, noise maps can be generated with the help of noise mapping software [6,7], based on numerical simulations that take into account estimated parameters (such as traffic flow, the type of road, rail, or vehicle data), emission models of transportation and industrial noise sources, noise propagation patterns, and the urban topology.

However, the END required that noise maps be based on empirical measures. Moreover, in 2006, the EC working group "Assessment of Exposure to Noise" (WG-AEN) published a document called "Good Practice Guide for Strategic Noise Mapping and the Production of Associated Data on Noise Exposure" [8], which strongly recommended obtaining accurate and real data on noise levels.

For this task, professionals have traditionally carried out measurements using instruments for noise collection and processing, called sound level meters, placed in a mesh pattern in the area to be mapped. They measure the noise using the A-weighting equivalent continuous sound pressure level, the L_{AeqT} indicator [9].

However, this procedure has a series of disadvantages inherent to the technology used, such as the impossibility of making continuous measurements for long periods of time (weeks/months), the lack of knowledge of the situation in real time, and the inability to take preventive or corrective actions in real time. This traditional method also presents many technical difficulties for complying with some regional legislations [10].

To solve, in part, these problems and inconveniences, in recent years, different studies have proposed the use of Internet of Things (IoT)-based technologies [11].

The potential applications of IoT are numerous and diverse. In the EC documents relating to IoT [12–14], 65 IoT scenarios were identified and presented, grouped into 14 domains. One of these domains is the so-called smart city, defined as "a place where traditional networks and services are made more efficient with the use of digital and telecommunication technologies for the benefit of its inhabitants and business" [14]. One of the trendiest scenarios in smart cities is identified as noise urban maps—sound monitoring in bar areas and centric zones in real time.

Wireless acoustic sensor networks (WASNs) [15] play a key role in this scenario of a smart city. Over the last few years, many researchers have devoted their attention to the design of these types of networks to monitor the real data of continuous and precise noise levels, and create noise maps in real time and space. Many research works and patents have been published [16–31], but very few real projects have been developed based on WASN approaches [32–37].

In the literature [37,38], the authors present two reviews of the most relevant WASN-based approaches developed to date focused on environmental noise monitoring in smart cities. In the literature [37], WASNs have been classified according their data quality, scale, longevity, affordability, and accessibility. On the other hand, in the literature [38], another classification is presented, where the sensor nodes are divided into three main categories, according to their measurement accuracy, cost, and computational capacity.

Although WASNs are becoming a reality in smart cities, in the literature [38], the authors argue that very few projects have been deployed around the world, and they conclude that further research should be conducted to improve the performance of WASNs in real-life operation conditions. They highlight the project DYNAMAP, the objective of which is the deployment of a low-cost WASN in two different cities, Milan and Rome [36], to monitor road traffic noise.

In this work, we present the design and implementation of a complete low-cost system for a WASN deployed in the city of Linares (Jaén), Spain, which has been running continuously for ten months. The complete system covers the hardware of the sensor nodes, signal processing for noise monitoring in the sensor nodes, network topology design, protocols, and the design of a private cloud platform with an intuitive graphical user interface to show clear and comprehensible information to the general public.

As a result, a complete system has been obtained to provide the information, shown in Table 1, for each of the locations where the nodes are deployed.

Table 1. Information provided by the system.

Parameter	Description
L_{AeqT}	A-weighting equivalent continuous sound pressure level
L_{den}	Day–evening–night level
L_{day}	A-weighted average sound level over the daytime period 07:00–19:00
$L_{evening}$	A-weighted average sound level over the evening period 19:00–23:00
L_{night}	A-weighted average sound level over the night period 23:00–07:00
L_{Amax}	Maximum A-weighted noise level during the measurement period
F_{eqmax}	Predominant frequency (Hz) of the noise
L_{A01T}, L_{A10T}, L_{A50T}, L_{A90T}, and L_{A99T}	Percentile noise levels, L_{AnT}, which are defined as the A-weighted sound level that is exceeded n% of the measurement time interval

In addition, along with this information, a map using the Google Maps platform Application Programming Interface (API) is also displayed, representing the L_{Aeq} in each location.

Based on the two classifications presented in the literature [37,38], the system is characterized by the following: (a) high data quality; (b) easily scalable to a large number of nodes; (c) can work continuously for long periods of time; (d) affordability due to its low-cost equipment; (e) data accessibility through a cloud web server; (f) the capacity to perform spectral analysis calculations, compute L_{AeqT}, and conduct real-time signal processing; and (g) high computational capacity and low-cost equipment.

The remainder of the paper is structured as follows. Section 2 describes the complete system for the WASN deployed in the city of Linares. The experimental results are provided in Section 3. Some conclusions and future works are presented in Section 4.

2. The Design and Implementation of the Deployed WASN

This section describes all of the elements that make up the complete system for noise monitoring in the city of Linares (Jaén)—the network topology design, the hardware and software of the sensor nodes, protocols, and a cloud web server platform.

2.1. Design Considerations

The City Council of Linares, through the area of urban planning, established those locations of the city that were considered the most critical from the point of view of noise pollution. Specifically, eight locations were established, which mostly covered the entire downtown area. To these locations, we decided to add one more, considered as noncritical. Therefore, in total, nine low-cost nodes have been installed as measuring points. Table 2 and Figure 1 show the exact locations of these points.

Table 2. The critical places to be monitored.

Id	Location	Id	Location
1	Andalucia Avenue	6	Cervantes Street
2	Ayuntamiento Square	7	Julio Burell Street
3	Isaac Peral Street	8	Ubeda Street
4	Santa Margarita Square	9	Noruega Street
5	San Francisco Square		

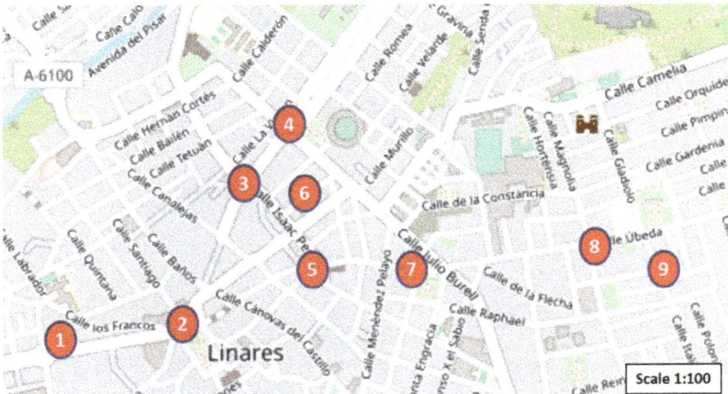

Figure 1. The critical locations identified for the measurement of acoustic noise in the city of Linares (Jaén).

The City Council of Linares specified the existence of a corporate Wi-Fi network deployed in the center area of the city, so that it was possible for the sensor nodes to transmit data. In addition, there was the possibility of using a power supply permanently at all of the measuring points.

2.2. Distribution of the Sensor Nodes

The topology design of a data network determines the connections between the nodes or between a node and a server. Because of the design considerations, we designed a network topology where each sensor node can send the measurements directly to a central server, which is a cloud web server in our case.

Because of the existence of the corporate Wi-Fi network that the City Council of Linares deployed in the city, as well as the absence of power supply restrictions, we proposed using this Wi-Fi network in all of the locations where this was possible. After analyzing the coverage, it was detected that, in seven of the nine locations, it was possible to use said Wi-Fi network. However, in two locations (nodes two and nine), there was no coverage. For these two locations, we decided to use 3G and Sigfox technologies, respectively. The network topology for the proposed complete system is shown in Figure 2.

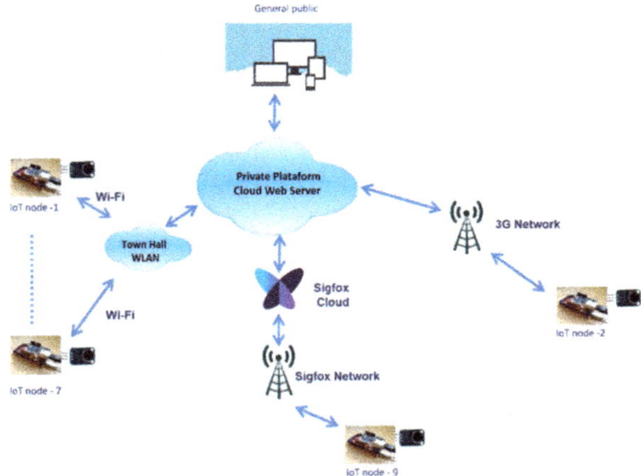

Figure 2. The network topology proposed for the wireless acoustic sensor network (WASN) deployed in the city.

Sigfox [39] is a reliable, low-power solution based on a dedicated radio-based network to connect sensors and devices, and it needs to continuously be on and emitting small amounts of data.

2.3. Hardware IoT Sensor Nodes

Typical IoT devices have constrained sensor resources, an actuator capacity, and local information processing, and they are able to communicate data with servers on the Internet cloud platform.

With the design considerations indicated above, when it is possible to have a continuous power supply, the chosen device is a standard hardware model (i.e., commercial sensor node) of the Arduino platform. Specifically, it is the Arduino Due device [40], which is based on a 32-bit ARM core microcontroller, and is an open-source platform designed for the development of solutions related to sensor networks. The choice of this device is mainly due to its technical specifications, in terms of the processor and the memory, which allow for the execution of a frequency domain-based algorithm to calculate L_{AeqT} in real time. This is not possible on other devices of the Arduino platform. However, any other device with similar or better characteristics to the Arduino Due could be used, such as Raspberry Pi.

The Arduino Due has the following technical specifications: Atmel SAM3X8E ARM Cortex-M3 processor (32-bit, clock speed of 82 MHz, 96 Kb of SRAM, and 512 Kb of flash memory), 54 I/O digital ports, 12 input analog ports with a 12-bit resolution, and two output analog ports. Arduino Due hardware uses standard components, and its software is based on C/C++.

Related to the communication hardware of the sensor nodes, we used the following:

- For sensor nodes one and three through seven: Arduino Ethernet Shield [41] and an antenna MikroTiK SXT 2 [42]. We used this external antenna to ensure the existence of wireless coverage for the sensor nodes, which were connected through a UTP cable to a RJ45 female connector installed in the enclosure box. The power consumption was approximately 180 mA.
- For sensor node two: Arduino Ethernet Shield, a 3G router (model TL-MR3020 [43]), and a 3G USB modem with an outdoor antenna. In this case, the power consumption was approximately 900 mA.
- For sensor node nine: an 868-MHz Sigfox module for Arduino [44], a communication shield, and a 4.5-dBi antenna. The power consumption was approximately 125 mA.

All of the nodes were powered through passive Power over Ethernet (PoE), using 12-V 1-A power adapters and PoE injectors. The electrical plugs were at a maximum distance of 20 m, and for that distance, the voltage drop in the UTP cable was 0.9 V, so there was 11.1 V to power the Arduino, enough for its operation.

The microphone used is based on a commercial design [45]. Each sensor node is equipped with an electret microphone, 20–20 kHz (Figure 3). It is much more than just a microphone, because it is integrated with an operational Maxim MAX4466 specifically designed for acoustic solutions (it amplifies and filters the noise). The gain is adjustable via an integrated potentiometer. Moreover, the microphones have a miniature foam windshield ball [46]. For the outdoor enclosure for the nodes, we used IP66-rated outdoor aluminum enclosures for wireless platforms, such as StationBox ALU RF elements [47]. Figure 3 shows some sensor nodes.

To the best of our knowledge, this system has one of the lowest economic costs per node, and this aspect is very important when implementing WASNs with a large number of nodes. Table 3 shows the costs of each node (tax not included). Costs can be more reduced by using compatible materials.

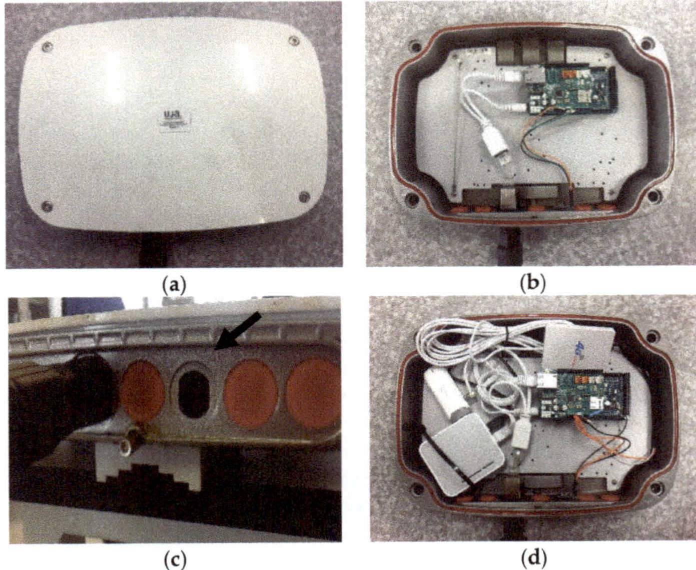

Figure 3. The sensor nodes: (**a**) enclosure box; (**b**) Wi-Fi sensor node; (**c**) microphone; (**d**) 3G sensor node.

Table 3. The cost of the sensor nodes.

Description	Original Material Cost (€)	Compatible Material Cost (€)
Wi-Fi Sensor Node	45	23
Wi-Fi Sensor Node + Antenna MikroTiK SXT2	99	47
Sigfox Sensor Node	109	49
3G Sensor Node	90	48

2.4. Software Implemented in the Sensor Nodes for Noise Monitoring

For the measurement of acoustic noise and to integrate the calculated L_{AeqT} into the sensor nodes, it is necessary to design and implement an algorithm that runs on these nodes. In the previous work [48], we presented a frequency domain-based algorithm to calculate L_{AeqT} in real time adapted to resource-constrained devices, such as wireless acoustic sensor nodes.

In this work, we improved the algorithm by introducing a new module to determine the frequencies with a higher energy and their degree of importance with respect to background noise or less significant frequencies. In this manner, we obtained the information from the predominant frequency of the noise. The optimized architecture used by the algorithm consists of four functional blocks, as shown in Figure 4.

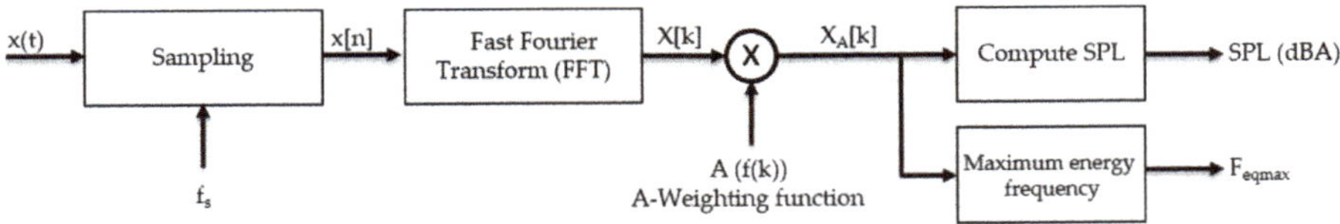

Figure 4. The algorithm's functional blocks. SPL—sound pressure level.

The sampling block is responsible for sampling the acoustic signal x(t). The IEC 60651 Type-2 SLM acoustic standard [49], superseded by IEC 61672 [50], established the measurement of environmental noise between 0–8 kHz, and the SLMs specified in the literature [50] are intended to measure sounds generally in the range of human hearing. As is well known, in urban areas, the acoustic signal energy is concentrated in a low-frequency region (<10 kHz). Based on this, we configured the Arduino Due with a frequency-sampling rate, f_s, of 33 kHz, by a software function with a resolution of 12 bits. Thus, it is not possible to measure frequencies higher than 16.5 kHz. Related to the time-constant of the integration or time capture, there are two time-weightings that have been internationally standardized, namely: (a) slow response (S) of one second; and (b) fast response (F) of 125 ms.

The second block receives the audio samples from the first block. The algorithm is based on a frequency analysis, which uses the discrete Fourier transform (DFT) to determine the frequency spectrum of a segment of audio samples. Let us denote $X[k]$ as the DFT of a windowed signal $x[n]$, at the digital frequencies $2\pi k/N$ radians, where N denotes the number of samples and $k = 0, \ldots, N-1$. To determine the samples of the DFT, we use the following equation:

$$f(k) = \frac{f_s \cdot k}{N}. \quad (1)$$

The difference between two consecutive samples is given by the following expression:

$$\Delta f = \frac{f_s}{N}. \quad (2)$$

Regarding the CPU and memory requirements, this block is the most demanding. For a computationally efficient implementation, we used the fast Fourier transform (FFT) to evaluate the DFT. Some analyzers have been designed to determine the optimal FFT length and thus achieve efficient implementation. Additionally, an exhaustive analysis has been performed using software functions to the reduce memory requirements and the execution time. The FFT length depends on the frequency-sampling rate chosen and the time capture (T_w), as follows:

$$N = f_s \cdot T_w \quad (3)$$

Table 4 shows the FFT length for the two time-weightings that have been standardized and for the frequency-sampling rate of 33 kHz.

Table 4. The sample length for a frequency-sampling rate of 33 kHz.

Frequency Sampling Rate	Type of Response	Time Capture	Sample Length
33 kHz	Fast	125 ms	4125
33 kHz	Slow	1 s	33,000

Taking into account that the FFT is more runtime efficient if a base-two sample length is selected, to reduce the execution time, we propose a length of 4096 samples (instead of 4125) for the fast response and 32,768 samples (instead of 33,000) for the slow response. This means a slight decrease in the time window at 124.1 ms for the fast response, and at 0.993 s for the slow response. Our proposal is to perform an FFT calculation for the fast response only, which uses a time window of 124.1 ms. This approximation of the time window produces an error of 0.7% (29 samples less than using a time window of 125 ms), which could be considered as negligible. Larger FFT sizes provide a higher spectral resolution, but take more resources to compute. A smaller window size means a shorter runtime, and, therefore less resource consumption.

Once the frequency components of the acoustic signal are available, an A-weighted filtering is implemented. This filtering stage allows for weighting of the different frequency components of the acoustic signal, consistent with a typical human ear response. The mathematical function used to obtain the value of the attenuation depends on the frequency, and is given by the normative IEC 61672 [50], as follows:

$$(f) = 10\log_{10}\left[\frac{1.562339^2 \cdot f^4}{(f^2+107.65265^2)(f^2+737.86223^2)}\right] + \\ + 10\ log_{10}\left[\frac{2.242881 \cdot 10^{16} \cdot f^4}{(f^2+20.598997^2)(f^2+12194.22^2)}\right] \tag{4}$$

In the above equation, f is the frequency and $A(f)$ is the associated attenuation. However, some resource-constrained devices have an anomalous behavior for math operations with large numbers. Therefore, we propose the use of an equivalent expression to reduce the complexity of the mathematical operations [51].

$$A(f) = 2 + 20\ log_{10}(R_A(f)) \tag{5}$$

$$R_A(f) = \frac{12200^2 \cdot f^4}{(f^2 + 20.6^2)\sqrt{(f^2 + 107.7^2)(f^2 + 737.9^2)}(f^2 + 12200^2)}$$

Using Equations (1) and (2), we can determine the frequency for each sample of the FFT. After applying the filter, the obtained signal is as follows:

$$X_A[k] = A(\Delta f \cdot k) \cdot X[k] \tag{6}$$

Finally, the last block computes the total energy of the weighted frequency components to obtain the sound pressure levels (SPLs) in dBA. Using the Parseval's relation [52], the total energy of the waveform can be summed across all of its frequency components, as follows:

$$\epsilon_x = \frac{1}{N}\sum_{k=0}^{N-1} |X[k]|^2 \tag{7}$$

Regarding the properties of the FFT, the samples have symmetry because they are complex conjugates, as follows:

$$X\left[\frac{N}{2} + k\right] = X^*\left[\frac{N}{2} - k\right]\ 1 \le k \le \left(\frac{N}{2}\right) - 1 \tag{8}$$

The input samples in the time domain are real values. In the frequency domain, they are symmetrical from the sample $N/2$ (N is even). Therefore, we can calculate the energy of the signal using only the first $N/2 + 1$ samples (filtered spectrum with A-weighting filter). The expression is the following:

$$\epsilon_x \approx \frac{2}{N}\sum_{k=1}^{(\frac{N}{2})-1} |X_A[k]|^2 + |X_A[0]|^2 + |X_A[N/2]|^2 \tag{9}$$

Equation (9) is used to determine the total energy of the signal in the time capture. Taking T_w seconds, the instantaneous average power of the signal is given by the following:

$$P_x \approx \frac{\epsilon_x}{T_w} \tag{10}$$

To obtain the *SPL* in dBA, we applied the following expression:

$$SPL\ (dBA) = 10 \cdot log_{10}(P_x) + C \tag{11}$$

where C is the calibration constant, which will be calculated in the calibration process.

Calibration and Test Results

Before deploying the sensor nodes in the urban area, some tests were carried out in the lab and in a street to verify the quality of the noise measurements. Figure 5 presents the lab scenario where we used an Arduino Due with the microphone, a commercial Sound Level Meter PCE-353 (SLM) [53], and one laptop with speakers. The SLM and the sensor node were connected to the laptop using a USB connection. The sensor node and the SLM were deployed closely; the distance from the speakers to the devices was 0.5 m.

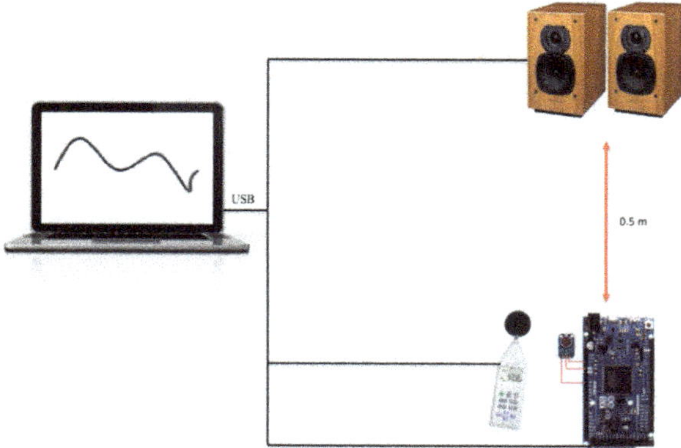

Figure 5. The scenario for indoor measurement tests.

First, the commercial SLM was calibrated using the Class-2 Sound Level Meter Calibrator PCE-SC 42, at one kHz and for an SPL of 94 dB. Later, to calibrate the sensor node, an acoustic signal of a 1-kHz tone was created using math software. The volume of the speakers was raised until the SLM measured 94 dB, and the microphone gain was adjusted to give that measurement. Once both devices were calibrated, an acoustic signal composed of white noise (30 Hz–20 kHz) was generated with three noise levels of acoustic intensity: 60, 70, and 85 dBA. For each level, the L_{Aeq} indicator was calculated after repeating the experiment 30 times. The duration of the acoustic signal was five seconds for each test. Figure 6 shows the results for this experiment.

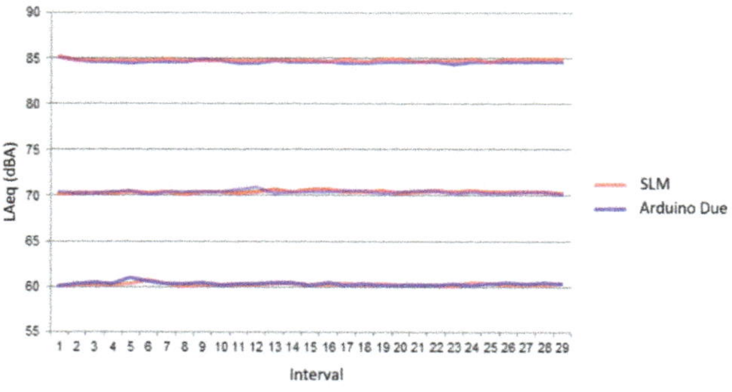

Figure 6. The white noise lab tests.

Table 5 shows the average value of the 30 measurements of the L_{Aeq} indicator for each level of acoustic intensity, and the absolute error (Diff) between the Arduino Due and the commercial SLM measurements.

Table 5. The Arduino Due measurements in the lab.

Intensity	L_{Aeq} SLM (dBA)	Diff Arduino Due (dBA)
Intensity 1	60.5	0.13
Intensity 2	70.3	0.14
Intensity 3	86.8	0.16

As can be observed, the differences between the Arduino Due and the SLM were lower than 0.2 dBA. For another test, we deployed the SLM and Arduino Due devices in an urban street. The distance between the devices and street was approximately eight meters. The measurements were made during one hour in daytime. Both devices calculated the SPLs each second, for four intervals of 15 min each. Table 6 shows the L_{Aeq} and the absolute error (Diff) between the SLM and Arduino Due measurements. Figure 7 shows the location where the sensor and the SLM were located.

Table 6. The SLM and Arduino Due measurements in an urban street.

Interval	L_{Aeq} SLM (dBA)	L_{Aeq} Arduino (dBA)	Diff (dBA)
Interval 1	60.59	60.71	0.12
Interval 2	60.01	60.90	0.89
Interval 3	60.85	61.44	0.59
Interval 4	61.15	61.70	0.55

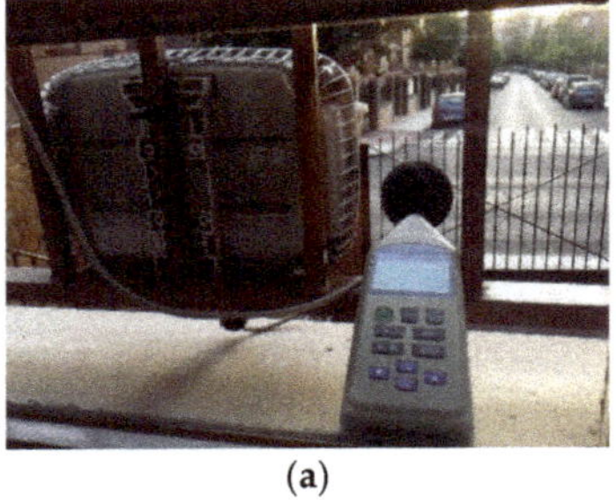

(a)

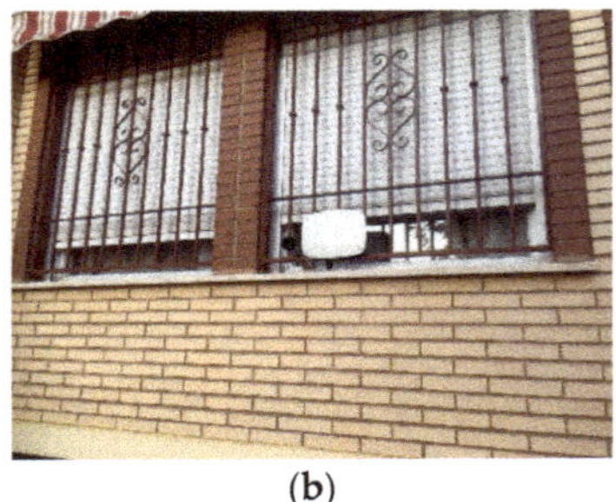

(b)

Figure 7. The SLM and Arduino Due deployed in a street for the tests: (**a**) back view; (**b**) front view.

In this case, the differences between the Arduino Due and the SLM were lower than 0.9 dBA, which represents a very good agreement.

Finally, we deployed the SLM and Arduino Due devices in the same urban street, but this time for a full day. The results are shown in Table 7.

Table 7. The SLM and Arduino Due measurements in an urban street for a full day.

Device	L_{Aeq} (dBA)	L_{day} (07:00–19:00)	$L_{evening}$ (19:00–23:00)	L_{night} (23:00–07:00)
SLM	56.90	57.60	58.90	53.10
Arduino Due	57.96	57.83	61.31	53.83

The differences between the Arduino Due and the SLM were approximately 1 dBA in the L_{Aeq} measured. The maximum difference was in the period $L_{evening}$, being 2.41 dBA. This is because the dynamic range of the sensor is 44–105 dBA, and therefore, it cannot measure noise levels below 44 dBA. Alternately, the L_{Amax} measured with the SLM was 88.1 dBA, while that with the Arduino Due was 90.5 dBA. The results of the previous tests indicate that the software designed for the Arduino Due has a good performance when we compare the differences between the acoustic measurements calculated by the Arduino and those of the commercial SLM.

2.5. Protocols and Platform Cloud Web Server

Many protocols have been specifically designed for communication between IoT devices, namely: Message Queue Telemetry Transport (MQTT) [54], Constrained Application Protocol (CoAP) [55], Advanced Message Queuing Protocol (AMQP) [56], Data Distribution Service (DDS) [57], etc. However, the commonly used protocol for the Internet, Hyper Text Transfer Protocol (HTTP), is used in most cases for IoT devices when they need to publish a considerable amount of data.

In fact, most of the commercial platforms that currently exist in the cloud and intend to provide services to IoT devices allow for communication from these devices through HTTP. Some examples are the Amazon AWS IoT Core platform, Microsoft Azure, Google Cloud IoT Core, and Thingworx. Each platform offers developers a series of application programming interfaces (APIs) and software development kits (SDKs) that make it possible to establish communication between the IoT devices and the cloud platform.

The use of the services of these platforms has advantages and disadvantages. Among the drawbacks of Azure, in addition to the cost involved in its use, are that the implementation of real-time data visualization systems is sometimes complex, and there is incompatibility with the Safari web browser. Google IoT Core Cloud does not support the MQTT protocol. In the AWS IoT Core, the use of services and functions is complex in some cases, and it is the most expensive option for many services.

Therefore, in our case, we decided to design and implement our own platform cloud web server using the infrastructure of the University of Jaén. The cloud web server is based on a model–view–controller (MVC) software architecture, which separates the application data, the user interface, and the control logic into three distinct components. For frontend technologies, we used HTML5, CSS3, Bootstrap, and JavaScript, and for the backend technology, we used PHP. For the database, a MySQL relational database was designed. In addition, for the representation of the L_{Aeq} values of the locations on a map with different colors, the Google Maps platform API was used.

In all of the sensor nodes, in addition to the acoustic noise monitoring software, the communication software was implemented to send data to the Platform Cloud Web Server. To send the data, we decided to use the HTTP protocol. Therefore, a web client was implemented on each sensor, except for on sensor node nine. Node nine was programmed using the Sigfox API for sending data, and a callback was configured in the Sigfox backend, so that an HTTP request with the GET method was made to our cloud web server.

As shown in Figures 8 and 9, all of the sensor nodes calculate the SPL every second. Every 30 s, sensor nodes one–eight send the following data to the Platform Cloud Web Server:

1. Id sensor node
2. L_{Aeq} calculated for these 30 s
3. L_{Amax} in the period of 30 s
4. Average predominant frequency of the noise during 30 s (F_{eq})
5. Higher predominant frequency of the noise during 30 s (F_{eqmax})

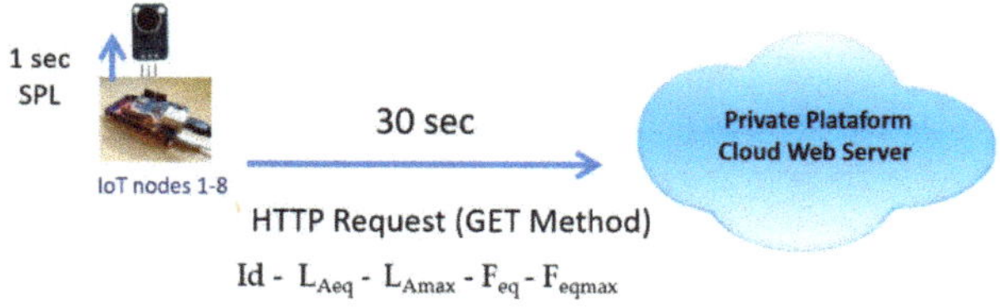

Figure 8. The Hyper Text Transfer Protocol (HTTP) request—GET method.

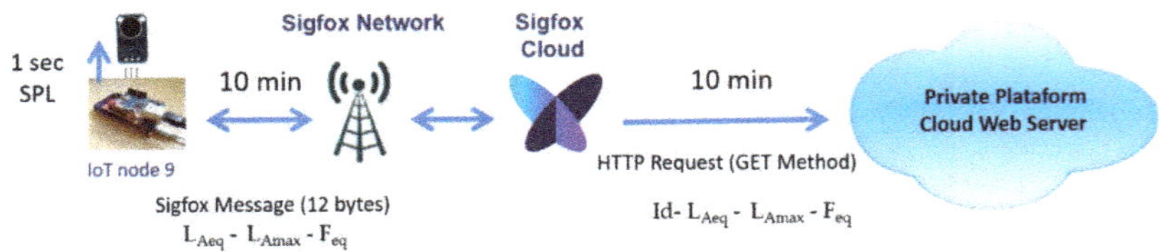

Figure 9. The Sigfox message. HTTP request—GET method.

In the case of sensor node nine, Sigfox messages can carry a payload (user data) of 12 bytes, and 140 messages are permitted per day, at most (although we have verified that this limit can actually be exceeded a bit). Therefore, as shown in Figure 9, sensor node nine sends a message every 10 min with the following parameters:

1. L_{Aeq} (four bytes) calculated for these 10 min
2. L_{Amax} (four bytes) in the period of 10 min
3. Average predominant frequency (four bytes) of the noise during 10 min (F_{eq})

3. Results

As a main result, we can say that an experimental wireless acoustic sensor network for real-time noise monitoring has been installed in the city of Linares (Jaén), Spain, and it has been running continuously for 10 months.

3.1. Sensor Nodes Deployed around the City

Figure 10 shows some locations where the sensor nodes were installed. From the data sent to the cloud web server by the sensor nodes, a huge amount of information has been obtained, with which it would be possible to characterize the city in terms of its activity. In this section, we will show some of the results obtained, but it is impossible to show all of the cases and events that have occurred in all of the locations.

3.2. The Information Offered by the Cloud Web Server

The cloud web server offers the possibility of selecting a date to display a map using Google Maps, where L_{Aeq} is represented by a color in each location measured during the 24 h of the previous day. Figure 11 shows an example of city noise for the day of 26 November 2016.

For each one of the locations, the temporal evolution of the noise during the selected time interval can be visualized, as well as the noise indicators. Figures 12 and 13 show an example of how this information is displayed. Figure 12 shows the acoustic noise measured by the sensor of San Francisco Square from 00:00 to 23:59 on 30 June 2017. Figure 13 shows the temporal evolution of acoustic noise for a full week at the Andalucia Avenue location. It can be seen that the acoustic noise has a similar pattern every day.

Figure 10. The sensor nodes deployed in the city. (**a**,**b**) Id 8: Ubeda Street; (**c**,**d**) Id 5: San Francisco Square; (**e**,**f**) Id 1: Andalucia Avenue.

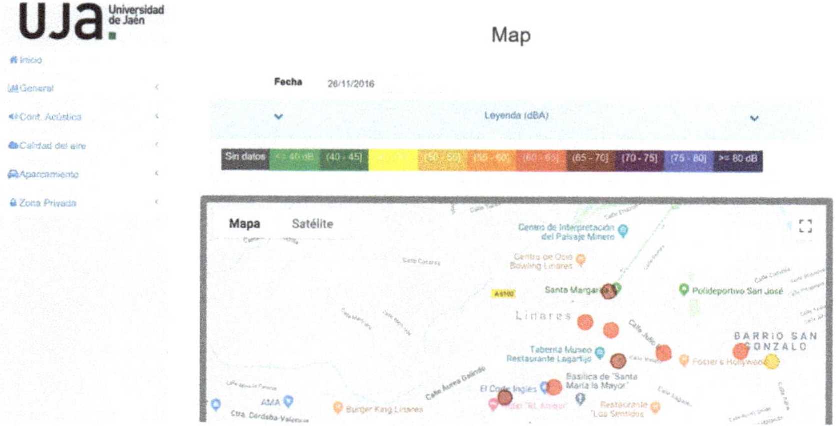

Figure 11. The map with the noise in the sensor locations on 26 November 2016.

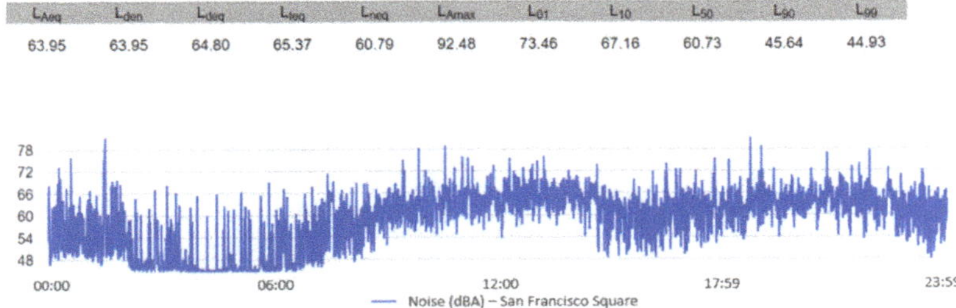

L_{Aeq}	L_{den}	L_{deq}	L_{eeq}	L_{neq}	L_{Amax}	L_{01}	L_{10}	L_{50}	L_{90}	L_{99}
63.95	63.95	64.80	65.37	60.79	92.48	73.46	67.16	60.73	45.64	44.93

Figure 12. The temporal evolution of the noise from 00:00 to 23:59 on 30 June 2017 in San Francisco Square.

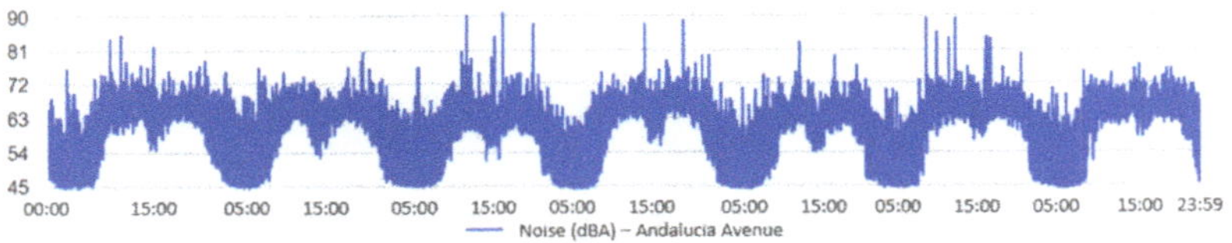

Figure 13. The temporal evolution of the noise in Andalucia Avenue location for a full week in the month of October.

It is possible to set the query for a certain period of time. For example, if L_{Amax} is 92.48 dBA, a query can be made to visualize the L_{Amax} that occurred in each one of the 30-s intervals, as shown in Figure 14.

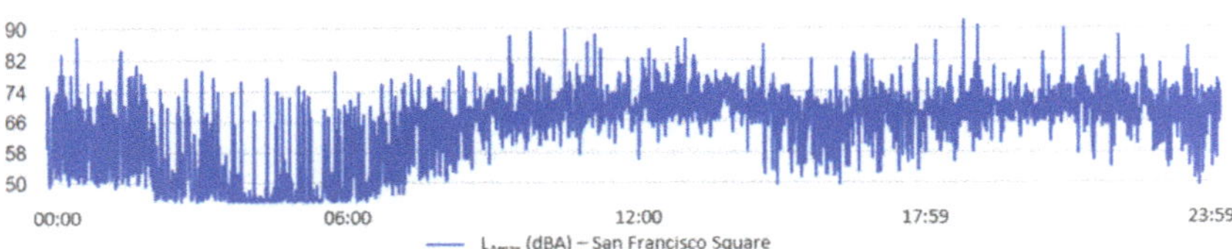

Figure 14. The L_{Amax} measured at each 30-s interval from 00:00 to 23:59 on 30 June 2017 in San Francisco Square.

It can be observed that the L_{Amax} of that day was produced in the strip from 17:30 to 19:30, so if we only consult that period, we can know exactly when the L_{Amax} occurred, which was at 18:44, as seen in Figure 15.

Figure 15. The L_{Amax} measured at each 30-s interval from 17:30 to 19:30 on 30 June 2017 in San Francisco Square.

A query can also be established for the predetermined day–evening–night periods, as shown in Figure 16.

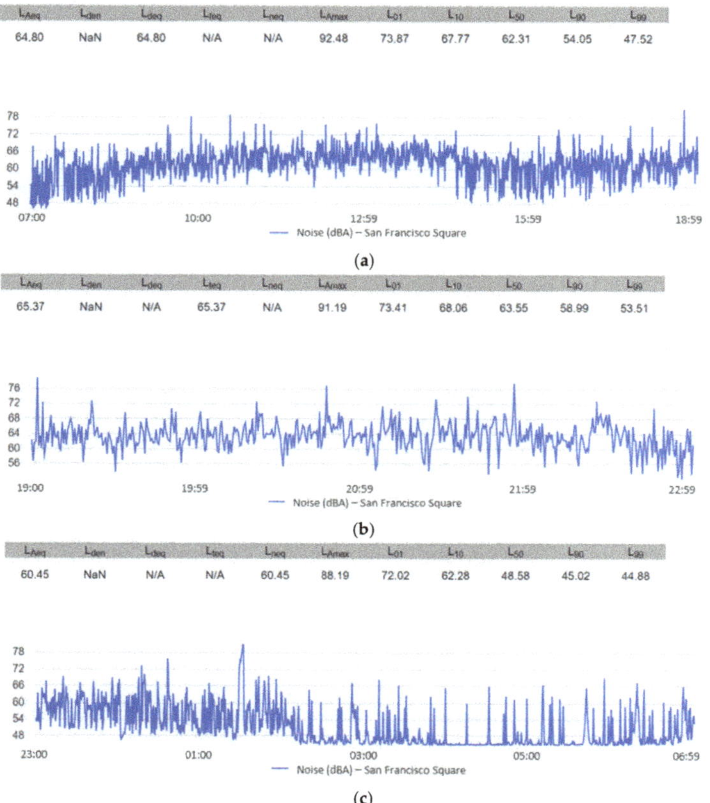

(**a**)

(**b**)

(**c**)

Figure 16. The temporal evolution of the noise on 30 June 2017 in San Francisco Square for the periods: (**a**) daytime 07:00–19:00; (**b**) evening 19:00–23:00; (**c**) night 23:00–07:00.

Appendix A contains different events that show examples of the activity of the city, namely: the garbage collection truck, the noise from construction work in a street, and the noise derived from a leisure activity—a bar.

4. Conclusions and Future Work

We have presented the design and implementation of a complete low-cost system, composed of nine sensors nodes, for a WASN deployed in the city of Linares (Jaén), Spain, which has been working continuously for ten months. The complete system has covered the network topology design, hardware and software of the sensor nodes, protocols, and a cloud web server platform. The information provided for the system for each location where the nodes have been deployed is as follows: L_{Aeq} for a given period of time; some noise indicators indicated in the END (L_{den}, L_{day}, $L_{evening}$, and L_{night}), the percentile noise levels (L_{A01T}, L_{A10T}, L_{A50T}, L_{A90T}, and L_{A99T}); a temporal evolution representation of the noise levels; and the predominant frequency of the noise. Moreover, a map using the Google Maps platform API has been displayed, representing the L_{Aeq} in each location.

Before deploying the sensor nodes in the city, different experiments were conducted to verify the performance of the Arduino Due hardware, together with the software implemented for the acoustic noise monitoring. The results were compared to the measurements acquired using a commercial SLM, which proved that the sensor nodes have a very good performance. However, the dynamic range of the sensor nodes was 44–105 dBA, and therefore, they cannot measure noise levels below 44 dBA. This can distort the results for measurements in non-noisy environments and especially during the night measurement period, from 23:00–07:00.

The performance of the Arduino Due is very good; the sensor nodes have been running continuously for 10 months, monitoring noise every second and sending the parameters to the cloud web server every 30 s. Some sensor nodes have presented problems derived from power outages and the subsequent restart of the devices (i.e., they did not restart properly). To solve this issue, it was necessary to mobilize technicians for a manual restart. This is a clear inconvenience if frequent power outages occur. There have also been problems with sending the data to the private cloud web server when the corporate Wi-Fi network of the city council did not work. In addition, there have been multiple hacker attacks on the cloud web server, with attempts to make insertions to the MySQL database, which succeeded in some cases. Therefore, we must increase the level of security in the cloud server.

Alternately, the amount of data received and stored has been enormous, given that every 30 s, each node sends the measured parameters. This approach is fine for analyzing acoustic noise over a considerable period of time, but it is not feasible to maintain it sustainably for many months or years.

An analysis of the results obtained from the acoustic noise in the city indicates that the variability of the acoustic noise in a specific location is very low, and therefore, a continuous measurement for one month is more than enough to characterize the noise in that location.

Although much work remains in order obtain accurate maps of noise levels in smart cities using WASN, the proposed system presented in this work can serve as an excellent starting point for this task.

Future research should aim to improve the dynamic range of the sensor nodes to be able to monitor acoustic noise below 44 dBA, and design and incorporate a fog computing platform between the sensor nodes and the cloud so as to avoid data loss due to the lack of a connection to the cloud. Security should be included in the protocols for sending the data, for example, HTTP. In addition, because the noise perception is affected by subjective factors, and there is not a direct correlation between the indicators and the subjective perception of the noise, the implementation of a module that is capable of evaluating the subjective impact of the noise annoyance is also proposed as a future work.

Author Contributions: Conceptualization, J.C.-B. and M.-A.G.-M.; investigation, J.-A.F.-P. and J.C.-B.; methodology, J.-A.F.-P. and M.-A.G.-M.; resources, J.-A.F.-P. and M.-A.G.-M.; software, J.-A.F.-P. and J.C.-B.; supervision, J.-A.F.-P. and J.C.-B.; validation, J.-A.F.-P. and M.-A.G.-M.; visualization, J.C.-B. and M.-A.G.-M.; writing (original draft), J.-A.F.-P.; writing (review and editing), J.C.-B. and M.-A.G.-M. All authors have read and agreed to the published version of the manuscript.

Funding: This research was partially funded by the Centro de Estudios Avanzados en Tecnologias de la Información y Comunicación, University of Jaén (project CEATIC-2013-001), and the University of Jaén (project UJA2014/06/16).

Acknowledgments: We would like to thank the support provided by Excmo. Ayuntamiento de Linares, especially for the technical support given by the ICT department.

Conflicts of Interest: The authors declare no conflict of interest.

Appendix A

Appendix A.1 Event 1—The Garbage Collection Truck

During the night period of the Figure 16, we appreciated an event of greater acoustic noise at approximately 01:30. As shown in Figure A1, if we visualize the period from 01:15–02:30 during some days of the week, we observed that the noise occurred periodically on all of the days, and lasted approximately five minutes. When we moved to the location to find the source of this noise, we could see that it was the garbage collection truck.

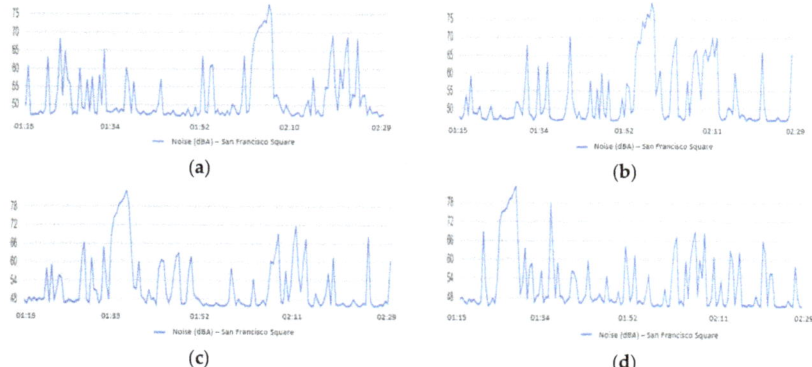

Figure A1. The temporal evolution of the noise in San Francisco Square for the period of 01:15–02:30 on the following days: (**a**) Monday 26 June 2017; (**b**) Tuesday 27 June 2017; (**c**) Wednesday 28 June 2017; (**d**) Thursday 29 June 2017.

Appendix A.2 Event 2—The Noise in the Works under Construction in a Street

Another event that we would like to highlight is how the sensor node, located in Julio Burell Street, detected the acoustic noise due to some works under construction, which were carried out for several days. Figure A2 shows the temporal evolution of noise on that street during two days, in the absence of works under construction and days where such works were carried out.

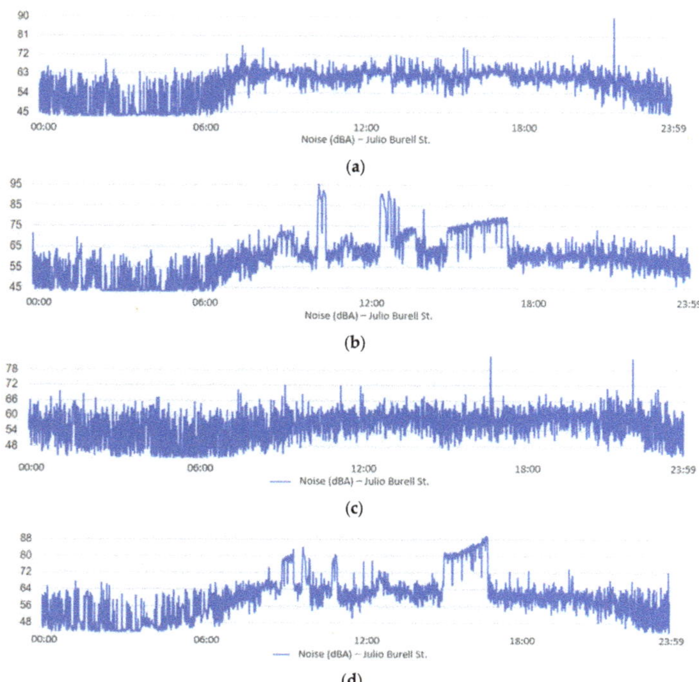

Figure A2. The temporal evolution of the noise at the Julio Burell St. location: (**a**) 15 November 2016, day without works under construction; (**b**) Friday 18 November 2016, starting the works under construction; (**c**) Sunday 20 November 2016, stopping the works (no labor on this day); (**d**) Monday 21 November 2016, restarting the works.

Figure A3 shows the maximum noise during 22 November 2016, for the daytime period (07:00–19:00) at Julio Burell Street. It can be seen that there is a permanent noise of approximately 92 dB from 13:00 to 15:00. At that time, the noise subsided, probably due to the break for the workers' lunch, and the activity restarted at 16:00. We could say that the workers had an hour's break and finished their workday at 17:30.

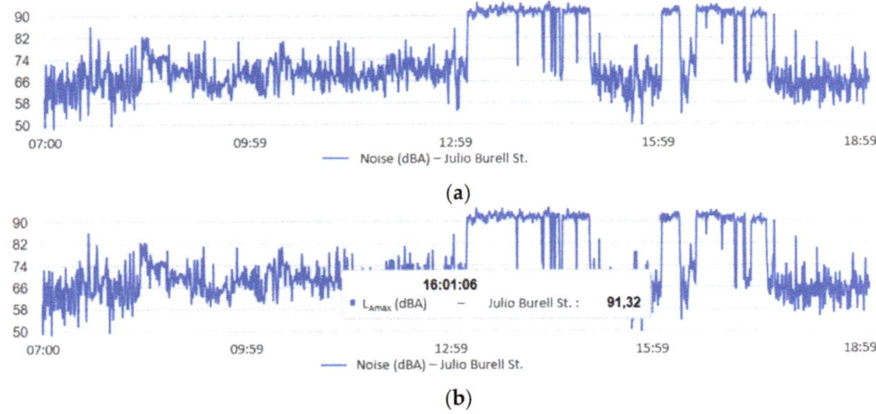

(a)

(b)

Figure A3. (a) The temporal evolution of the maximum noise at the Julio Burell St. location for the daytime period on 22 November 2016; (b) instant where the work resumes at 16:01.

If we analyze the predominant frequency of noise during a full day without construction work, we can see that the frequency is lower than 100 Hz for most of the day, as shown in Figure A4.

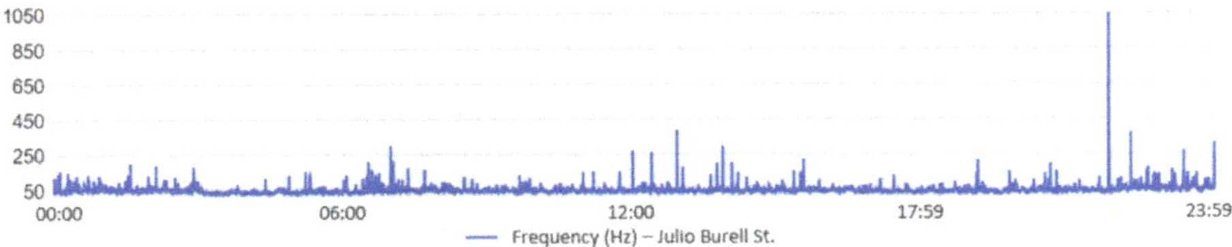

Figure A4. The predominant frequency during 15 November 2016, at Julio Burell Street.

However, if we now show the frequency on a day where construction work is being performed, we can see that when there is an increase in noise due to construction work, the predominant frequency increases, which could be due to some hammering. Figure A5 shows the predominant frequency against the maximum noise during 22 November 2016, during a daytime period (07:00–19:00) on Julio Burell Street.

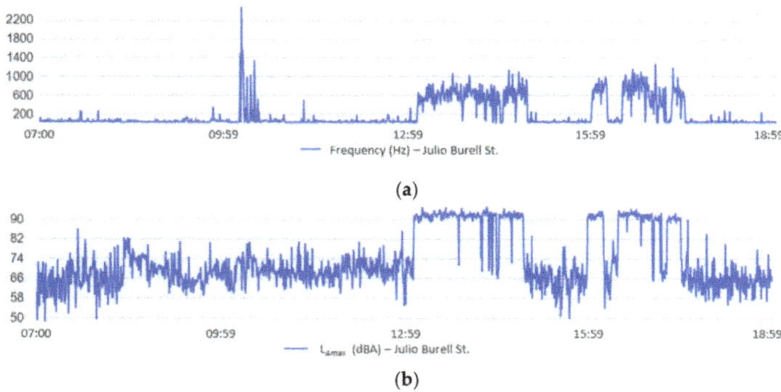

Figure A5. (**a**) The predominant frequency measured by the sensor node located at Julio Burell St. during 22 November 2016, in a daytime period. (**b**) Maximum noise measured in Julio Burell St. location for the daytime period on 22 November 2016.

Appendix A.3 Event 3—The Noise Derived from a Leisure Activity (A Bar)

Finally, we show the measurements made by the sensor with Id eight located on Ubeda Street. As seen in Figure 10b, the sensor was located just above a terrace of a bar where people usually go to eat and drink outdoors in good weather. Figure A6 shows the temporal evolution of acoustic noise during a weekday (Wednesday, 9 November 2016) and during a weekend day (Sunday, 13 November 2016).

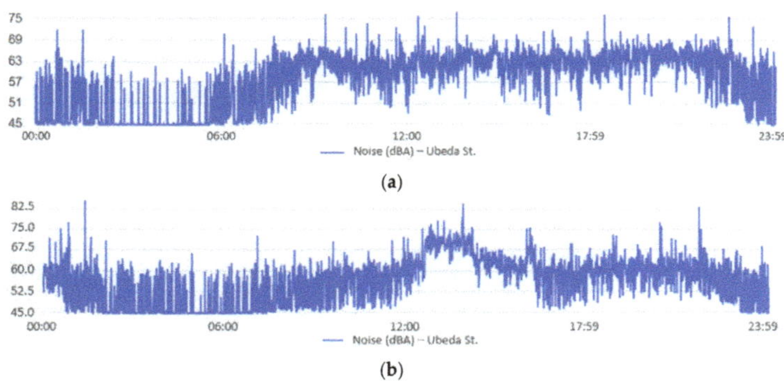

Figure A6. The temporal evolution of the noise in Ubeda Street during: (**a**) a weekday, 9 November 2016; (**b**) a weekend day, 13 November 2016.

We can appreciate how the noise on a weekday began to increase at approximately 07:30, and remained at almost 63 dB until 21:00, at which point it began to decrease. However, on the weekend, it can be seen that the noise began to increase at approximately 12:45, and remained here until approximately 16:15, because it was a sunny day, and many people probably went to eat at that bar terrace.

This noise pattern is always repeated for weekdays and weekend days. Figure A7a,b shows the temporal evolution of several weekdays in different months (Thursday, 18 May 2017 and Tuesday, 11 April 2017), while Figure A7c,d shows the temporal evolution of the noise on two weekend days (Sunday 3 May 2017 and Sunday 16 April 2017). It should also be noted that during many Saturday nights, there was also noise from people, which can be inconvenient for neighbors, as shown in Figure A7d. The noise did not begin to decrease until 01:00 on Sunday 16 May 2017.

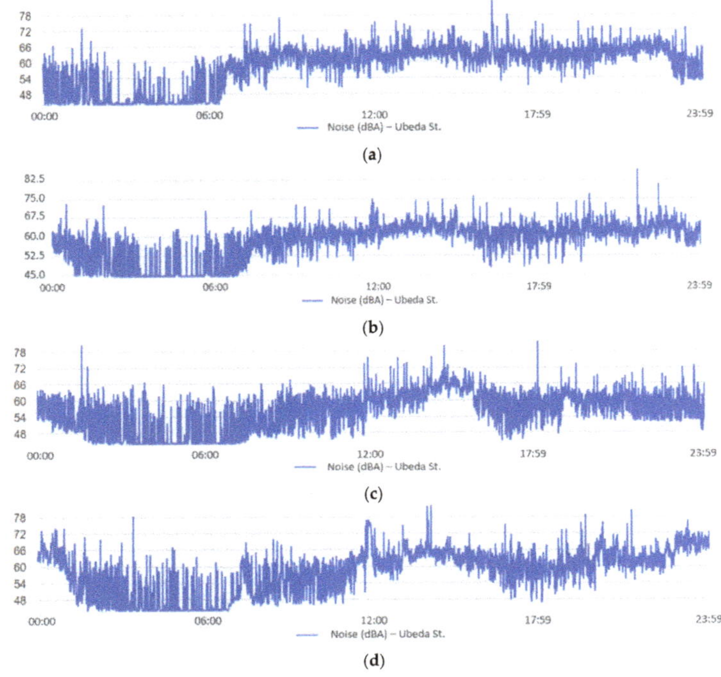

Figure A7. The temporal evolution of the noise in Ubeda Street during: (**a**) a weekday, 18 May 2017; (**b**) a weekday, 1 April 2016; (**c**) a weekend day, 3 May 2017; (**d**) a weekend day, 16 April 2017.

If the predominant frequency of the noise is shown, we can verify that the frequency is higher in that time slot because of the presence of people, as shown in Figure A8.

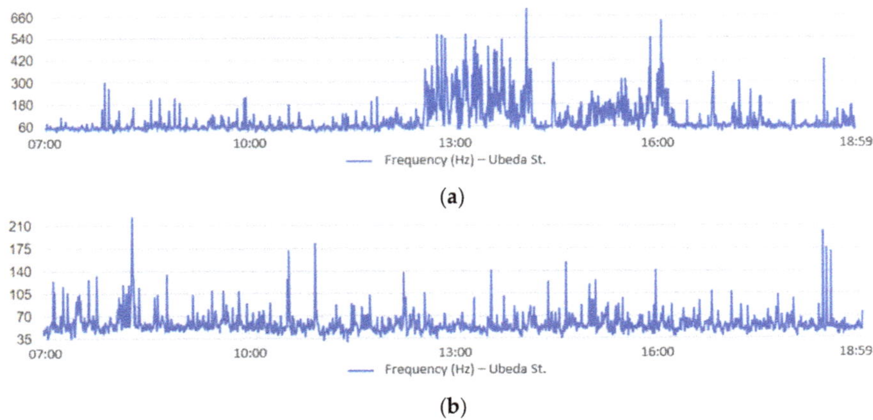

Figure A8. (**a**) The predominant frequency measured by the sensor node located in Ubeda Street on a weekend day, 13 November 2016, with the presence of people; (**b**) the predominant frequency measured by the sensor node on a weekday, 9 November 2016, without the presence of people.

References

1. World Health Organization Regional Office for Europe and the Joint Research Centre of the European Commission. *Burden of Disease from Environmental Noise: Quantification of Healthy Life Years Lost in Europe*; WHO Regional Office for Europe: København, Denmark, 2011; ISBN 9789289002295.
2. Ecobarometro 2013. Available online: http://www.juntadeandalucia.es/medioambiente/portal_web/web/temas_ambientales/educacion_y_voluntariado_ambiental/invest_soc_y_ma/ecobarometro/EBA_Sintesis_2013.pdf?lr=lang_es (accessed on 1 October 2019).
3. The Fifth EC Environmental Action Programme. Available online: https://ec.europa.eu/environment/archives/action-programme/env-act5/pdf/5eap.pdf (accessed on 1 October 2019).
4. European Commission. *Future Noise Policy: European Commission Green Paper. Com (96) 540 Final*; European Commission: Brussels, Belgium, 1996.
5. European Union. Directive 2002/49/EC of the European Parliament and of the Council of 25 June 2002 relating to the Assessment and Management of Environmental Noise. *Off. J. Eur. Communities* **2002**, *189*, 2002.
6. Bocher, E.; Guillaume, G.; Picaut, J.; Petit, G.; Fortin, N. NoiseModelling: An Open Source GIS Based Tool to Produce Environmental Noise Maps. *ISPRS Int. J. Geo-Inf.* **2019**, *8*, 130. [CrossRef]
7. İlgürel, N.; Akdağ, N.; Akdag, A. Evaluation of noise exposure before and after noise barriers, a simulation study in Istanbul. *J. Environ. Eng. Landsc.* **2016**, *24*, 293–302. [CrossRef]
8. European Commission Working Group Assessment of Exposure to Noise. Good Practice Guide for Strategic Noise Mapping and the Production of Associated Data on Noise Exposure, January 2006. Available online: http://sicaweb.cedex.es/docs/documentacion/Good-Practice-Guide-for-Strategic-Noise-Mapping.pdf (accessed on 1 October 2019).
9. European Environment Agency. State of the Art of Noise Mapping in Europe. Internal Report. 2005. Available online: http://egra.cedex.es/EGRA-ingles/I-Documentacion/Urban_Strategic_Noise_Maps/2005_eea-eionet-urbanareas.pdf (accessed on 1 October 2019).
10. Boletín Oficial de la Junta de Andalucía. Decreto 6/2.012, de 17 de Enero, por el que se aprueba el Reglamento de Protección contra la Contaminación Acústica en Andalucía. *Actual. Juríd. Ambient.* **2012**, *11*, 13–15.
11. Vermesan, O.; Friess, P. *Internet of Things—From the Research and Innovation to Market Deployment*; Vermesan, O., Friess, P., Eds.; Rivers Publisers: Aalborg, Denmark, 2014.
12. Vermesan, O.; Friess, P.; Furness, A. *The Internet of Things 2012 New Horizons*; Smith, I.G., Ed.; IERC-European Research Cluster: Halifax, UK, 2012.
13. Strategic Research Agenda. Internet of Things: Strategic Research Roadmap. 2009. Available online: http://www.internet-of-things-research.eu/pdf/IoT_Cluster_Strategic_Research_Agenda_2009.pdf (accessed on 1 October 2019).
14. Europen Comission. Smart Cities: Cities Using Technological Solutions to Improve the Management and Efficiency of the Urban Environment. Available online: https://ec.europa.eu/info/eu-regional-and-urban-development/topics/cities-and-urban-development/city-initiatives/smart-cities_en (accessed on 1 October 2019).
15. Bertrand, A. Applications and trends in wireless acoustic sensor networks: A signal processing perspective. In Proceedings of the 18th IEEE Symposium on Communications and Vehicular Technology in the Benelux (SCVT 2011), Ghent, Belgium, 22–23 November 2011; pp. 1–6.
16. Polastre, J.; Szewczyk, R.; Culler, D. Telos: Enabling ultra-low power wireless research. In Proceedings of the Fourth International Symposium on Information Processing in Sensor Networks, Los Angeles, CA, USA, 15–15 April 2005; pp. 24–27.
17. Santini, S.; Vitaletti, A. Wireless Sensor Networks for Noise Pollution Monitoring. In Proceedings of the 6th GI/ITG Workshop on Sensor Networks, Aachen, Germany, 16–17 July 2007.
18. Santini, S.; Ostermaier, B.; Vitaletti, A. First Experiences Using Wireless Sensor Networks for Noise Pollution Monitoring. In Proceedings of the Workshop on Real-World Wireless Sensor Networks, Glasgow, UK, 1 April 2008.
19. Santini, S.; Ostermaier, B.; Adelmann, R. On the use of sensor nodes and mobile phones for the assessment of noise pollution levels in urban environments. In Proceedings of the 6th International Conference on Networked Sensing Systems, Pittsburgh, PA, USA, 17–19 June 2009.

20. Hakala, I.; Kivela, I.; Ihalainen, J.; Luomala, J.; Gao, C. Design of low-cost noise measurement sensor network: Sensor function design. In Proceedings of the First International Conference on Sensor Device Technologies and Applications, Venice, Mestre, Italy, 18–25 July 2010.
21. Kivelä, I.; Gao, C.; Luomala, J.; Hakala, I. Design of Noise Measurement Sensor Network: Networking and Communication Part. In Proceedings of the Fifth International Conference on Sensor Technologies and Applications, Nice, France, 21–27 August 2011.
22. Bhusari, P.; Asutkar, G.M.; Tech, M. Design of Noise Pollution Monitoring System Using Wireless Sensor Network. *IJSWS* **2013**, 55–58.
23. Singla, T.; Manshahia, M.S. Wireless Sensor Networks for Pollution Monitoring and Control. *IJSRCSEIT* **2013**, *2*, 217–222.
24. Gubbi, J.; Marusic, S.; Rao, A.S.; Law, Y.W.; Palaniswami, M. A Pilot Study of Urban Noise Monitoring Architecture using Wireless Sensor Networks. In Proceedings of the International Conference on Advances in Computing, Communications and Informatics, Mysore, India, 22–25 August 2013.
25. Peckens, C.; Porter, C.; Rink, T. Wireless sensor networks for long-term monitoring of urban noise. *Sensors* **2018**, *18*, 3161. [CrossRef] [PubMed]
26. Noise Monitoring System. Patent CN202024817U, 2 November 2011.
27. Environment Noise Wireless Real-Time Monitoring System Based on ZigBee Technology. Patent CN102183294A, 13 March 2013.
28. A Construction Site Remote Real-Time Monitoring and Flowing Dust and Noise Monitoring System Based on an Internet of Things. Patent CN106919065A, 24 December 2015.
29. Mietlicki, C.; Mietlicki, C. Medusa: A new approach for noise management and control in urban environment. In Proceedings of the EuroNoise, Crete, Greece, 27–31 May 2018.
30. Camps-Farrés, J.; Casado-Novas, J. Issues and challenges to improve the Barcelona noise monitoring network. In Proceedings of the EuroNoise, Crete, Greece, 27–31 May 2018.
31. Gloaguen, J.R.; Can, A.; Lagrange, M.; Petiot, J.F. Road traffic sound level estimation from realistic urban sound mixtures by non-negative matrix factorization. *Appl. Acoust.* **2019**, *143*, 229–238. [CrossRef]
32. Filipponi, L.; Santini, S.; Vitaletti, A. Data Collection in Wireless Sensor Networks for Noise Pollution Monitoring. In Proceedings of the 4th IEEE International Conference on Distributed Computing in Sensor Systems, Santorini Island, Greece, 11–14 June 2008.
33. Nencini, L.; De Rosa, P.; Ascari, E.; Vinci, B.; Alexeeva, N. SENSEable Pisa: A wireless sensor network for real-time noise mapping. In Proceedings of the EuroNoise, Prague, Czech Republic, 10–13 June 2012.
34. Domínguez, F.; Dauwe, S.; Cuong, N.T.; Cariolaro, D.; Touhafi, A.; Dhoedt, B.; Botteldooren, D.; Steenhaut, K. Towards an environmental measurement cloud: Delivering pollution awareness to the public. *IJDSN* **2014**, *10*. [CrossRef]
35. Bartalucci, C.; Borchi, F.; Carfagnim, M. The smart noise monitoring system implemented in the frame of the Life MONZA project. In Proceedings of the EuroNoise 2018, Crete, Greece, 27–31 May 2018.
36. Bellucci, P.; Peruzzi, L.; Cruciani, F.R. Implementing the DYNAMAP system in the suburban area of Rome. In Proceedings of the Inter-Noise, Hamburg, Germany, 21–24 August 2016.
37. Mydlarz, C.; Sharma, M.; Lockerman, Y.; Steers, B.; Silva, C.; Bello, J.P. The Life of a New York City Noise Sensor Network. *Sensors* **2019**, *19*, 1415. [CrossRef] [PubMed]
38. Alías, F.; Alsina-Pagès, R.M. Review of Wireless Acoustic Sensor Networks for Environmental Noise Monitoring in Smart Cities. *J. Sens.* **2019**, *2019*, 13. [CrossRef]
39. Sigfox. Available online: https://www.sigfox.com/en (accessed on 8 October 2019).
40. Arduino Due. Available online: https://store.arduino.cc/due (accessed on 8 October 2019).
41. Arduino Ethernet Shield. Available online: https://store.arduino.cc/arduino-ethernet-shield-2 (accessed on 8 October 2019).
42. MikroTik SXT2. Available online: https://mikrotik.com/product/RBSXTG-2HnDr2-168 (accessed on 8 October 2019).
43. Portable 3G/4G Wireless N Router TL-MR3020. Available online: https://www.tp-link.com/en/home-networking/3g-4g-router/tl-mr3020/ (accessed on 8 October 2019).
44. Sigfox Module. Available online: https://www.cooking-hacks.com/sigfox-module-for-arduino-waspmote-raspberry-pi-intel-galileo-868-mhz-7184 (accessed on 8 October 2019).

45. Adafruit Electret Microphone. Available online: https://www.adafruit.com/products/1063 (accessed on 8 October 2019).
46. Fonestar Foam Windshield Ball YS-11. Available online: https://fonestar.com/EN/YS-11/Foam-windshield-ball (accessed on 8 October 2019).
47. Station Box ALU Enclosure. Available online: https://rfelements.com/products/integration-platforms/stationbox-alu/stationbox-s-carrier-class-2/ (accessed on 8 October 2019).
48. Mariscal-Ramirez, J.A.; Fernandez-Prieto, J.A.; Cañada-Bago, J.; Gadeo-Martos, M.A. A new algorithm to monitor noise pollution adapted to resource-constrained devices. *Multimed. Tools Appl.* **2015**, *74*, 9175–9189. [CrossRef]
49. *International Standard IEC 60651 Sound Level Meters*; International Electrotechnical Commission: Geneva, Switzerland, 2001.
50. *International Standard IEC 61672 Electroacoustic—Sound Level Meters*; International Electrotechnical Commission: Geneva, Switzerland, 2003.
51. Standard ASA S1.42-2001 (R2016). *Design Response of Weighting Networks for Acoustical Measurements*; Standard by American National Standards of the Acoustical Society of America: Melville, NY, USA, 2001.
52. Oppenheim, A.V.; Willsky, A.S.; Hamid Nawab, S. *Signals & Systems*, 2nd ed.; Prentice-Hall: Upper Saddle River, NJ, USA, 1996.
53. PCE-353 Sound Level Meter. Available online: http://www.industrial-needs.com/technical-data/datalogging-leq-sound-level-meter.htm (accessed on 11 October 2019).
54. MQTT. Available online: http://mqtt.org/ (accessed on 14 October 2019).
55. IETF. The Constrained Application Protocol (CoAP). Available online: https://tools.ietf.org/html/rfc7252 (accessed on 14 October 2019).
56. AMQP. Available online: https://www.amqp.org/ (accessed on 14 October 2019).
57. DDS. Available online: https://www.dds-foundation.org/ (accessed on 14 October 2019).

Article

A Digital Signal Processor Based Acoustic Sensor for Outdoor Noise Monitoring in Smart Cities

Juan Manuel López *, Jesús Alonso, César Asensio, Ignacio Pavón, Luis Gascó and Guillermo de Arcas *

Grupo de Investigación en Instrumentación y Acústica Aplicada, Universidad Politécnica de Madrid, 28040 Madrid, Spain; jalonsof@i2a2.upm.es (J.A.); casensio@i2a2.upm.es (C.A.); ignacio.pavon@upm.es (I.P.); luis.gasco@i2a2.upm.es (L.G.)

* Correspondence: juanmanuel.lopez@upm.es (J.M.L.); g.dearcas@upm.es (G.d.A.); Tel.: +34-910-673-306 (J.M.L.); +34-910-678-952 (G.d.A.)

Received: 9 December 2019; Accepted: 18 January 2020; Published: 22 January 2020

Abstract: Presently, large cities have significant problems with noise pollution due to human activity. Transportation, economic activities, and leisure activities have an important impact on noise pollution. Acoustic noise monitoring must be done with equipment of high quality. Thus, long-term noise monitoring is a high-cost activity for administrations. For this reason, new alternative technological solutions are being used to reduce the costs of measurement instruments. This article presents a design for a versatile electronic device to measure outdoor noise. This device has been designed according to the technical standards for this type of instrument, which impose strict requirements on both the design and the quality of the device's measurements. This instrument has been designed under the original equipment manufacturer (OEM) concept, so the microphone–electronics set can be used as a sensor that can be connected to any microprocessor-based device, and therefore can be easily attached to a monitoring network. To validate the instrument's design, the device has been tested following the regulations of the calibration laboratories for sound level meters (SLM). These tests allowed us to evaluate the behavior of the electronics and the microphone, obtaining different results for these two elements. The results show that the electronics and algorithms implemented fully fit within the requirements of type 1 noise measurement instruments. However, the use of an electret microphone reduces the technical features of the designed instrument, which can only fully fit the requirements of type 2 noise measurement instruments. This situation shows that the microphone is a key element in this kind of instrument and an important element in the overall price. To test the instrument's quality and show how it can be used for monitoring noise in smart wireless acoustic sensor networks, the designed equipment was connected to a commercial microprocessor board and inserted into the infrastructure of an existing outdoor monitoring network. This allowed us to deploy a low-cost sub-network in the city of Málaga (Spain) to analyze the noise of conflict areas due to high levels of leisure noise. The results obtained with this equipment are also shown. It has been verified that this equipment meets the similar requirements to those obtained for type 2 instruments for measuring outdoor noise. The designed equipment is a two-channel instrument, that simultaneously measures, in real time, 86 sound noise parameters for each channel, such as the equivalent continuous sound level (Leq) (with Z, C, and A frequency weighting), the peak level (with Z, C, and A frequency weighting), the maximum and minimum levels (with Z, C, and A frequency weighting), and the impulse, fast, and slow time weighting; seven percentiles (1%, 5%, 10%, 50%, 90%, 95%, and 99%); as well as continuous equivalent sound pressure levels in the one-third octave and octave frequency bands.

Keywords: outdoors noise; sound level meter; digital signal processing; multirate filters

---

1. Introduction

According to the United Nations, 55% of the world's population currently resides in urban areas, with this percentage projected to reach 66% by 2050 [1]. This rapid urban growth has caused environmental impacts, including environmental noise exposure to citizens. Noise pollution is one of the most important environmental health concerns around world. Environmental noise is produced by a variety of sources and is widely present in urban environments. Among the adverse effects produced by environmental noise exposure are those that threaten the well-being of human populations, deteriorate health, and decrease the ability of children to learn properly at school, leading to high economic costs for society [2].

In 2018, the World Health Organization (WHO) Regional Office for Europe published Environmental Noise Guidelines for the European Region. Compared to previous WHO guidelines on noise, there are some significant developments in the new version, among which the following should be highlighted: the inclusion of leisure noise in addition to noise from transportation (aircraft, rail and road traffic), and the use of long-term average noise exposure indicators to better predict adverse health outcomes compared to short-term noise exposure measures [3,4]. Both leisure noise and long-term average noise exposure indicators are issues to be considered in the management of noise in urban environments.

The Environmental Noise Directive (END) provides mechanisms for annoyance and sleep disturbance assessment, which if exceeded require action plans to be drawn that are designed to reduce exposure and protect areas not yet polluted by noise [5]. One of the most important evaluation mechanisms of the END is strategic noise mapping. At present, only industrial noise sources and noise from means of transport are taken into account (roads, railways, and airports) for strategic noise mapping. Meanwhile, there are many other sources of noise within urban environments not covered by strategic noise maps, such as citizen behavior, festive and cultural events, public works, urban maintenance and cleaning, and leisure noise, including night-life activities.

In order to carry out a comprehensive assessment of all the noise sources present in an area, one of the options that many cities usually use is environmental noise monitoring networks [6]. Environmental noise monitoring systems consist of a network of discrete sensor stations, usually integrated with an averaging sound level meter using an outdoor microphone. One of the main advantages of environmental noise monitoring systems lies in their ability to use the required time evolution data. On the other hand, the main drawback of environmental noise monitoring systems is that the discrete number of points implies weaknesses in the representativeness of spatial data. Different approaches have been proposed in recent times to solve the problems related to spatial representativeness. Examples of this are proposals to perform environmental noise measurements based on smart phones [7–9] and mobile monitoring networks using means of transport [10,11]. As a general rule, the lower the cost of the sensor, the more sensors can be used and the more spatially accurate the data will be.

Traditionally, professional systems used for noise measurement are designed to comply with very high-quality measurement requirements and are manufactured under strict international standards, such as IEC 61672 and IEC 61260 [12–17]. This situation makes this kind of equipment unsuitable for creating wide grids of measurement points in smart cities due to its high cost, large size, and other factors.

Recent technological developments related to the availability of cheaper and smaller equipment and innovations in communication networks and acoustic signal processing have led to the emergence of low-cost environmental noise sensor networks [18]. In recent years, several projects based on low-cost environmental noise sensor networks have been developed [19–22].

These solutions show different approaches for implementing noise monitoring systems in smart acoustic sensor networks. In each reference, we can see that different open topics in noise measurement are covered. However, there are several common elements. These solutions use commercial hardware to produce a test concept of the proposed architecture. Not all references use exhaustive tests to measure quality characterization, and when these tests are implemented, they are only implemented from an acoustic point of view; no electrical tests are used, as suggested by the standard IEC 61672-3. In addition, these solutions measure only a few acoustic parameters, mainly the equivalent continuous sound level (Leq). Although some of these solutions point toward future trends in the application of algorithms for source identification, the equipment used is mono-channel, which reduces the possibility of implementing algorithms for source localization, one of the open topics for future noise measurement in cities.

To address these points, in this paper, a low-cost instrument to measure outdoor noise is presented. The equipment has been designed to form an electronics–microphone set, which allows it to be seen and used as a sensor device to be connected to microprocessor systems, thereby increasing its versatility and ease of use. The electronic device has been designed keeping in mind some of the challenges to be covered by future environmental noise monitoring networks. For this reason, the equipment incorporates two measurement channels, which together with a digital signal processor (DSP) will allow the future implementation of algorithms for detecting and locating noise sources. Likewise, noise measurement algorithms have been designed to meet the measurement requirements of type 1 instruments, so in the future, different types of microphones can be connected, thereby covering the different degrees of precision needed.

The equipment has been designed with fully digital implementation capability to increase its quality and reliability and to reduce its cost. The system has been designed around a Texas Instruments C5000 DSP, due to its low power consumption and high performance. Our instrument implements two fully functional measuring channels, having the basic functionalities of a sound level meter together with the ability to perform octave-level and one-third octave-level frequency analyses.

One of the key elements in sound level meters is a condenser microphone, which has exceptional characteristics for measuring noise, such as a flat frequency response, a large dynamic range, high precision, and repetitiveness [23]. However, these characteristics make this type of instrument more expensive, hindering the widespread use of this type of microphone in the implementation of monitoring networks with a large number of nodes in smart cities.

One of the open challenges in environmental noise sensor networks is to build mixed networks where low-cost instruments can be used without the overall quality of the network measures being substantially affected. To comply with these requirements, the use of condenser microphones must be heavily restricted. However, the equipment used must be subjected to the tests indicated by the standard for sound level meters to assure the quality of their measurements [12–17]. For this reason, and to characterize the quality and usability of the equipment, a set of exhaustive tests have been performed. These tests were mostly laboratory tests similar to those applied to sound level meters, including electric and acoustic tests [14,17]. One of the main objectives of the designed equipment is its capability to be used as a sensor in any acoustic sensor network. For this reason, the device has a simple interface for connection to microprocessor systems. The designed equipment was connected to a NRG2 panStamp wireless module. This module is based on the CC430F5137 system-on-chip (SoC) design, which provides a communication radio channel at free industrial, scientific and medical (ISM) bands (868 MHz). With this configuration, a set of eight instruments were built to be connected to an existing acoustic sensor network in the city of Málaga in Spain. Málaga is a city that is very concerned about the quality of life of its citizens [24] and has the infrastructure to carry out an outdoors test with the units we built.

2. System Description

Figure 1 shows a block diagram of the designed instrument and the form factor of the implemented module. The set formed by the microphone and the designed card was implemented as an original equipment manufacturer (OEM) module, allowing it to be easily integrated into any monitoring network. The core of the system is a digital signal processor that digitally implements all acoustic functions in order to achieve a more robust, economical, and adjustment-free architecture. Power consumption is a very important issue in systems designed for monitoring purposes. Usually, these systems are installed in remote or isolated places and are powered by solar panels. Thus, a family of fixed-point and low power consumption DSPs has been chosen. The use of a DSP allows us to implement many parameters without an increase in hardware cost, requiring only firmware upgrades. This system is capable of measuring, in real time, the following parameters for both channels simultaneously: the equivalent continuous sound level (Leq) (with Z, C, and A frequency weightings); the peak level (with Z, C, and A frequency weightings); the maximum and minimum levels with (Z, C, and A frequency weightings); the impulse, fast, and slow time weightings; seven percentiles (1%, 5%, 10%, 50%, 90%, 95%, and 99%); as well as continuous equivalent sound pressure levels in the one-third octave and octave frequency bands.

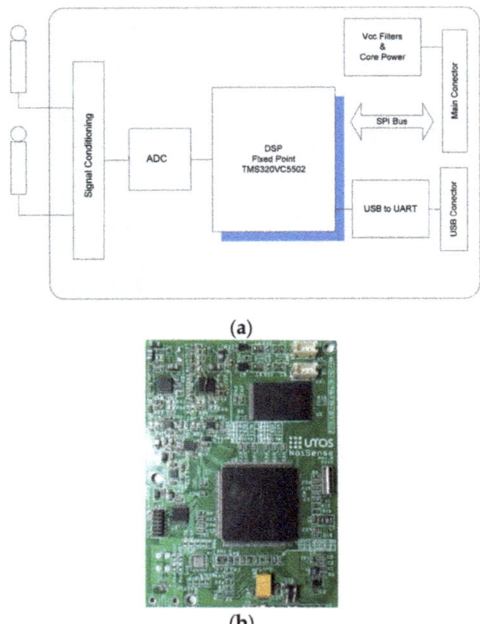

Figure 1. (**a**) Block diagram of the original equipment manufacturer (OEM) module for noise measurement; (**b**) final circuit implementation of 87 × 62 mm. DSP = digital signal processor; SPI bus = serial peripheral interface; ADC = analog to digital converter; UART = universal asynchronous receiver-transmitter.

2.1. Hardware Solution

One of the most import elements in a noise measurement system is the microphone, which determines the overall quality of the equipment. In order to build cost-effective solutions, condenser microphones must be avoided due to their high price. The microphone used is an electret Panasonic WM 63-PR [25], whose frequency response is shown in Figure 2. The frequency response and dynamic range of the microphone suggest that it can be used to design an instrument with characteristics similar to type 2 sound level meters, according to the IEC 61672 standard [13], with a frequency range of 63 HZ to 8 kHz.

The first block in Figure 1a shows the signal conditioning stage, which is used to adapt the signal from the microphone to the analog to digital converter (ADC). This stage is formed by two active filters. One of them is a high-pass filter with a cutoff frequency of 5 Hz, and the other is a low-pass filter with a cutoff frequency of 22.7 kHz; both feature a unity gain. To adapt the microphone's signal level to the dynamic input range of the ADC, a gain stage was added. To minimize the noise and protect the equipment against electrostatic discharge (ESD) and electromagnetic interference (EMI), a front end circuit was added between the microphone and the first filter. Figure 3 shows this circuit.

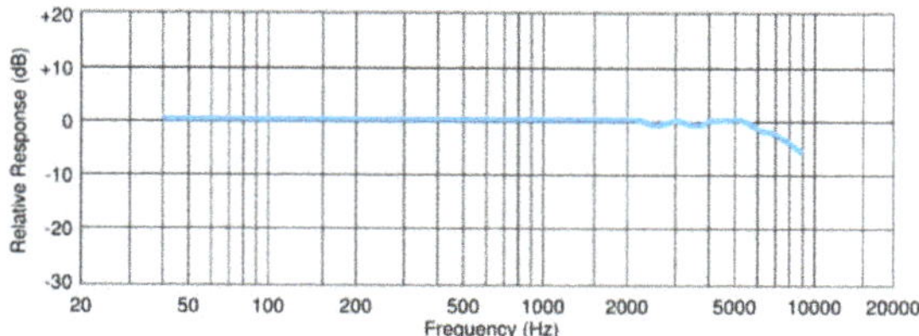

Figure 2. Panasonic WM 63-PR microphone frequency response.

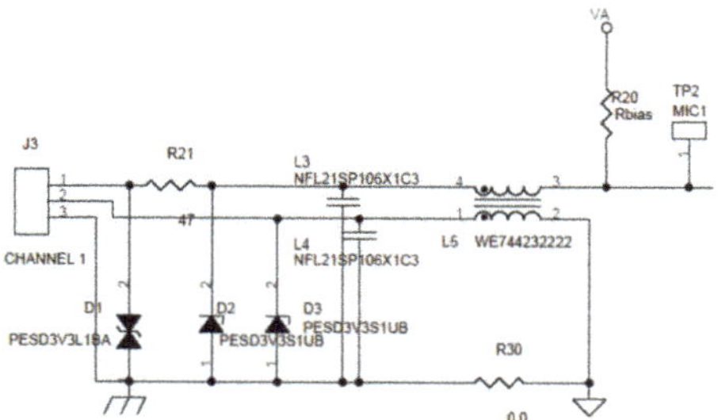

Figure 3. Filter stage for noise minimization and electrostatic discharge (ESD) and electromagnetic interference (EMI) protection.

The ADC used is a CS5344 from Cirrus Logic [26], whose main features are a 24-bit sigma–delta converter, a sample rate up to 108 kHz, a dynamic range of 98 dB at 5V, a power consumption less than 40 mW at 3.3V, and a single power supply. The ADC is connected to the DSP using an multichannel buffered serial port (MCBSP) interface and left-justified with 256x speed. The transfers are managed using a direct memory access (DMA) channel; the block transfer includes 2048 samples for the two channels. Figure 4 shows the ADC connection.

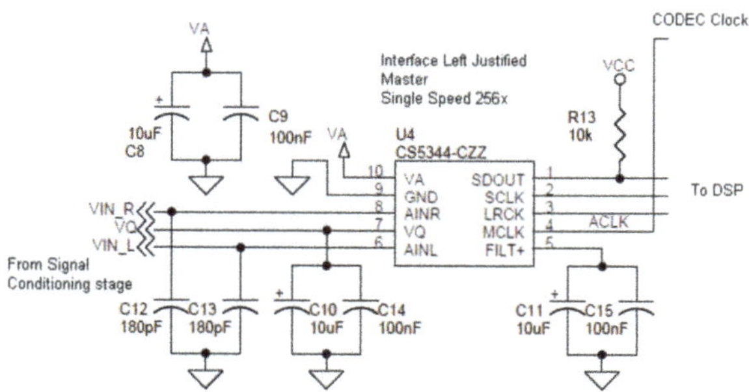

Figure 4. ADC connection.

The core of the system is a Texas Instruments fixed-point DSP (TMS320VC5502) in a low-profile quad flat package (LQFP) [27]. To optimize the system, all algorithms are executed in the internal RAM of the DSP, so only flash memory is connected to the DSP itself, which is used to store the code with the algorithms. This code is downloaded to the RAM at boot time. All algorithms are implemented through cycle optimization and have been coded in assembly language to reduce the code size and improve run time. In this way, the code can be allocated in the internal RAM of the DSP, and a lower main frequency can be used for the DSP, thereby reducing the overall power consumption [28].

The system is designed to be used similarly to an OEM module to measure environmental noise; it can be managed by an external control unit using a serial peripheral interface (SPI) interface. This equipment can be connected to a microprocessor system to manage it using a simple interface. In this interface, there are four SPI signals, four control lines (these lines are used for the microprocessor system to reset the instrument, to send a configuration command, to start the measurements, and for when the equipment sends data to the microprocessor), and two lines for the power supply.

2.2. Software Implementation

Software for non-time-critical tasks, such as system initialization, commands parsing, and flow control, were implemented in the C language. However, the audio signal processing was coded in assembly language to improve its run time. Figure 5 shows the main flowchart of the code.

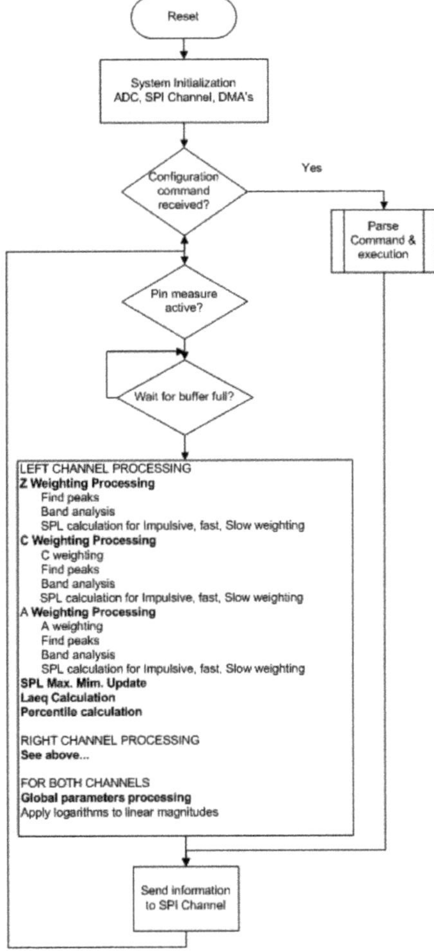

Figure 5. Software flowchart.

For the time weighting processing (impulsive, fast, and slow), the following formula is used:

$$L_m[n] = \frac{1}{Fs \cdot T_m}\left[(Fs \cdot T_m - 1)L_m[n-1] + x^2[n]\right] \quad (1)$$

where T_m is 0.125 seconds for fast weighting, 1 second for slow weighting, and 0.035 seconds for impulse weighting [12]. As divisions are computationally expensive for fixed point DSPs, the values (Fs·Tm − 1) and 1/(Fs·Tm) are pre-calculated to optimize run time. In the few places where a division cannot be pre-computed, the Newton–Raphson method is used to quickly obtain the reciprocal and perform a multiplication instead of a division. The impulse time weighting detector also requires an additional filter at the output, with a maximum drop slope of 2.9 dB.

The frequency weighting filters A and C were calculated using MATLAB [29]. First, the analog filters have been designed using the poles shown in Table 1 [12]. Then, the analog filters were transformed into digital filters using the impulse invariance method [30].

Table 1. Poles used to design the analog filters.

Weighting	Poles
C	Two real poles at 20.6 Hz Two real poles at 12,200 Hz
A	Poles in C and One pole at 107.7 Hz One pole at 737.9 Hz

The IEC 61260-1 standard allows two methods for design of the octave and one-third octave filters in base 10 (preferred) and in base 2 (allowed) modes [15]. For optimization and computational efficiency, the filters were implemented in base 2. A multirate filter structure was used for the design of the third octave filter bank. This allowed us to simplify the design process; instead of designing 30 filters, it was only necessary to design the three corresponding to the octave of greatest frequency, with a central frequency of 16 kHz, in addition to the anti-aliasing filter for the decimator. Figure 6 shows the base block used to build the bank filter. This block consists of a down sampler (by 2), an anti-aliasing filter, and three filters with normalized center frequencies of 0.388, 0.488, and 0.615 [31].

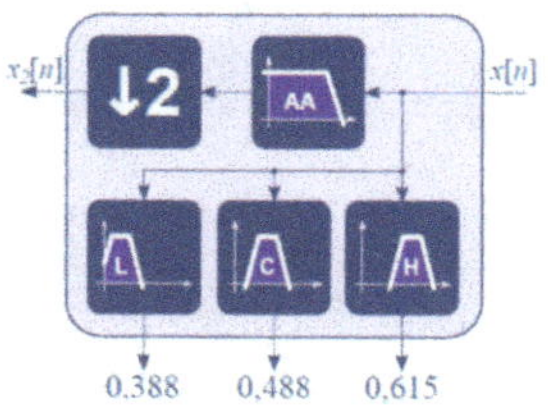

Figure 6. Base block used to build the bank filter. Frequencies are normalized to sampling frequency (Fs/2).

The final frequencies obtained in the real domain have slight differences compared to the nominal values. However, this situation is understood and allowed in the IEC 61260 standards [15–17]. Figure 7 shows the complete filter bank that was implemented. This kind of implementation reduces the workload because only four filters are designed. This methodology also has two additional advantages: avoiding the cutoff frequencies to get close to the limit frequency's minimum and maximum (0 and 32,768 Hz), and reducing the processing power required by the DSP. Due to the successive decimation performed at each stage, the processing load required to compute the whole filter bank is about twice that required to calculate the basic building block shown in Figure 6.

To obtain the response of the octave band filter bank from the one-third octave band filter bank, it is sufficient to add (in linear magnitude, not in dB) the values obtained from the three adjacent one-third octave filters in the octave of interest.

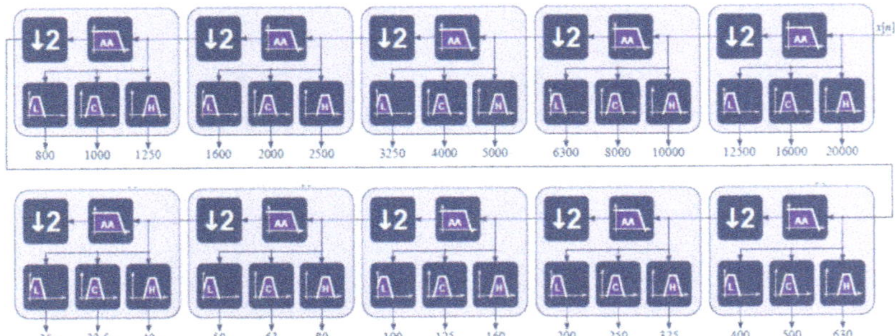

Figure 7. Filter bank for the one-third octave. The frequencies are shown in Hz.

3. Implemented System Verification

In order to verify the designed system, the equipment underwent different tests to analyze the acoustic and electrical properties. The electrical tests were used to verify the quality of the digital implementation (digital filters, algorithms, etc.) and the analog input chain, while the acoustic tests were used to verify the microphone's behavior. Both tests together were used to verify the overall features of the equipment.

3.1. Electrical Test

The third section of the international standard IEC 61672 specifies the tests required to verify the frequency weightings implemented in a sound level meter [14], which should be determined relative to the response at 1 kHz using steady sinusoidal electrical input signals. At the reference level range, and for each frequency, the 1 kHz input signal should be adjusted to yield an indication that is 45 dB less than the upper limit for the sound level meter [14]. The tolerance limits are specified in IEC 61672-1 [12]. For class 1 SLM, the test should be performed with nine frequencies at nominal octave intervals from 63 Hz to 16 kHz. For a class 2 SLM, eight frequencies should be used from 63 Hz to 8 kHz. Table 2 shows the results obtained. Here, the equipment fully fits the requirements of a class 1 SLM.

Table 2. Results obtained for the frequency weighting test.

	Frequency (Hz)	Frequency Weights (dB)	Correction (dB)	Read Level (dB)	Expected Level (dB)	Deviation (dB)	U(uncertainty) (dB)	Positive Tolerance (dB)	Negative Tolerance (db)
101.20	63	−26.2	0	75.2	75.0	0.20	0.18	1.5	−1.5
91.10	125	−16.1	0	75.1	75.0	0.10	0.18	1.5	−1.5
83.60	250	−8.6	0	74.9	75.0	−0.10	0.18	1.4	−1.4
78.20	500	−3.2	0	75.0	75.0	0.00	0.18	1.4	−1.4
75.00	1000	0	0	75.0	-	-	-	-	-
73.80	2000	1.2	0	75.0	75.0	0.00	0.18	1.6	−1.6
74.00	4000	1	0	74.9	75.0	−0.10	0.18	1.6	−1.6
76.10	8000	−1.1	0	75.0	75.0	0.00	0.18	2.1	−3.1
86.10	16,000	−6.6	0	75.2	75.0	0.26	0.18	3.5	−17

For the level linearity test, the IEC 61672-3 states that the test must be performed using a sinusoidal electrical signal at a frequency of 8 kHz that varies its amplitude for the SLM linear measurement range [14]. The linearity will be measured in steps of 5 dB until it reaches 5 dB before the extreme limits of the linear range. Then, the steps will be 1 dB increments until the limits are reached. Table 3 shows the results obtained for the linearity test. This table only shows the central values and extremes for clarity. With these results, we can establish the linear dynamic range of the equipment in 80 dB.

Table 3. The results obtained for the linearity test. Values marked with " ... " are omitted for clarity (the device passed the test). SLM = sound level meter.

Applied SPL (dB)	Frequency (Hz)	SLM Level (dB)	Expected Level (dB)	Deviation (dB)	U (dB)	Positive Tolerance (dB)	Negative Tolerance (db)	Test Result
122.1	8000	118.9	122.0	−2.1	0.14	1.1	−1.1	ERROR
121.1	8000	118.9	121.0	−1.1	0.14	1.1	−1.1	ERROR
120.1	8000	118.5	119.0	−0.5	0.14	1.1	−1.1	PASS
...	...	...	...	...	...	...	...	...
105.1	8000	104.0	104.0	0.0	0.14	1.1	−1.1	PASS
100.1	8000	99.0	99.0	0.0	0.14	1.1	−1.1	PASS
95.1	8000	94.0	-	-	-	-	-	
85.1	8000	84.0	84.0	0.0	0.14	1.1	−1.1	PASS
80.1	8000	79.0	79.0	0.0	0.14	1.1	−1.1	PASS
...	...	...	...	...	...	...	...	...
39.1	8000	38.9	38.0	0.9	0.14	1.1	−1.1	PASS
38.1	8000	38.1	37.0	1.1	0.14	1.1	−1.1	ERROR
37.1	8000	37.4	36.0	1.4	0.14	1.1	−1.1	ERROR

The instrumentation equipment used for the electrical tests included:

- Multifunction acoustic calibrator Brüel & Kjaer B&K 4226;
- Signal Generator Stanford DS360;
- Multimeter Keithley 2015-P.

The reference conditions were Temperature = 23 °C ± 2 °C; Relative humidity = 50% ± 20%; Atmospheric pressure = 95 kPa ± 10 Pa.

3.2. Acoustic Test

A set of tests was performed using a multifunction acoustic calibrator (B&K 4226) in a calibration laboratory. The use of a multifrequency calibrator allowed us to determine the microphone's influence on the equipment, thus removing the effects of diffraction and refraction that appear when the microphone is inside an acoustic field. Thirteen different devices were tested, and Figure 8 shows the results obtained. The individual measurements of each of the devices are shown in light blue, and the measurement corresponding to a type 1 reference sound level meter is shown in gray, adjusted to the level offered by the 1 kHz reference microphone. Also, the mean ($\overline{\mu}_{mic}$) and the typical deviation with a 95% confidence interval ($\overline{\mu}_{mic} \pm 2\sigma$) are shown in solid and dotted dark blue, respectively.

The variability of the thirteen units with their microphones is low and constant for frequencies between 31.5 Hz and 4000 Hz (the predominant bandwidth in city noise). This deviation increases for higher frequencies. Table 4 shows the values of the typical deviation.

Table 4. Standard Deviation for the frequency bands.

Frequency (Hz)	31.5	63	125	250	500	1000	2000	4000	8000	12,500	16,000
Standard deviation (dB)	1.3	1.3	1.3	1.3	1.3	1.3	1.3	1.4	1.8	2.4	2.6

Figure 9 shows the difference between the mean of the thirteen units and the measures of the reference equipment (type 1). This difference is less than 3 dB at all frequencies, except at 16,000 Hz, where the difference is 10 dB.

Figure 10 shows the response of the designed equipment in reference to a tone of 1 kHz and a 94 dB sound pressure level, the reference commonly used in the characterization of this type of measurement instrument. At 1 kHz, all frequency responses are about 0 dB, and the variability is visible with respect to the other frequency bands. For this equipment, the variation of the response is reduced (almost flat) to 4 kHz, thereby worsening its frequency response noticeably from 8 kHz.

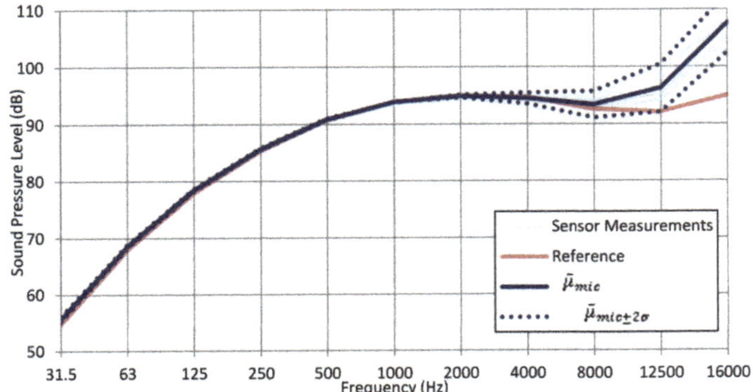

Figure 8. Tests results using a multifrequency calibrator.

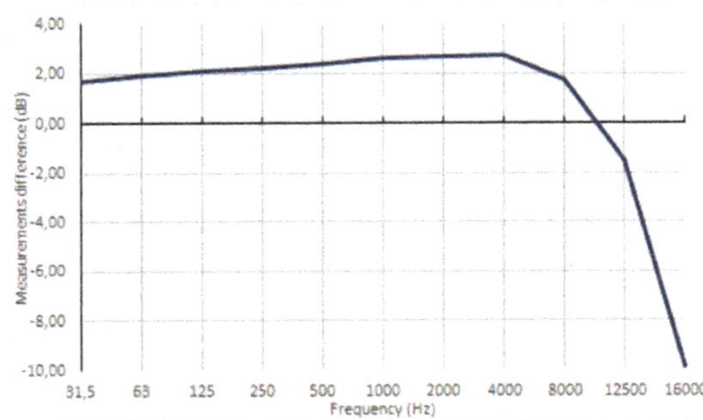

Figure 9. Difference in the average value between the type I reference equipment and the mean of the measures of the equipment designed for each frequency band.

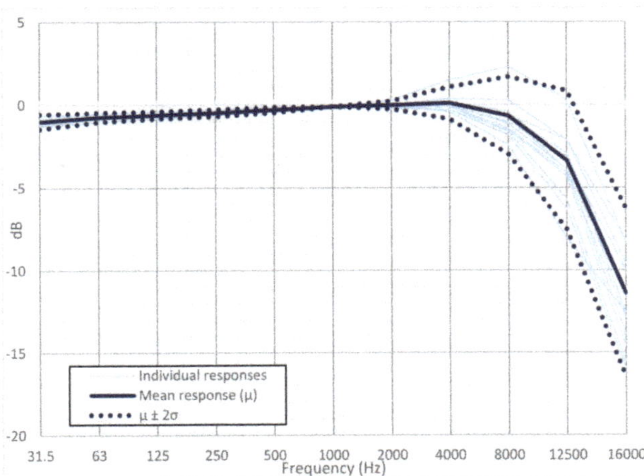

Figure 10. Frequency response for the thirteen units, including the mean and standard deviation for 2 σ.

Finally, the closest equipment to the average was chosen to be tested according to the IEC-61672-3 standard [14]. Table 5 shows the measurements and their comparisons with the requirements specified in the aforementioned standard. At 8 kHz, the equipment does not meet the specifications of a type 1 instrument. The values here fully fit with those of type 2 equipment [12].

Table 5. The obtained results. At 8000 Hz, we can see that the microphone response deviation falls outside the positive tolerance for a type I instrument (marked in grey).

Applied SPL (dB)	Frequency (Hz)	Frequency Weights (dB)	Correction (dB)	Read Level (dB)	Expected Level (dB)	Deviation (dB)	U (dB)	Tolerance (dB)
93.96	63	−26.2	0.0	66.45	65.90	0.55	0.23	1.5; −1.5
93.95	125	−16.1	0.0	76.40	75.99	0.41	0.20	1.5; −1.5
93.95	250	−8.6	0.0	83.70	83.49	0.21	0.20	1.4; −1.4
93.94	500	−3.2	0.0	88.95	88.88	0.07	0.23	1.4; −1.4
93.96	1000	0	0.1	92.00	–	–	–	–; -
93.97	2000	1.2	0.4	92.80	92.91	−0.11	0.20	1.6; −1.6
93.96	4000	1	1.6	92.35	91.50	0.85	0.23	1.6; −1.6
93.90	8000	−1.1	2.9	91.05	88.04	3.01	0.24	2.1; −3.1

3.3. *System Verification Conclusions*

Once the different tests had been carried out, we concluded that from the perspective of digital signal processing, this equipment is capable of complying with the specifications of type 1 instruments up to 16 kHz. However, the use of a low-cost electret microphone means that the acoustic behavior can only fully fit a type II instrument when tests are done using a multifrequency calibrator for a frequency range up to 8 kHz.

4. Results: Deployment in Málaga City

Once it was verified that the equipment offers sufficient quality to measure the noise in cities, the next step was to test the OEM module design in a real environment. The Spanish city of Málaga had optimal conditions to carry out this experiment. Málaga is a modern city concerned about the problem of outdoor noise; it has an operational data monitoring platform with the ability to insert new nodes, and it is possible to access these data to analyze them. For these reasons, eight units were deployed in the city to verify their behavior under real operating conditions over three months. The noise monitoring systems were located on the façades of residential buildings, in areas affected by night-time leisure noise located on pedestrian streets, and near bars, pubs, restaurants, and terraces [32]. To help to verify the quality of the measurements of these units, the equipment was installed in places where type 1 monitors were available. To keep the power consumption as low as possible, the equipment under study was configured to provide data every minute to reduce the power used for communication. In this way, the equipment used batteries, making the installation process easier. Unlike the network of type I monitors, where the equipment sends data directly to the central server through a general packet radio service (GPRS) connection, the terminals under study send their data via radio (868 MHz) to several access points, which then send the data to the same central server through a GPRS connection. Figure 11 shows the architecture used. The implemented OEM module was connected to a NRG2 panStamp-based CPU, and the code was developed using Arduino. This CPU obtained data from the OEM module and sent it by radio to the access point, which transformed it into frames for the existing architecture.

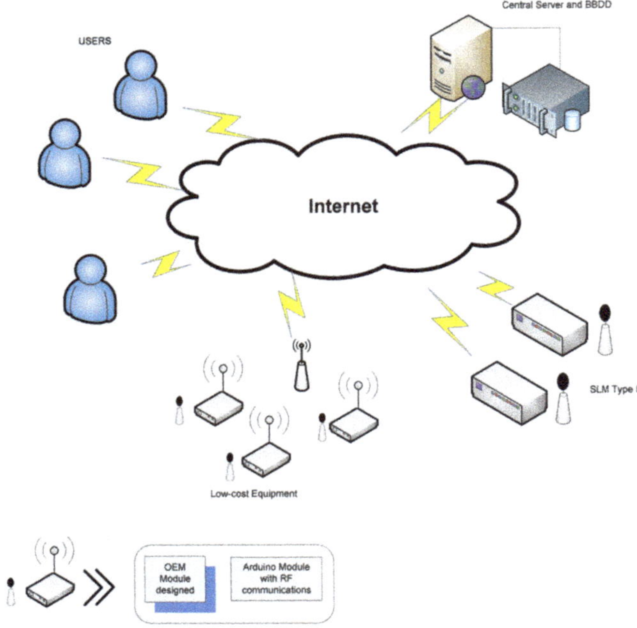

Figure 11. Network architecture and details of the low-cost monitors.

In addition, to verify the evolution of the equipment, three tests measurements were carried out in situ to determine the evolution of the operation of the units. One was performed after the outdoor installation using an acoustic calibrator (94 db). Table 6 shows the results for the eight devices and their locations.

Table 6. Verification values after installation outdoors (94 dB applied).

Monitor Location street	Value (dB) at Calibration	Monitor Location street	Value (dB) at Calibration
Andromeda, 9	93.6	Plutarco, 20	93.2
Angel	93.7	Plutarco, 57	93.8
Capitán	93.6	Velázquez	93.4
M. Vado Maestre, 4	93.6	M. Vado Maestre, 6	93.4

The second verification was carried out a month later; in this case, a type 1 instrument was placed next to the microphone of the equipment under study to measure the same noise at the same time. A three-step cycle of about ten minutes was performed. First, the background noise was measured for three minutes. Then, a five-minute stationary measurement was taken using a speaker placed above the vertical axis of the microphones with white noise. Then, the background noise was measured again for three minutes. Finally, at the end of the measurement period, another comparison was made using an acoustic calibrator for the eight devices used. After analyzing the data from these three tests, two of the units presented important errors, so they were removed for the final analysis under suspicion that they had been damaged during the measurement period.

The data from the central server can be remotely consulted and downloaded using a web browser. After three months of the campaign, an exhaustive analysis of the data stored on the central server was performed. The downloaded data were analyzed and a report was made. Figure 12 shows the information for each monitor. The top part shows a view of the equipment installation, along with its localization on the city map. The equivalent noise levels for the day (Ld), evening (Le), and night (Ln) were then calculated. Using a semaphore color code, the obtained noise levels were classified. A figure with a calendar shows the percentage of measurements received for each day. Days that received less than 30% data were deleted for the calculation.

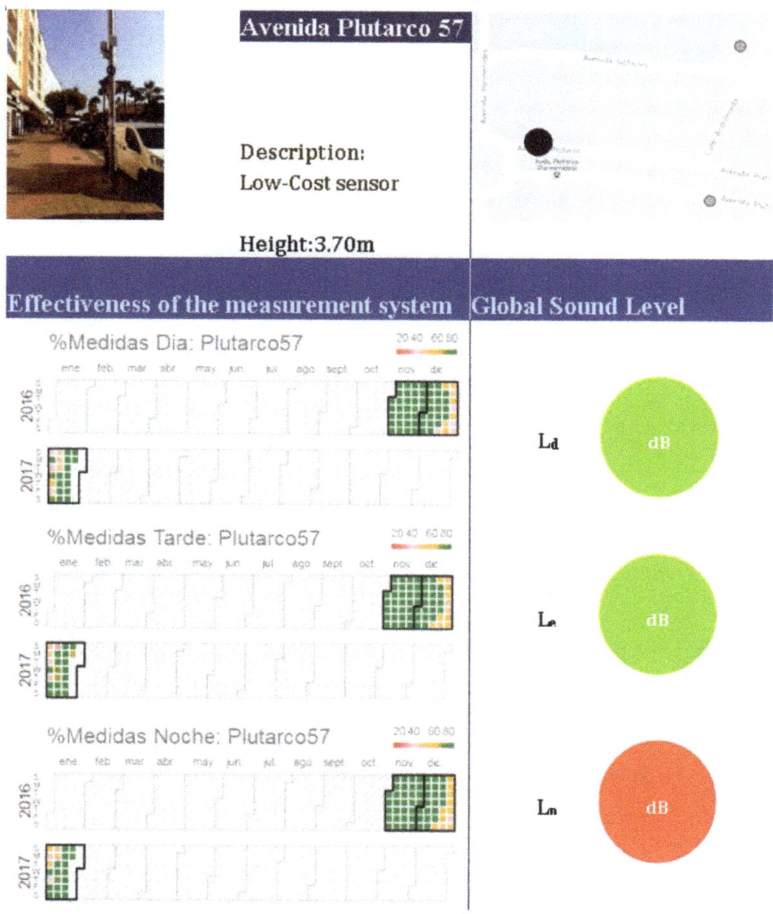

Figure 12. Information template used for each measurement point.

The report also shows box plot figures with the measurement difference between a type 1 instrument and a low-cost instrument. Figure 13 shows a diagram for the equipment placed on Plutarco Street. We can see that there is a greater dispersion in sound levels at night, which is due to the minimum value of the dynamic input range of the instrument. The figure also shows the outlier values, where one can see the low outlier numbers over three months.

Table 7 shows the average values (day, afternoon, and night) measured during the 84 campaign days for type 1 and low-cost instruments.

The device installed on Plutarco Street 57 shows a great difference in the night's equivalent noise level. After a deeper analysis of the data, this deviation was found to be due to the measurements made during the Christmas season. The Christmas lights installed near the devices produced some interference in the measurements.

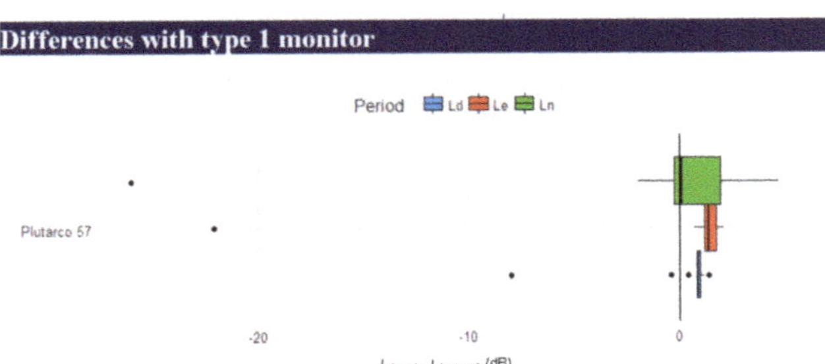

Figure 13. Box plot diagram to compare the measurements between instruments.

Table 7. Ld, Le, and Ln levels. The left values are from low-cost monitors and the right values are from the type 1 monitors used as a reference.

Monitor's Location	L_d (dB)	L_e (dB)	L_n (dB)
Andromeda, 9	62/63	62/63	57/59
Angel	65/67	69/71	71/70
Capitán	69/74	78/79	74/76
M. Vado Maestre, 4	65/67	66/71	68/71
Plutarco, 20	61/63	62/64	58/59
Plutarco, 57	65/65	67/64	71/62

5. Discussion

When the design started, the main objective was to obtain a noise measurement device with the following features:

- Quality and characteristics comparable to the sound level meters that are used today to measure this type of magnitude, according to the standards and regulations that apply to them;
- Low cost required to build a large-scale monitoring network;
- Processing capabilities for the detection of target sources and their spatial localization;
- Easy integration into sensor networks.

All these objectives, also mentioned by other authors, have been covered [19], and the developed equipment was conceived as a generic OEM card. The unit can be controlled by two serial interfaces: SPI or USB. These interfaces are widely used in microprocessor-based systems, which allows this device to be easily integrated, and thus can be used to build noise measurement equipment in a simple way. The developed equipment has the ability to measure two channels simultaneously and implement the basic functions of an integrating sound level meter, carrying out measurements simultaneously for A, C, fast, slow, and impulsive weights. It also allows one to obtain the maximum and minimum values, perform an analysis in the octave and one-third octave frequency bands, and calculate the percentiles of the measurements, resulting in eighty-six parameters being measured for each channel in real time. All these parameters are calculated digitally; the code optimization allows the DSP to compute all these parameters for both channels in real time using a clock frequency of 60 MHz. As the DSP allows one to raise the clock speed to 300 MHz, further algorithms for tasks such as noise source identification and pattern recognition of acoustic footprints can be implemented in the future using this equipment after a software upgrade.

The electrical tests applied to the equipment show that the electronics and algorithms implemented have the behavior of a type 1 sound level meter for an input frequency range spanning from 10 Hz to 20 KHz and a linear dynamic range of 80 dB. However, the use of an electret microphone instead of a condenser microphone reduces the electric performance of the equipment, limiting its characteristics to those of a type 1 instrument for a frequency range up to 8 kHz (Table 5). For type 2 instruments, the tolerance allowed at 8 kHz increases to ± 5.6 dB. Thus, the equipment complies with the requirements of a type 2 instrument according to IEC61672-3 [14]. The use of the electret microphone also reduces the acoustic features of the equipment. However, this error can be assumed for the application of acoustic noise measurement in cities, as has been shown with the correlation analysis using the measurements of type 1 equipment installed as a reference in the measurement sites.

Eight units were deployed in the city of Málaga from November to January to verify their operation under real conditions of outdoor noise measurement. Some very interesting conclusions were obtained when analyzing the data uploaded by the units compared to the type 1 equipment used to contrast the results. Finally, only six devices could be used for the final results, since two of them were damaged and suffered large deviations in their measurements after the three months of the measurement campaign. The results obtained for four of the devices were very good; however, two of them presented a deviation in their data right at the limit of what is acceptable for a measurement, with quality similar to that provided by type 2 instruments. A detailed study of these data showed that the devices were more influenced by meteorological phenomena, such as rain, than type 1 monitors. In addition, it was possible to verify how—coinciding with the installation and lighting of the streets for Christmas—some measurements were affected by electromagnetic interference.

To summarize, the designed equipment can easily be integrated by third party users into any wireless sensor network due to the available digital interface. The device is designed to have SLM functionality and is capable of measuring 86 parameters by channel, including frequency analysis. This creates new possibilities for low-cost sensor networks, where traditionally only the equivalent continuous sound level (Leq) is measured. The designed instrument had an input linear range of 80 dB and passed the electrical test according to the standards applied to a type 1 SLM, although the use of an electret microphone reduced the quality of the measurements to those indicated for type 2 instruments. However, this is an important point. If the device was to be built with a better microphone (at present, microelectromechanical (MEMS) microphones possess good features similar to those of traditional condenser microphones [19,32]), the instrument could be a very good solution for creating low-cost sensor networks with high measurement quality.

The results have shown that there are two aspects to need to be improved in the proposed solution. One is the vulnerability that some units showed to electromagnetic interference. The case used to house the equipment was a standard plastic case. To improve this feature, a metal case or plastic cases with anti-EMI treatments must be used. The other major change would be the use of new technological advances in MEMS microphones to achieve better measurement quality. Thus, the next step will be to modify the signal conditioning stage to use MEMS microphones [33].

Author Contributions: J.M.L., G.d.A., and J.A. performed the electronics design and firmware implementation. J.A. and C.A. performed the acoustic tests, and L.G. and I.P. performed the data analysis. All authors have read and agreed to the published version of the manuscript.

Acknowledgments: The authors want to thank the company Vatia S.L. for providing use of and access to the web platform used to store the measurements and use the data for the objectives of this paper, and to the Málaga city council for allowing the installation and use of the data acquired.

Conflicts of Interest: Declare conflicts of interest or state "The authors declare no conflict of interest."".

References

1. United Nations. Statistical Papers—United Nations (Ser. A), Population and Vital Statistics Report. Available online: https://doi.org/10.18356/e59eddca-en (accessed on 3 December 2019).
2. European Environmental Agency. Environmental Topics, Environment and Health, Noise. Available online: https://www.eea.europa.eu/themes/human/noise (accessed on 3 December 2019).
3. *Environmental Noise Guidelines for the European Region 2018*; World Health Organization, Regional Office for Europe: København, Denmark, 2018.

4. Jarosińska, D.; Héroux, M.-È.; Wilkhu, P.; Creswick, J.; Verbeek, J.; Wothge, J.; Paunović, E. Development of the WHO Environmental Noise Guidelines for the European Region: An Introduction. *Int. J. Environ. Res. Public Health* **2018**, *15*, 813. [CrossRef] [PubMed]
5. European Commission. *Directive 2002/49/EC of the European Parliament and of the Council of 25 June 2002 Relating to the Assessment and Management of Environmental Noise*; European Commission: Brussels, Belgium, 2002.
6. Garrido Salcedo, J.C.; Mosquera Lareo, B.M.; Echarte Puy, J.; Sanz Pozo, R. Management Noise Network of Madrid City Council. In Proceedings of the Internoise 2019—Noise Control for a Better Environment, Madrid, Spain, 16–19 June 2019.
7. Zamora, W.; Calafate, C.T.; Cano, J.-C.; Manzoni, P. Accurate Ambient Noise Assessment Using Smartphones. *Sensors* **2017**, *17*, 917. [CrossRef] [PubMed]
8. Zuo, J.; Xia, H.; Liu, S.; Qiao, Y. Mapping Urban Environmental Noise Using Smartphones. *Sensors* **2016**, *16*, 1692. [CrossRef] [PubMed]
9. Murphy, E.; King, E.A. Smartphone-based noise mapping: Integrating sound level meter app data into the strategic noise mapping process. *Sci. Total Environ.* **2016**, *562*, 852–859. [CrossRef] [PubMed]
10. Alsina-Pagès, R.M.; Hernandez-Jayo, U.; Alías, F.; Angulo, I. Design of a Mobile Low-Cost Sensor Network Using Urban Buses for Real-Time Ubiquitous Noise Monitoring. *Sensors* **2017**, *17*, 57. [CrossRef] [PubMed]
11. Quintero, G.; Aumond, P.; Can, A.; Balastegui, A.; Romeu, J. Statistical requirements for noise mapping based on mobile measurements using bikes. *Appl. Acoust.* **2019**, *156*, 271–278. [CrossRef]
12. IEC 61672-1. *Electroacoustics-Sound Level Meters. Part 1: Specifications*; International Electrotechnical Commission: Geneva, Switzerland, 2013.
13. IEC 61672-2. *Electroacoustics-Sound Level Meters. Part 2: Patten Evaluation Test*; International Electrotechnical Commission: Geneva, Switzerland, 2013.
14. IEC 61672-3. *Electroacoustics-Sound Level Meters. Part 3: Periodic Test*; International Electrotechnical Commission: Geneva, Switzerland, 2013.
15. IEC 61260-1. *Electroacoustics-Octave-Band and Fractional-Octave-Band Filters—Part 1: Specifications*; International Electrotechnical Commission: Geneva, Switzerland, 2014.
16. IEC 61260-2. *Electroacoustics-Octave-Band and Fractional-Octave-Band Filters—Part 2: Patten Evaluation Test*; International Electrotechnical Commission: Geneva, Switzerland, 2016.
17. IEC 61260-3. *Electroacoustics-Octave-Band and Fractional-Octave-Band Filters—Part 3: Periodic Test*; International Electrotechnical Commission: Geneva, Switzerland, 2016.
18. Alías, F.; Alsina-Pagès, R.M. Review of Wireless Acoustic Sensor Networks for Environmental Noise Monitoring in Smart Cities. *J. Sens.* **2019**, *2019*, 13. [CrossRef]
19. Mydlarz, C.; Salamon, J.; Bello, J.P. The implementation of low-cost urban acoustic monitoring devices. *Appl. Acoust.* **2017**, *117*, 207–218. [CrossRef]
20. Maijala, P.; Zhao, S.; Heittola, T.; Virtanen, T. Environmental noise monitoring using source classification in sensors. *Appl. Acoust.* **2018**, *129*, 258–267. [CrossRef]
21. Sevillano, X.; Socoró, J.C.; Alías, F.; Bellucci, P.; Peruzzi, L.; Radaelli, S.; Coppi, P.; Nenci, L.; Cerniglia, A.; Bisceglie, A.; et al. DYNAMAP—Development of low cost sensors networks for real time noise mapping. *Noise Mapp.* **2016**, *3*, 172–189. [CrossRef]
22. Peckens, C.; Porter, C.; Rink, T. Wireless sensor networks for Log-term Monitoring of Urban Noise. *Sensors* **2018**, *18*, 3161. [CrossRef] [PubMed]
23. Wong, G.S.K.; Embleton, T.F.W. Principles of Operation of Condenser Microphones. In *AIP Handbook of Condenser Microphones. Theory, Calibration, and Measurements*; American Institute of Physics: New York, NY, USA, 1995.
24. Asensio, C.; Gasco, L.; de Arcas, G.; López, J.M.; Alonso, J. Assessment of Residents' Exposure to Leisure Noise in Málaga (Spain). *Environments* **2018**, *5*, 134. [CrossRef]
25. Panasonic WM 63-PR.Datasheet. Available online: https://www.alldatasheet.com/datasheet-pdf/pdf/534460/PANASONIC/WM-63PR.html (accessed on 9 December 2019).
26. Cirrus Logic CS5343/4 Datasheet (Rev. F5). Available online: https://statics.cirrus.com/pubs/proDatasheet/CS5343-44_F5.pdf (accessed on 9 December 2019).
27. Texas Instrument. TMS320VC5502 Fixed-Point Digital Signal Processor Datasheet (Rev. K). Available online: https://www.ti.com/lit/ds/sprs166k/sprs166k.pdf (accessed on 9 December 2019).

28. Texas Instrument. TMS320C55x DSP CPU Reference Guide (Rev. F). Available online: http://www.ti.com/lit/ug/spru371f/spru371f.pdf (accessed on 9 December 2019).
29. Matlab. Trademark of MathWhorks. Available online: https://www.mathworks.com/products/matlab.html (accessed on 9 December 2019).
30. Oppennheim, A.V.; Schafer, R. Filter Design Techniques. In *Book Discrete-Time Signal Processing*; Prentice Hall: Upper Saddle River, NJ, USA, 1989.
31. Alonso, J. Discriminación del Estado de la Carretera Mediante Procesado Acústico en Vehículo. Ph.D. Thesis, Universidad Politécnica de Madrid, Madrid, Spain, 2014.
32. González, D.M.; Morillas, J.M.B.; Gozalo, G.R.; Moraga, P.A. Microphone position and noise exposure assessment of building façades. *Appl. Acoust.* **2020**, *160*, 107157. [CrossRef]
33. Peña-García, N.N.; Aguilera-Cortés, L.A.; González-Palacios, M.A.; Raskin, J.-P.; Herrera-May, A.L. Design and Modeling of a MEMS Dual-Backplate Capacitive Microphone with Spring-Supported Diaphragm for Mobile Device Applications. *Sensors* **2018**, *18*, 3545. [CrossRef] [PubMed]

Article

Stabilization of a *p-u* Sensor Mounted on a Vehicle for Measuring the Acoustic Impedance of Road Surfaces

Francesco Bianco [1], Luca Fredianelli [2], Fabio Lo Castro [3], Paolo Gagliardi [2], Francesco Fidecaro [2] and Gaetano Licitra [4,*]

1 iPOOL S.r.l., via Cocchi 7, 56121 Pisa, Italy; francesco.bianco@i-pool.it
2 Physics Department, University of Pisa, Largo Bruno Pontecorvo 3, 56127 Pisa, Italy; fredianelli@df.unipi.it (L.F.); paolo.gagliardi@df.unipi.it (P.G.); francesco.fidecaro@unipi.it (F.F.)
3 CNR-INM Section of Acoustics and Sensors O.M. Corbino, via del Fosso del Cavaliere 100, 00133 Rome, Italy; fabio.locastro@cnr.it
4 Environmental Protection Agency of Tuscany Region, via Vittorio Veneto 27, 56127 Pisa, Italy
* Correspondence: g.licitra@arpat.toscana.it; Tel.: +39-055-530-5353

Received: 29 January 2020; Accepted: 15 February 2020; Published: 25 February 2020

Abstract: The knowledge of the acoustic impedance of a material allows for the calculation of its acoustic absorption. Impedance can also be linked to structural and physical proprieties of materials. However, while the impedance of pavement samples in laboratory conditions can usually be measured with high accuracy using devices such as the impedance tube, complete in-situ evaluation results are less accurate than the laboratory results and is so time consuming that a full scale implementation of in-situ evaluations is practically impossible. Such a system could provide information on the homogeneity and the correct laying of an installation, which is proven to be directly linked to its acoustic emission properties. The present work studies the development of a measurement instrument which can be fastened through holding elements to a moving laboratory (i.e., a vehicle). This device overcomes the issues that afflict traditional in-situ measurements, such as the impossibility to perform a continuous spatial characterization of a given pavement in order to yield a direct evaluation of the surface's quality. The instrumentation has been uncoupled from the vehicle's frame with a system including a Proportional Integral Derivative (PID) controller, studied to maintain the system at a fixed distance from the ground and to reduce damping. The stabilization of this device and the measurement system itself are evaluated and compared to the traditional one.

Keywords: p-u sensor; p-p sensor; noise; Adrienne; stabilization; damping; acoustic impedance; road surfaces

---

1. Introduction

Transport infrastructures continue to be a source of noise, representing a serious issue in modern society. Among the sources that should be monitored according to European Environmental Noise Directive (END) [1], roads are the one that reach most citizens, and even if the noise produced is not the most disturbing, they contribute the most to the overall noise exposure. According to the 2017 END's review [2], nearly 100 million European citizens are exposed to road traffic *Lden* higher than 55 dB(A) and among them 32 million are exposed to *Lden* higher than 65 dB(A). Many studies show that prolonged exposure to noise can induce cardiovascular disease [3,4], alterations in blood pressure [5,6], respiratory diseases [7], hypertension [8], learning impairment [9,10], annoyance [11,12], and sleep disturbance [13–15].

In order to avoid the aforementioned health issues, the END set the instruments in mandatory action plans for big infrastructures and urban agglomerations [16]. More detailed noise maps are mandatory for roads with traffic flows greater than 3 million vehicles per year and mitigation should be applied when limits' exceedances emerge. In the last decades, economic efforts have been spent in mitigation actions for road traffic noise, with rigid barriers used as an optimal choice. However, it has been proven that this solution suffers from several problems, such as diffraction at the edges, reflection of sound energy in the opposite direction, and complaints of citizens due to the reduction of fields of view, natural light, and air flow [17].

Thus, researchers are recently looking for innovative and possibly "green" solutions, such as live and integrated monitoring systems [18,19], sustainable metamaterial absorbers [20,21], sonic crystals noise barriers made of recycled materials [22], electric cars, and car sharing [23].

The overall noise produced by a vehicle is given by the sum of propulsion, tire-pavement interaction, and aerodynamic contribution, depending on the vehicle speed. As electric vehicles become more widespread, the contribution of engines to overall noise is decreasing; however, it is only relevant at low speeds. Noise produced by tire–road interaction is the responsible for most of the noise emitted by roads and is a complex phenomenon resulting from the combination of aerodynamic and vibro-dynamic phenomena [24,25]. The former is related to compression of the air trapped within the tread of the rolling tire [26] (known as air pumping) and cause noise at frequencies higher than 1 kHz. In addition, pipe and Helmholtz resonances due to the coupling of a vibrating mass of air within the tread acting as a cavity contribute to the aerodynamic noise. Vibro-dynamic noise covers frequencies lower than 1 kHz and is due to tire vibrations caused by the impact of the tire against irregularities of the road surface and by non-linear effects, such as the stick-and-slip and stick-and-snap mechanisms [25].

The knowledge of noise generation mechanisms is vital for improving mitigation actions [27] applied to the two main actors in the phenomenon: Tires have been optimized [28], and mostly, low noise pavements have recently spread, some of which have even been developed with recycled materials [29,30]. Since tire/road noise depends on structural parameters of road surfaces [31], the European Union has recently prescribed acoustic tests of newly laid surfaces [32] by means of the Close Proximity Index (CPX) method [33] during the first three months from laying date. Therefore, knowing the acoustic performance of a pavement plays an extremely important role [34,35].

Even after the laying, monitoring a pavement over time is important to keep track of the ageing effect, which dramatically reduces the acoustic performance of surfaces. Traffic and weather cause voids in the surface to clog with detritus, increasing noise emitted even by 5 dB [36,37].

CPX measurements are generally performed to evaluate the acoustic emission properties of a pavement, while other measurement methods are needed for a wider evaluation of the acoustic properties. In order to assess the effects of noise on the receptors, for example, other factors are involved besides the emission, such as acoustic propagation, which in turn depends on the absorption coefficient of the road surface. The evaluation of the absorption coefficient is also an important parameter related to the state of the installation. For this reason, being able to measure the absorption yields a broader characterization of the pavement. In fact, the surface texture and volumetric of the friction course of a pavement are the main parameters influencing rolling noise, also related to absorption coefficient [35–39].

Different methods are generally used to evaluate the acoustic properties of a pavement, in order to derive information about their use in practical applications: From the CPX method [40], that also evaluates the emission properties, to the absorption measurement performed with an impedance tube using standing wave ratio (ISO 10534-1) [41] in-lab or on-site (ISO 13472-2) [42], and the extended surface method, usually called "Adrienne method" (ISO 13472-1) [43,44]. A recent study [45,46] showed that Adrienne method achieves good estimates for pavement's characteristics, whereas the impedance tube scores the best accuracy in-lab, at the expense of time consumption, and the on-site Kundt tube can give an acceptable estimate of porosity.

Models of surface impedance of the materials currently present in the literature (e.g., Allard and Atalla [47]) are based on parameters related to the geometric properties of the interconnected porous structure and therefore to the effective structure of the medium. The sound absorption is often used as a proxy of the impedance, however, knowledge on the impedance represents a more complete information and allows a better characterization of surfaces from an acoustical point of view. Thus, in the last decade, several authors explored the possibility of measuring the surface impedance, defined as the ratio between the local pressure p and particle velocity u. This measurement is usually carried out by using a pressure-velocity (*p-u*) probe. The convenience of using a p-u probe is twofold: On one hand, it provides a local and simultaneous measure of both pressure and particle speed, while on the other hand it allows a direct measure of the impedance. A direct measurement of velocity is not possible with other known methods, such as the *p-p* sensors (pressure – pressure) reported below but can be achieved indirectly. In this way, the two quantities are related to the same incident and reflected field, with consequently simpler and error-free calculations. The Adrienne method, for example, requires evaluating a direct field that requires subsequent subtraction from the measured one, but is clearly not stable over time. Another important advantage of using the p-u probe to measure the impedance/absorption coefficient is the broadband "figure of eight" directivity of the particle velocity sensor, which significantly restricts background noise and allows the sensor to be used in-situ.

Most of the studies used the probe in laboratory conditions [48,49], while, to the best of the authors' knowledge, only few approaches have tried in-situ road surface impedance measurement [50–52]. Nevertheless, Tijs and de Bree's concept [53] can be considered a precursor of the present work, as it uses a small source attached to the bumper and close to the road surface and a p-u probe, operating in near field conditions and using a spherical wave.

The methodology reported in the paper uses a p-u sensor that measures the sound intensity by means of a pressure transducer combined with a particle velocity transducer, instead of the more common *p-p* sensor combining two pressure microphones and then applying a finite-difference approximation to the pressure gradient. Random errors can be overcome by repeating the measurement, while bias errors can occur for multiple causes, such as phase mismatch between the two measurement channels, errors in the scanning procedure, errors due to airflow, influence of the pressure equalization vents of the microphones, non-stationary external sources, reflections from the operator as they move around the source, influence of the environment on the sound power output of the source, and the absorption of the source itself [54]. These reasons contribute to the choice of the *p-u* probe over a *p-p* one. Furthermore, background noise from sources outside the measurement surface can increase phase mismatch in *p-p* sensors, but not in p-u ones [55]. On the contrary, strongly reactive sound fields can increase *p-u* phase mismatch and have no influence with the *p-p* ones. Another advantage of p-u probes over *p-p* ones is their reduced size; however, *p-u* sensors are not easily calibrated.

In the framework of the NEREiDE LIFE project, a new measurement system based on a pressure-velocity (*p-u*) probe has been implemented on a mobile laboratory. The present paper describes an instrument mounted on a mobile laboratory able to measure the absorption coefficient of a pavement in a continuous way. A mobile laboratory allows in-situ measurements of long road sections for a more precise assessment of the quality of asphalt pavements, which can vary along the installation [56]. It also allows to make measurements in safe conditions and without necessarily closing the road to traffic.

After addressing the choice of the sensor used, this work tries to solve the primary issue regarding these measurements, which is the stabilization of the instrumentation respect to the road surface. The new approach has been derived from the Adrienne method and can measure the acoustic absorption in a contactless way, thus resulting in a wider frequency bandwidth and a larger studied portion of the road surface. A proper device, consisting of an actuator driven by a laser distance sensor, has been developed for stabilizing and damping the measuring system. This task is extremely important for maintaining the stability and correctness of the impedance evaluation and reducing the measuring errors. Its effects have been simulated and measured in order to validate its functioning.

2. Life NEREIDE Project

The present work is part of the LIFE15 ENV/IT/000268 NEREiDE project, which aims improve porous asphalt pavements and low noise surfaces made of recycled asphalts and tires. A warm mixture was produced at a high temperature giving birth to pavements that improve:

- safety in urban areas by better draining;
- the reduction of waste materials and virgin materials;
- acoustical performances and a significant reduction of noise emitted;
- asphalt laying procedure, thus reducing air pollution emissions.

Crumb rubber from scrap tires were used in order to improve the road surfaces from elastics, soundproofing and toughening characteristics points of view, while recycled asphalts were used to reduce virgin aggregates and bitumen. The project characterizes raw materials searching for the best that fit the manufacturing processes of crumb rubber in order to modify the vulcanized recycled rubber and to make it suitable for incorporation in the bitumen. The last was the binder and matrix of the new composite material, with structural and waterproofing functions that should avoid the leaching of chemicals. Laboratory tests had already selected the correct composition, grading curve and the structural characteristics, the percentage of crumb rubber, and the methods to realize them.

The new pavements had been laid in two urban areas in Tuscany and one of the major tasks of the project was the evaluation of their effectiveness. Different approaches are currently used to test their effectiveness; from surface characteristics, to acoustical properties, and surveys submitted to the exposed population. The effectiveness was evaluated through a comparison of the surface acoustical properties prior and after the laying, but also by comparing the results with measurements over standard porous asphalt. This led to another objective, which is not less important than the main one: To improve the reliability of the results by suggesting a new evaluation technique for the performances of new pavements. The method should work on-site using techniques also applicable in urban context. Then, the effectiveness of new asphalts was monitored, also considering the subjective response to noise ante and post-opera. Roadside noise measurements were used to validate outputs from a noise model in order to assign noise levels to nearby residents who were also interviewed with questionnaires that compared their response [57].

Both Close Proximity Index and statistical pass-by (SPB) methods [58] were used to measure the noise produced by tires rolling on new and reference pavements. Unfortunately, standard SPB is applicable under environmental conditions that are almost never realized in the urban context. Therefore, an urban statistical pass by method was developed adding a monitoring station placed roadside at a known distance, in addition to traffic counters. In this way, the SPB has been modified to be used in urban areas by reducing the confounding factors. The method for deriving pass by values in urban contexts can define the index of vehicles noisiness on the specific pavement and place, thus allowing a comparison of efficiency of new pavements in terms of noise at roadside.

The project also aimed at the development of a new on-site acoustical absorption measuring system built on a mobile laboratory. For this purpose, a car was equipped with several instruments in order to perform moving and static measures. A preliminary data analysis was set up and described in the present paper. The system was used to monitor acoustic properties of new surfaces realized on experimental sites and was able to give real time results and a fine spatial acoustic characterization of the road surface. The results were linked to the microscopic and macroscopic properties of the asphalt, with parameters such as porosity and tortuosity, but also to macroscopic discontinuities or structural problems. They also provide an evaluation of the current state of a whole pavement stretch, thus giving useful information on the homogeneity of results along the whole installation. This allows for the evaluation of the goodness of the mixture and its laying. In the rest of the paper, a description of the measurement's methodology will be carried out, focusing on the difficulties regarding the stabilization of the system over a moving vehicle.

3. Methodology

Measuring the absorption coefficient with a moving vehicle is best achieved using a contactless method. The system used in this work is the sum of the following parts, each separately tested and calibrated both in controlled conditions and on the actual mobile laboratory:

(1) p-u probe, also known as intensity probes: A sensor measuring the local value of acoustic pressure (p) and particle velocity (u). With these two parameters, the probe can easily evaluate the acoustic impedance;

(2) inertial damper: An electro mechanic actuator driven by a laser distance measure between the car and the pavement surface. This tool is needed to maintain constant the distance between the probe and the asphalt and to reduce vibrations from the moving vehicle;

(3) acoustic emitter: The sound emitted by the source is measured by the probe after being reflected by the pavement;

(4) A pc and an acquisition board acting as instrumentation control system.

The p-u probe can simultaneously and directly detect local pressure and particle velocity. It consists of a miniaturized pre-amplified microphone for pressure measurements and hot wires for the particle velocity evaluation [59]. This instrument is sensitive to specific velocity direction. As reported in Figure 1, the velocity sensor measures the perpendicular component respect to the pavement. An older method for measuring the particle's velocity consisted of using two matched microphones and calculating the pressure difference, but problems in low pressure variation occurred when considering the microphones dimension and the spacing between them. The introduction of a technique based on a hot wire overcome this issue and the particle velocity could then be evaluated with a wire whose electric resistance changes with the temperature, and hence with the speed of the flow of the air that flows over the wire. The velocity u was then evaluated with Equation (1) [60].

$$u = \left(\frac{E^2 - a}{b}\right)^{\frac{1}{n}} \tag{1}$$

where E is the voltage difference at the two ends of the wire resistance, a, b, and n are three constants evaluated during calibration. The voltage across the resistance is given by Equation (2) [61].

$$E = (R_0 + \Delta R)i \tag{2}$$

where R_o is the value of the resistance at the temperature T_o and the resistance variation ΔR is given by Equation (3), where α is thermal resistance coefficient and T the temperature of the hot wire.

$$\Delta R = R_0\alpha(T - T_0) \tag{3}$$

The model used in this work considers a monopole source placed over the ground at h_s and a receiver with height h_r from the ground, as shown in Figure 1. The receiver acquires both p and u [62]. An innovative stabilization system is used to reduce the variation of the height of the source and the receiver, and it will be described in the next chapter.

The acoustic impedance was calculated with Equation (4) starting from the ratio of the complex pressure to complex velocity amplitudes measured by the probe.

$$Z_r(r,\omega) = \frac{p_r(r,\omega)}{u_r(r,\omega)} \tag{4}$$

The pressure and velocity values were calculated with Equations (5) considering a spherical wave of radius r:

$$p(r,t) = \frac{A}{r}e^{-kri}e^{i\omega t} \quad u(r,t) = \frac{A}{\rho cr}\left(1 - \frac{i}{kr}\right)e^{-kri}e^{i\omega t} \tag{5}$$

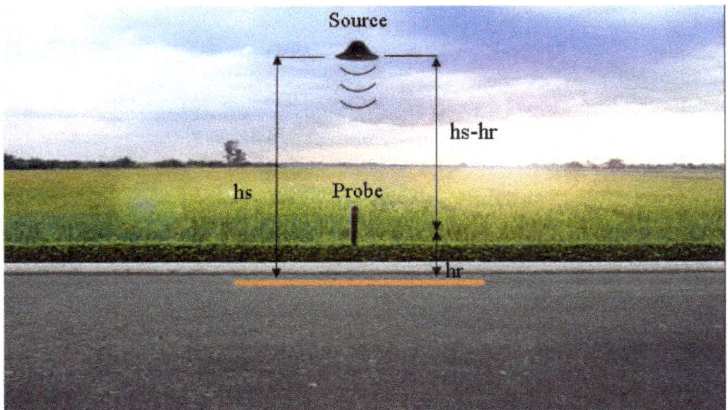

Figure 1. Set-up of the measurement system.

A is the wave amplitude, ρ is the air density, c is the speed of sound, $\omega = 2\pi f$ is the angular frequency, $k = \frac{2\pi}{\lambda} = \frac{2\pi f}{c}$ is the acoustic wave number, λ is the wavelength.

The absorption coefficient α is defined as $\alpha = 1 - \Gamma$, where Γ is the reflection coefficient defined in Equation 6 as the ratio of the reflected power P_r and incident power P_i.

$$\Gamma = \frac{P_r}{P_i} \cong \frac{|p_r|^2}{|p_i|^2} = |R|^2 \tag{6}$$

R is the sound pressure reflection factor. p is the sum of the direct and reflected pressure wave. Similarly, u is the sum of direct and reflected velocity wave. However, since they are vectors, the negative sign must be taken into account. The impedance can then be reported in Equation (7).

$$\begin{aligned} Z_r(r,t) &= \frac{p(h_s-h_r,t)+p(h_s+h_r,t)}{u(h_s-h_r,t)-u(h_s+h_r,t)} = \\ &= \frac{\frac{A}{d_1}e^{-kd_1 i}e^{i\omega t}+R\frac{A}{d_2}e^{-kd_2 i}e^{i\omega t}}{\frac{A}{\rho c d_1}\left(1-\frac{i}{kd_1}\right)e^{-kd_1 i}e^{i\omega t}-R\frac{A}{\rho c d_2}\left(1-\frac{i}{kd_2}\right)e^{-kd_2 i}e^{i\omega t}} = \\ &= \rho c\frac{\frac{1}{d_1}e^{-kd_1 i}+R\frac{1}{d_2}e^{-kd_2 i}}{\frac{1}{d_1}\left(1-\frac{i}{kd_1}\right)e^{-kd_1 i}-R\frac{1}{d_2}\left(1-\frac{i}{kd_2}\right)e^{-kd_2 i}} \end{aligned} \tag{7}$$

where for simplicity $d_1 = (h_s - h_r)$ and $d_2 = (h_s + h_r)$. According to Equation (7), the explicit time dependency disappears and $Z_r(r,t)$ becomes in $Z_r(r)$ in the following. From the previous equation, R can be obtained Equations (8) and (9).

$$R(r) = -\frac{(h_s + h_r)\left(\frac{Z_r[k(h_r-h_s)+i]}{\rho c k(h_r-h_s)} - 1\right)}{(h_r - h_s)\left(\frac{Z_r[k(h_s+h_r)-i]}{\rho c k(h_s+h_r)} + 1\right)}e^{-2kh_r i} \tag{8}$$

$$|R|^2 = \left[\left(\frac{h_s + h_r}{h_s - h_r}\right)\left|\frac{\frac{Z_r[-k(h_r-h_s)-i]}{\rho c k(h_r-h_s)} + 1}{\frac{Z_r[k(h_r+h_s)-i]}{\rho c k(h_r+h_s)} + 1}\right|\right]^2 \tag{9}$$

In order to know which parameter mostly affects the absorption coefficient, according to the theory of propagation of uncertainty, the propagation error for a multivariable function is derived for each component from the partial derivative of the function itself. Thus, the sensitivity of α respect to h_s or h_r can be calculated with Equations (10).

$$\frac{\partial \alpha}{\partial h_r} = \frac{\partial |R|^2}{\partial h_r} \qquad \frac{\partial \alpha}{\partial h_s} = \frac{\partial |R|^2}{\partial h_s} \tag{10}$$

4. Results

4.1. Instrumentation Height

Since it derives from the Adrienne method, the new method is dependent on the height from the plane. Considering that the height of the vehicle from the road surface changes due to its acceleration and to superficial irregularities of the surface, the very first part of the analysis are in-lab simulations aimed to optimize receiver's position from the ground, which is the one leading to the smallest error in the calculation of the absorption coefficient. The results are compared with the Adrienne ones.

In these calculations, the ideal source and receiver were used and the variation of receiver to source height brought by the running vehicle were considered.

The sensitivity of α respect to the variation of h_s or h_r were studied in the Adrienne method with Equations (11).

$$\frac{\partial \alpha}{\partial h_r} = \frac{\partial |R|^2}{\partial h_r} = -\frac{4h_s(h_s - h_r)}{(h_s + h_r)^3}\left|\frac{P_r(f)}{P_i(f)}\right|^2, \qquad \frac{\partial \alpha}{\partial h_s} = \frac{\partial |R|^2}{\partial h_s} = \frac{4h_r(h_s - h_r)}{(h_s + h_r)^3}\left|\frac{P_r(f)}{P_i(f)}\right|^2 \tag{11}$$

From Equations (11), is possible to derive Equations (12), which shows that the ratio of the sensitivities of the absorption coefficients is equal to the ratio of h_s and h_r. Therefore, the sensitivity to variation of height of receiver is higher in module than that respect to the variation of the height of source.

$$\frac{\frac{\partial \alpha}{\partial h_r}}{\frac{\partial \alpha}{\partial h_s}} = -\frac{h_s}{h_r} \tag{12}$$

In the new method with a p-u probe as receiver, the sensitivities ratio is different and changes with frequency, pavement impedance, receiver, and source height. The sensitiveness of the absorption coefficient values obtained using the Adrienne method is far greater than the new method proposed. The absorption coefficient sensitivity to the variation of the height of the probe also results higher in the Adrienne method than in the new one.

The height of the probe corresponding to the lowest error is at 0.16 m from the pavement for the new method, while for Adrienne the height of the microphone is 0.25 m, as suggested by the standard. However, the error for the Adrienne method is higher than the new method around ±0.05 to the minimum of both curves.

The main features of Adrienne and the new method are summarized in Table 1. The new method offers a wider frequency band width and is less sensitive to variations of the receiver's height, which makes it more appropriate for a running vehicle.

Table 1. Comparison of in-situ measurement methods.

	Adrienne Method	Adopted Method Based On *P-U* Probe
In-Situ Measurement	✓	✓
Contactless Measurement	✓	✓
Frequency Bandwidth	250 Hz ÷ 4 kHz	315 Hz ÷ 10 kHz
Exposed Area Diameter	≈ 1.4 m	≈ 1.4 m
Height of The Sound Source	1.25 m	1.5 m

Table 1. *Cont.*

	Adrienne Method	Adopted Method Based On *P-U* Probe
Height of The Sound Microphone / *P-U* Probe	0.25 m	0.16 m
Absorption Coefficient Sensitivity to Receiver Height Variation (F ≥ 315)	2.4 m^{-1}	2.1 m^{-1}
Absorption Coefficient Sensitivity to Source Height Variation (F ≥ 315)	0.5 m^{-1}	0.3 m^{-1}

4.2. Damping System

The mobile laboratory moves while acquiring data, therefore the height of the measurement instrument from the road pavement changes, as stated previously. The absorption measurement system was fastened to the frame of the vehicle and hence it suffered the same oscillations, shown in Figure 2. Figure 3 shows the vertical displacement of the rear of the vehicle, measured using an accelerometer and a laser distance meter. The accelerometer displacement was obtained through a double integration, which increased low frequency noise. On the contrary, the laser worked well over the whole band, but above a certain frequency the measurement was mainly influenced by the road texture (around 20 Hz). Red and blue curves in Figure 3 are similar within the range 2–20 Hz. Therefore, the spectrum of the searched displacement corresponds to the measurement performed with the laser up to that maximum identified frequency, while the following signal is attributable to the influence of road texture.

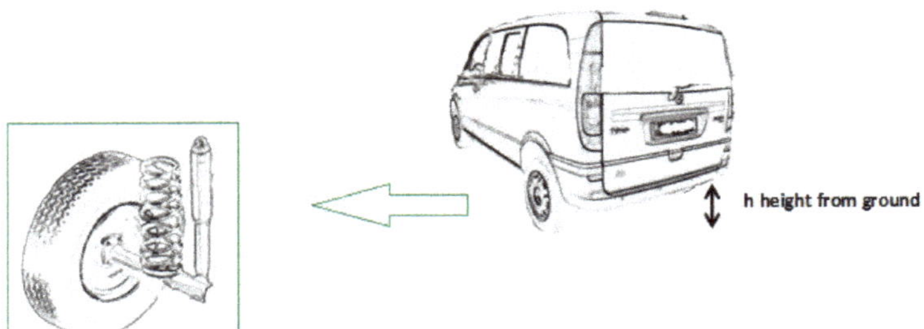

Figure 2. Example of vehicle where is installed the absorption measurement system, with a focus on the shock absorber.

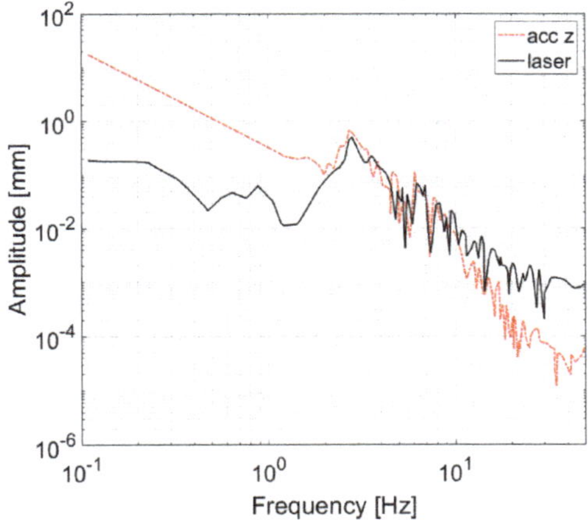

Figure 3. Example of vehicle vertical displacement spectrum while moving.

In order to reduce the influence of height variations on the measurement, the measurement instrumentation must be uncoupled from the vehicle frame, using a system capable of keeping a constant distance *h* from the pavement. The characteristics of this active damping system are hereby described.

As shown in Figure 4, an active controller on the distance *h* has been developed using a Proportional Integral Derivative (PID) controller with the following characteristics:

1. Maximum allowable displacement x_{max}: 100 mm;
2. Damping system with 1 degree of freedom placed along vertical axis;
3. Typical vehicle frame displacement (x′): ±20 mm;
4. Displacement stabilization range (error = h_{target} -h): 2 mm ÷ 5 mm;
5. Working frequency: 0 Hz ÷ 30 Hz;
6. Actuator load: 3 kg payload (loudspeaker, sensors, windscreen, laser distance sensor) plus damping system frame.

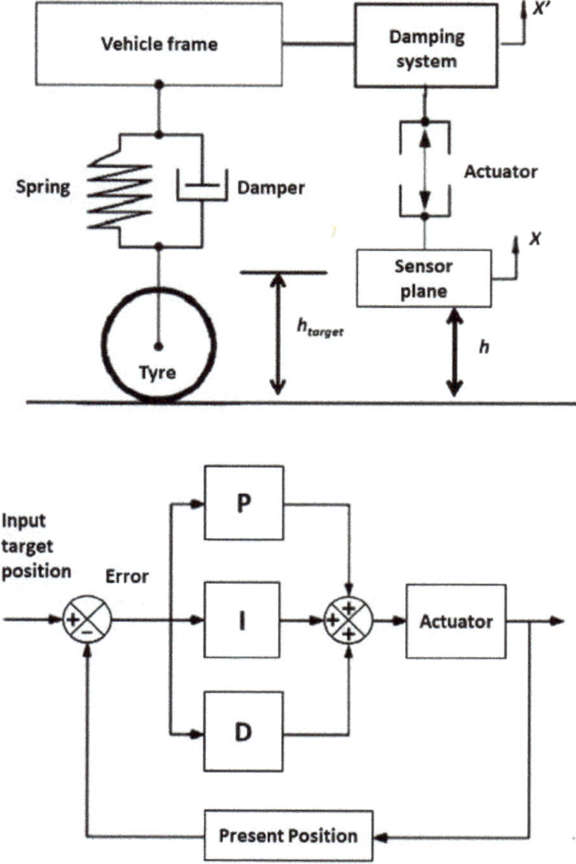

Figure 4. Damping system model and Proportional Integral Derivative (PID) controller of the height of the absorption measurement system.

The Proportional Integral Derivative (PID) controller acts on the error *e* equal to the difference of the target position h_{target} and the current position *h* of the measurement system with respect to the pavement. The target position h_{target} is defined as the almost fixed height of the system. Furthermore, it takes into account the error e itself through a proportional relation, its integral I and its derivative D. The error can be expressed with Equation (13).

$$e = h_{target} - h \tag{13}$$

The function f used to drive the actuator in order to stabilize the measurement system is defined in Equation (14).

$$f(e) = k_p e(t) + k_I \int_0^t e(t)dt + k_D \frac{\partial e(t)}{\partial t} \tag{14}$$

where k_p, k_I, k_D are constants to be calculated in accordance to the actuator system response and the acceptance error. In the present work, the accepted error was considered equal or less than 5 mm, corresponding to an absorption error due to the vehicle oscillation of around 0.01.

A simulation of the process is reported in Figure 5. The control was made through the PID controller which acts on a linear slide actuator that keeps the sensor at a fixed height h from the pavement. The actuator linear speed used in the simulation is 20 mm/s, while the road profile is taken from the on-field measurement, described below.

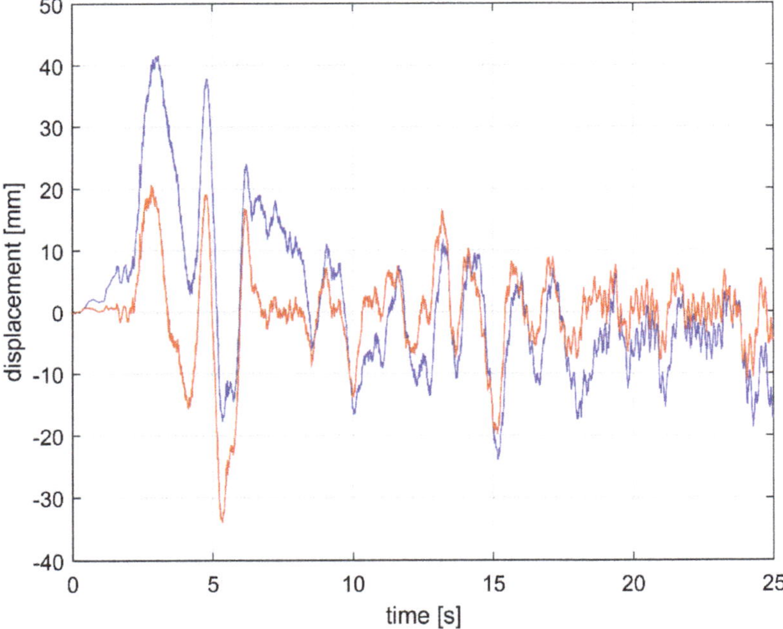

Figure 5. Displacement simulation of the sensor plate stabilized by a PID controller (red); acquired vehicle frame vibration used as input of the PID controller (blue).

A field test was carried out on a stretch of pavement in good condition, running on the same stretch with the stabilizer device off and on. The measurements were repeated several times at a speed of 30 km/h. Although it is not easy to repeat the exact condition of each run, the results showed good repeatability. Measurements with or without stabilizer are not contemporary because only one laser displacement sensor was available for measurements. Consequently, the two measurements do not correspond to the same exact profiles; however, it can be argued that the two profiles share the same mean properties since they derive from measurements of the same road surface.

Figure 6 shows the displacements measured. The bold lines represent a 1-second smoothing for improving the visibility of the average stabilization effect. The maximum excursion using the stabilizer is approximately 0.021 m at 30 km/h. The order of magnitude of the average displacement from the expected equilibrium position in standard operating conditions are reported in Table 2.

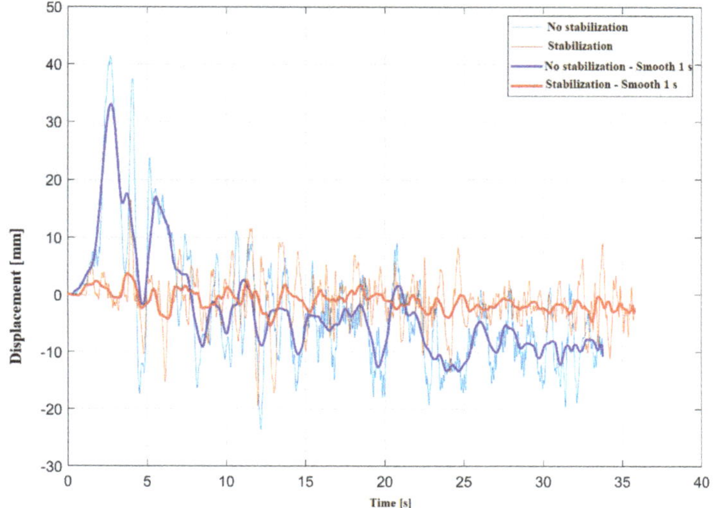

Figure 6. Displacement measured in-situ with stabilization (red) and without stabilization (blue), with smoothing (bold lines) and without smoothing (thin lines).

Table 2. RMS (Root mean square) of the average displacement measured at 30 km/h.

	RMS of Average Displacement [mm]
No stabilization	11.5
Stabilization	4.7

In order to evaluate how using the stabilizer affects the absorption measurement, an idealized case where the only effective variable was the position of the sensor/speaker group relative to the surface investigated was taken into account. Thus, considering a perfectly homogeneous surface, which has the same spectral trend for the reflection coefficient at each point. Furthermore, a high directionality of the sensor is assumed, so that the area actually investigated can be considered constant.

The subsequent calculations were carried out considering a hypothetical pavement with absorption from almost zero to almost one along the whole frequency range, a vehicle in motion at 30 km/h and the displacement trends with respect to an equilibrium position of 0.16 m from the pavement, as in Figure 5, both for the stabilized and non-stabilized system.

In order to simplify the application of Equation (7), it was assumed that the system remains stable, i.e., at a constant distance from the pavement, within the single measurement made by the laser distance meter. Since the instrument works at a frequency of 64 Hz, the measurements were carried out every 15.6 ms, enough time to resolve rather low signal frequencies. For example, up to about five cycles can be measured at 315 Hz.

A measurement of the p and v signals was simulated in successive windows of 15.6 ms length, calculating for each the signal $Z_{r,i}$ as the ratio of the respective spectra (i is the sections index). In order to do so, Equation (7) was used by imposing R, which was identical for each repetition, and the values of $d_{1,i}$ and $d2_{,i}$ related to the section. Using the calculated impedances but imposing a fixed distance of 0.16 m, the reflection coefficients R_i, representing an estimate of the error made in the absence of stabilization or with an imperfect one, were obtained. Calculating the relative error for each section becomes simple, since the actual value that R must assume is known.

Figure 7 shows a well-defined relationship between the measured absorption value and the error related to it, which is certainly influenced in absolute terms by the displacements with respect to the equilibrium.

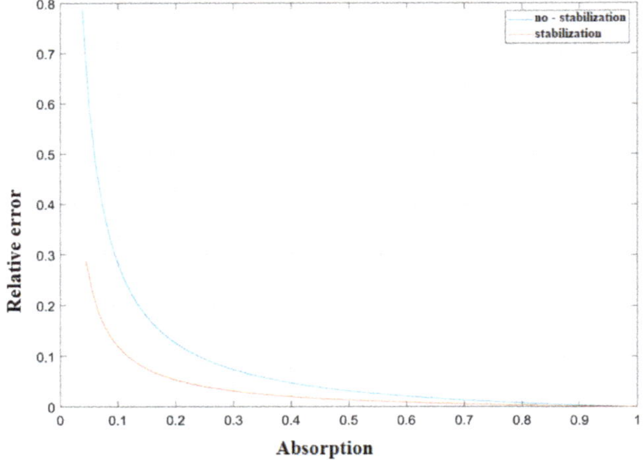

Figure 7. Simulated relative error vs the sound absorption value.

5. Discussions and Conclusions

The present paper reported a new measurement system for the acoustic absorption of a pavement developed inside the LIFE15 ENV/IT/000268 NEREiDE project. The new method is derived from the Adrienne one (ISO 13472-1), but the sensor used is a *p-u* probe mounted on a vehicle moving over the road surface. With this important changes, it is now possible to measure acoustic absorption coefficient on-site, in a contactless way along all the site. Furthermore, it has been shown that the resulting frequency bandwidth is wider.

The correct set-up of the instrumentation has been studied, with particular emphasis on the height of the instrumentation above the ground. It has been shown that the ideal height, at which the lowest measurement error is achieved, is 0.16 m, slightly different from the 0.25 m used in the Adrienne method.

The present paper mainly reported the solution applied to overcome the greater difficulty for this kind of measurements, i.e., the stabilization and damping of the system, required in a vehicle in motion. The instrumentation has been uncoupled from the vehicle's frame with a system studied to maintain constant the distance from the ground. As reported, a PID (Proportional Integral Derivative) controller was used to act on the error given by the difference of the target position and the real position of the measurement system with respect to the pavement. The test performed showed good results for keeping the instrumentation at a fixed height within an order of magnitude of 1 cm, which should be enough to achieve good accuracy in the evaluation of the absorption coefficient. For completeness, the extent of the influence is currently under study. The laboratory tests showed an excellent behavior of the stabilizer in dynamic conditions, perfectly within what was initially expected. Field measurements necessarily suffer from a greater uncertainty, which however allows to the use of the system within good margins for acoustic characterizations. In particular, the implemented possibility of obtaining the distance data from the system itself is rather useful, although it has not yet been tested. In the future, the authors expect to use the data for a possible real time correction to the calculations of the parameters that may depend on the distance.

This study improves on the work of Tijs and de Bree [53] by introducing a stabilization, but also allows the measurements at lower frequency by using a set-up similar to the Adrienne method. The required higher distance from probe to surface is safer for the instrumentation but also indirectly leads to a greater surface under study. However, this set-up implies a bigger source-pavement distance that allows a better approximation of the loudspeaker as point source on both the high and low frequencies. In addition, the methodology presented can be also considered as an improvement of the Adrienne method itself because it goes beyond the need to subtract the signals over time, resulting mathematically easier to operate and analyze.

At present, the system is designed to work at low speeds, within 30 km/h. Placed in the wake of the vehicle, the effect of the wind on the sensor is negligible, also because the sensor is equipped with a special windproof cap that gives it additional protection. In addition, the sensor is located inside an amplification horn, with the axis perpendicular to the ground, where a small windproof layer placed on the entrance and exit significantly reduces the disturbing effects generated by wind turbulence. In addition to this, the source power is sufficient to make other noise sources per se negligible. In particular, the speed sensor is highly directional, and its direction of measure is perpendicular to the road surface, thus further reducing the rolling noise produced by the wheel. Specific laboratory tests, not reported in this article, have shown effective independence of the probe measurement from sources placed outside the axis and of power comparable to the tire/road emission at the speeds investigated. While the calibration of a microphone can be performed with a suitable calibrator, no standard speed references to rely on exists for speed probes. For this purpose, the authors' actual efforts and future developments will investigate the development of a calibration system. This phase needs to measure the acoustic impedance, which is given by the ratio between the velocity values of the particles and pressure measured at the same point. The calibration measurement will be performed inside an anechoic chamber or inside a stationary wave tube (Kundt tube) [63]. Another method could be to perform a comparative measure with the Adrienne method, estimating the absorption coefficient of a material from its surface acoustic impedance.

Finally, the authors do not exclude the study of the new presented methodology also using a suitable *p-p* sensor, which can still be a reliable, fast, and convenient tool for on-site measurements.

Author Contributions: Conceptualization, F.L., G.L.; methodology, F.B.; software, F.B., F.L.C.; validation, G.L., F.F.; formal analysis, F.B., F.LC., P.G.; investigation, F.B., F.L.; resources, G.L., F.L.; data curation, F.L., P.G.; writing—original draft preparation, L.F.; writing—review and editing, L.F., F.B., P.G., F.L.C.; supervision, G.L., F.F.; project administration, F.F. All authors have read and agreed to the published version of the manuscript.

Funding: The authors thank the Tuscany Region for research grants financed for 50% with the resources of the POR FSE 2014-2020, falling within the scope of Giovanisì (www.giovanisi.it), the project of the Tuscany Region for the autonomy of young people.

Acknowledgments: The authors wish to thank Sandra Hill and Alessandro Del Pizzo for providing linguistic help and for their proof reading of the article. The authors also thank the LIFE15 ENV/IT/000268 NEREiDE project.

Conflicts of Interest: The authors declare no conflict of interest.

References

1. Directive (European Commission) Directive 2002/49/EC of the European parliament and the Council of 25 June 2002 relating to the assessment and management of environmental noise. *Off. J. Eur. Communities* **2002**, *189*, 2002.
2. European Commission. *Report from the Commission to the European Parliament and the Council on the Implementation of the Environmental Noise Directive in Accordance with Article 11 of Directive 2002/49/EC. COM/2017/015*; European Commission: Brussels, Belgium, 2017.
3. Héritier, H.; Vienneau, D.; Foraster, M.; Eze, I.C.; Schaffner, E.; Thiesse, L.; Brink, M. Transportation noise exposure and cardiovascular mortality: A nationwide cohort study from Switzerland. *Eur. J. Epidemiol.* **2017**, *32*, 307–315. [CrossRef]
4. Vienneau, D.; Schindler, C.; Perez, L.; Probst-Hensch, N.; Röösli, M. The relationship between transportation noise exposure and ischemic heart disease: A meta-analysis. *Environ. Res.* **2015**, *138*, 372–380. [CrossRef]

5. Dratva, J.; Phuleria, H.C.; Foraster, M.; Gaspoz, J.M.; Keidel, D.; Künzli, N.; Liu, L.J.S.; Pons, M.; Zemp, E.; Gerbase, M.W.; et al. Transportation noise and blood pressure in a population-based sample of adults. *Environ. Health Perspect.* **2011**, *120*, 50–55. [CrossRef] [PubMed]
6. Babisch, W.; Swart, W.; Houthuijs, D.; Selander, J.; Bluhm, G.; Pershagen, G.; Dimakopoulou, K.; Haralabidis, A.S.; Katsouyanni, K.; Davou, E.; et al. Exposure modifiers of the relationships of transportation noise with high blood pressure and noise annoyance. *J. Acoust. Soc. Am.* **2012**, *132*, 3788–3808. [CrossRef] [PubMed]
7. Recio, A.; Linares, C.; Banegas, J.R.; Díaz, J. Road traffic noise effects on cardiovascular, respiratory, and metabolic health: An integrative model of biological mechanisms. *Environ. Res.* **2016**, *146*, 359–370. [CrossRef] [PubMed]
8. Van Kempen, E.; Babisch, W. The quantitative relationship between road traffic noise and hypertension: A meta-analysis. *J. Hypertens.* **2012**, *30*, 1075–1086. [CrossRef] [PubMed]
9. Lercher, P.; Evans, G.W.; Meis, M. Ambient noise and cognitive processes among primary schoolchildren. *Environ. Behav.* **2003**, *35*, 725–735. [CrossRef]
10. Minichilli, F.; Gorini, F.; Ascari, E.; Bianchi, F.; Coi, A.; Fredianelli, L.; Licitra, G.; Manzoli, F.; Mezzasalma, L.; Cori, L. Annoyance judgment and measurements of environmental noise: A focus on Italian secondary schools. *Int. J. Environ. Res. Public Health* **2018**, *15*, 208. [CrossRef]
11. Lechner, C.; Schnaiter, D.; Bose-O'Reilly, S. Combined Effects of Aircraft, Rail, and Road Traffic Noise on Total Noise Annoyance—A Cross-Sectional Study in Innsbruck. *Int. J. Environ. Res. Public Health* **2019**, *16*, 3504. [CrossRef]
12. Basner, M.; Babisch, W.; Davis, A.; Brink, M.; Clark, C.; Janssen, S.; Stansfeld, S. Auditory and non-auditory effects of noise on health. *Lancet* **2014**, *383*, 1325–1332. [CrossRef]
13. Tiesler, C.M.; Birk, M.; Thiering, E.; Kohlböck, G.; Koletzko, S.; Bauer, C.P.; Berdel, D.; von Berg, A.; Babisch, W.; Heinrich, J. Exposure to road traffic noise and children's behavioural problems and sleep disturbance: Results from the GINIplus and LISAplus studies. *Environ. Res.* **2013**, *123*, 1–8. [CrossRef] [PubMed]
14. Onakpoya, I.J.; O'Sullivan, J.; Thompson, M.J.; Heneghan, C.J. The effect of wind turbine noise on sleep and quality of life: A systematic review and meta-analysis of observational studies. *Environ. Int.* **2015**, *82*, 1–9. [CrossRef] [PubMed]
15. Park, T.; Kim, M.; Jang, C.; Choung, T.; Sim, K.A.; Seo, D.; Chang, S. The Public Health Impact of Road-Traffic Noise in a Highly-Populated City, Republic of Korea: Annoyance and Sleep Disturbance. *Sustainability* **2018**, *10*, 2947. [CrossRef]
16. Licitra, G.; Ascari, E.; Fredianelli, L. Prioritizing process in action plans: A review of approaches. *Curr. Pollut. Rep.* **2017**, *3*, 151–161. [CrossRef]
17. Lee, J.; Kim, J.; Park, T.; Chang, S.; Kim, I. Reduction Effects of Shaped Noise Barrier for Reflected Sound. *J. Civil Environ. Eng.* **2015**, *5*, 1.
18. Wong, M.; Wang, T.; Ho, H.; Kwok, C.; Lu, K.; Abbas, S. Towards a smart city: Development and application of an improved integrated environmental monitoring system. *Sustainability* **2018**, *10*, 623. [CrossRef]
19. Merenda, F.G.; Praticò, R.; Fedele, R.; Carotenuto, F.G. Della Corte. A Real-time decision platform for the management of structures and infrastructures. *Electronics* **2019**, *8*, 1180. [CrossRef]
20. Gori, P.; Guattari, C.; Asdrubali, F.; de Lieto Vollaro, R.; Monti, A.; Ramaccia, D.; Bilotti, F.; Toscano, A. Sustainable acoustic metasurfaces for sound control. *Sustainability* **2016**, *8*, 107. [CrossRef]
21. Danihelová, A.; Němec, M.; Gergeľ, T.; Gejdoš, M.; Gordanová, J.; Ščensný, P. Usage of Recycled Technical Textiles as Thermal Insulation and an Acoustic Absorber. *Sustainability* **2019**, *11*, 2968. [CrossRef]
22. Fredianelli, L.; Del Pizzo, A.; Licitra, G. Recent developments in sonic crystals as barriers for road traffic noise mitigation. *Environments* **2019**, *6*, 14. [CrossRef]
23. Kim, D.; Ko, J.; Park, Y. Factors affecting electric vehicle sharing program participants' attitudes about car ownership and program participation. *Transp. Res. Part D Transp. Environ.* **2015**, *36*, 96–106. [CrossRef]
24. Kuijpers, A.; Van Blokland, G. Tyre/road noise models in the last two decades: A critical evaluation. In *Proceedings of INTER-NOISE and NOISE-CON Congress and Conference;* No. 2, 2494-2499; Institute of Noise Control Engineering: Washington, DC, USA, 2001.
25. Sandberg, U.; Ejsmont, J. *Tyre/Road Noise Reference Book;* INFORMEX: Kisa, Sweden, 2002.

26. Morgan, P.A.; Phillips, S.M.; Watts, G.R. *The Localisation, Quantification and Propagation of Noise from a Rolling Tyre*; TRL Limited: Berkshire, UK, 2007.
27. Ögren, M.; Molnár, P.; Barregard, L. Road traffic noise abatement scenarios in Gothenburg 2015–2035. *Environ. Res.* **2018**, *164*, 516–521. [CrossRef] [PubMed]
28. Regulation (European commission) No 1222/2009 of the European Parliament and of the Council of 25 November 2009 on the labelling of tyres with respect to fuel efficiency and other essential parameters. *Off. J. Eur. Union* **2009**, *342*, 59.
29. Kleizienė, R.; Šernas, O.; Vaitkus, A.; Simanavičienė, R. Asphalt Pavement Acoustic Performance Model. *Sustainability* **2019**, *11*, 2938. [CrossRef]
30. Berengier, M.C.; Stinson, M.R.; Daigle, G.A.; Hamet, J.F. Porous road pavements: Acoustical characterization and propagation effects. *J. Acoust. Soc. Am.* **1997**, *101*, 155–162. [CrossRef]
31. Losa, M.; Leandri, P.; Licitra, G. Mixture design optimization of low-noise pavements. *Transp. Res. Rec. J. Transp. Res. Board* **2013**, *2372*, 25–33. [CrossRef]
32. Garbarino, E.; Quintero, R.R.; Donatello, S.; Wolf, O. Revision of green public procurement criteria for road design, construction and maintenance. In *Procurement Practice Guidance Document*; European Union: Brussels, Belgium, 2016. [CrossRef]
33. ISO. *11819-2:2017 Acoustics—Measurement of the Influence of Road Surfaces on Traffic Noise—Part 2: The Close-Proximity Method*; ISO: Geneva, Switzerland, 2017.
34. Praticò, F.G. On the dependence of acoustic performance on pavement characteristics. *Transp. Res. Part D Transp. Environ.* **2014**, *29*, 79–87. [CrossRef]
35. Biligiri, K.P. Effect of pavement materials' damping properties on tyre/road noise characteristics. *Constr. Build. Mater.* **2013**, *49*, 223–232. [CrossRef]
36. Licitra, G.; Moro, A.; Teti, L.; Del Pizzo, A.; Bianco, F. Modelling of acoustic ageing of rubberized pavements. *Appl. Acoust.* **2019**, *146*, 237–245. [CrossRef]
37. Licitra, G.; Cerchiai, M.; Teti, L.; Ascari, E.; Fredianelli, L. Durability and variability of the acoustical performance of rubberized road surfaces. *Appl. Acoust.* **2015**, *94*, 20–28. [CrossRef]
38. Praticò, F.G.; Vaiana, R. A study on the relationship between mean texture depth and mean profile depth of asphalt pavements. *Constr. Build. Mater.* **2015**, *101*, 72–79. [CrossRef]
39. Praticò, F.G. Roads and loudness: A more comprehensive approach. *Road Mater. Pavement Des.* **2001**, *2*, 359–377. [CrossRef]
40. Licitra, G.; Teti, L.; Cerchiai, M.; Bianco, F. The influence of tyres on the use of the CPX method for evaluating the effectiveness of a noise mitigation action based on low-noise road surfaces. *Transp. Res. Part D Transp. Environ.* **2017**, *55*, 217–226. [CrossRef]
41. ISO. *EN ISO 10534-1, Determination of Sound Absorption Coefficient and Impedance in Impedance Tubes Part 1: Method Using Standing Wave Ratio*; ISO: Geneva, Switzerland, 2001.
42. ISO. *13472-1, Acoustics—Measurement of Sound Absorption Properties of Road Surfaces In Situ—Part 2: Spot Method for Reflective Surfaces*; ISO: Geneva, Switzerland, 2010.
43. ISO. *13472-1, Acoustics—Measurement of Sound Absorption Properties of Road Surfaces In Situ Part 1: Extended Surface Method*; ISO: Geneva, Switzerland, 2002.
44. Morgan, P.A.; Watts, G.R. A novel approach to the acoustic characterisation of porous road surfaces. *Appl. Acoust.* **2003**, *64*, 1171–1186. [CrossRef]
45. Praticò, F.G.; Fedele, R.; Vizzari, D. Significance and reliability of absorption spectra of quiet pavements. *Constr. Build. Mater.* **2017**, *140*, 274–281. [CrossRef]
46. Praticò, F.G.; Vizzari, D.; Fedele, R. Estimating the resistivity and tortuosity of a road pavement using an inverse problem approach. In Proceedings of the ICSV24 24th International Congress on Sound and Vibration, London, UK, 23–27 July 2017.
47. Allard, J.; Atalla, N. *Propagation of Sound in Porous Media: Modelling Sound Absorbing Materials 2e*; John Wiley & Sons: New York, NY, USA, 2009.
48. Liu, Y.; Jacobsen, F. Measurement of absorption with a p-u sound intensity probe in an impedance tube. *Acoust. Soc. Am. J.* **2005**, *118*, 2117–2120. [CrossRef]
49. Lanoye, R.; Vermeir, G.; Lauriks, W.; Kruse, R.; Mellert, V. Measuring the free field acoustic impedance and absorption coefficient of sound absorbing materials with a combined particle velocity-pressure sensor. *J. Acoust. Soc. Am.* **2006**, *119*, 2826–2831. [CrossRef]

50. Basten, T.G.; de Bree, H.E. Full bandwidth calibration procedure for acoustic probes containing a pressure and particle velocity sensor. *J. Acoust. Soc. Am.* **2010**, *127*, 264–270. [CrossRef]
51. Brandão, E.; Lenzi, A.; Paul, S. A review of the in situ impedance and sound absorption measurement techniques. *Acta Acust. United Acust.* **2015**, *101*, 443–463. [CrossRef]
52. Li, M.; van Keulen, W.; Tijs, E.; van de Ven, M.; Molenaar, A. Sound absorption measurement of road surface with in situ technology. *Appl. Acoust.* **2015**, *88*, 12–21. [CrossRef]
53. Tijs, E.; de Bree, H.E. An in situ method to measure the acoustic absorption of roads whilst driving. In Proceedings of the German Annual Conference on Acoustics, Rotterdam, The Netherlands, 23–26 March 2009.
54. Jacobsen, F. An overview of the sources of error in sound power determination using the intensity technique. *Appl. Acoust.* **1997**, *50*, 155–166. [CrossRef]
55. Jacobsen, F.; de Bree, H.E. A comparison of two different sound intensity measurement principles. *J. Acoust. Soc. America* **2005**, *118*, 1510–1517. [CrossRef]
56. Del Pizzo, A.; Teti, L.; Moro, A.; Bianco, F.; Fredianelli, L.; Licitra, G. Influence of texture on tyre road noise spectra in rubberized pavements. *Appl. Acoust.* **2020**, *159*, 107080. [CrossRef]
57. Lo Castro, F.; Iarossi, S.; De Luca, M.; Bernardini, M.; Brambilla, G.; Licitra, G. Life NEREiDE project: Preliminary evaluation of road traffic noise after new pavement laying. In *INTER-NOISE and NOISE-CON Congress and Conference Proceedings*; Institute of Noise Control Engineering: Washington, DC, USA, 2019; Volume 259, pp. 5494–5501.
58. ISO. *11819-1 Acoustics—Measurement of the Influence of Road Surfaces on Traffic Noise—Part 1: Statistical Pass-By Method*; ISO: Geneva, Switzerland, 1997.
59. De Bree, H.E. The Microflown: An acoustic particle velocity sensor. *Acoust. Aust.* **2003**, *31*, 91–94.
60. Davies, T.W. Modelling the response of a hot-wire anemometer. *Appl. Math. Model.* **1986**, *10*, 256–261. [CrossRef]
61. De Bree, H.E.; Leussink, P.; Korthorst, T.; Jansen, H.; Lammerink, T.S.; Elwenspoek, M. The μ-flown: A novel device for measuring acoustic flows. *Sens. Actuators A Phys.* **1996**, *54*, 552–557. [CrossRef]
62. Yntema, D.R.; Druyvesteyn, W.F.; Elwenspoek, M. A four particle velocity sensor device. *J. Acoust. Soc. Am.* **2006**, *119*, 943–951. [CrossRef]
63. Crocker, M.J.; Hanson, D.; Li, Z.; Karjatkar, R.; Vissamraju, K.S. Measurement of acoustical and mechanical properties of porous road surfaces and tire and road noise. *Transp. Res. Rec.* **2004**, *1891*, 16–22. [CrossRef]

Article

Characterization of Noise Level Inside a Vehicle under Different Conditions

Daniel Flor [1,*], Danilo Pena [2], Luan Pena [2], Vicente A. de Sousa and Jr. [1] and Allan Martins [2]

[1] Department of Communications Engineering, Federal University of Rio Grande do Norte, Natal 59078-970, Brazil; vicente.a.souza@gmail.com

[2] Department of Electrical Engineering, Federal University of Rio Grande do Norte, Natal 59078-970, Brazil; danilo@dca.ufrn.br (D.P.); luan.gppcom@gmail.com (L.P.); allan@dca.ufrn.br (A.M.)

* Correspondence: dlucenaflor@gmail.com; Tel.: +55-84-99132-5962

Received: 16 December 2019; Accepted: 20 February 2020; Published: 9 April 2020

Abstract: Vehicular acoustic noise evaluations are a concern of researchers due to health and comfort effects on humans and are fundamental for anyone interested in mitigating audio noise. This paper focuses on the evaluation of the noise level inside a vehicle by using statistical tools. First, an experimental setup was developed with microphones and a microcomputer located strategically on the car's panel, and measurements were carried out with different conditions such as car window position, rain, traffic, and car speed. Regression analysis was performed to evaluate the similarity of the noise level from those conditions. Thus, we were able to discuss the relevance of the variables that contribute to the noise level inside a car. Finally, our results revealed that the car speed is strongly correlated to interior noise levels, suggesting the most relevant noise sources are in the vehicle itself.

Keywords: noise sources; regression analysis; contribution analysis; vehicle interior noise

---

1. Introduction

Acoustic noise has been considered a crucial issue and one of the most important topics in sensing and communication systems in vehicles in the last years. General vehicle applications depend on the audio signal quality such as multimedia [1–4], security [5–9], and assistive [10–12] and autonomous vehicle applications [13,14]. In addition, vehicle noise is a concern to researchers due to its effects on human health and comfort, both inside and outside the vehicle. The influence of environmental noise in sleep and mental health [15–17] has been investigated. There is also research on the impact of noisy environments on the performance of students in school [18] and on the development of the cognitive processes in children [19]. Exposure to road and transportation related noise has been associated with a higher risk of ischemic heart disease [20], myocardial infarction [21], and diabetes [22]. These topics have motivated studies on better modeling and evaluation tools for acoustic noise with machine-learning based approaches [23,24]. Notably, a study uses an artificial neural network technique to model the sound quality of vehicle interior noise [25] and another one proposes a new sound quality metric for vehicle suspension shock absorber noise [26].

These interior noises result from a composition of different noise natures, such as wind noise, engine noise, and rolling noise. The understanding of the more relevant noise sources might indicate what are the key challenges in acoustic systems and what are the better mathematical models for them. Therefore, their evaluations are fundamental for anyone who is interested in vehicular acoustic signal analysis.

There are different noise source contributions to car environments and they can be separated into many different aspects. According to the literature [27], vehicle interior noise can be focused on vehicle subsystems and components, such as tire-road interaction noise [28,29] or aerodynamic noise. The tire-road interaction is usually refereed by two components, namely structure borne noise, contributing to low-frequency excitation (below 500 Hz), and air borne component with mid- and high-frequency excitation (above 500 Hz). Understanding the characteristics of this particular source of noise is essential in the development of low-noise road surfaces [30–32]. It has also been investigated the sound quality related to specifics noise phenomena in vehicles, such as closing doors [33,34], engine sound [35,36], and wind noise [37,38]. The authors of [39] presented the frequency characteristics and a model for wet road traffic noise, while the authors of [40,41] proposed wet road detection schemes based on acoustic measurements. However, to our knowledge, no efforts have been made to evaluate the effects of rain on noise inside the vehicle.

In the evaluations, the efforts are focused on finding better contributions to represent the general sound quality from interior and exterior vehicle noises. However, although many works show us the effect of different contributions to the acoustic noise in vehicles, little attention has been paid to statistical analysis of individual noise sources. For example, the authors of [42] showed the contribution of air-conditioner noise in sound quality analysis and more recently, the work in [43] establishes correlations between some features and the acoustic noise in cars.

One way to evaluate the acoustic noise contributions inside cars is to measure acoustic signals using microphones in real conditions. For this approach, a set of controlled or known relevant conditions is specified, and the other uncontrolled or unknown noise sources are spread in invariant random events, which ensure that these kinds of noise sources will not cause a disturbance in the analysis such as a bias. In this case, the uncontrolled or unknown variables will interfere with the same frequency that they happen in a real scenario. As a result, this experimental evaluation allowed us to investigate the degree of influence of sources into acoustic noise inside a car. Based on these criteria, the specifications were defined carefully considering the most relevant controlled noise sources in the experiment.

Even though studies considering all vehicle noise sources have many advantages, they have not been developed due to the amount of resources (time and cost) necessary in the analysis of the entire vehicle system. On the other hand, a reduced analysis of the main noise sources contributions in a vehicle makes the study more feasible and realistic. Thus, the variables with higher contributions to noise levels suggest where researchers should focus on when designing noise mitigation systems, such as filters and acoustic noise control.

Thus, one of the objectives of this study was to statistically analyze 212 acoustic noise measurements conducted on different known conditions. The procedure is described in two parts: measurements and evaluations. First, we planned the measures using the controlled and uncontrolled variables. The known or controlled variables were defined based on preliminary experiments, in which we evaluated their main contributions qualitatively. After that, we established the criteria, relationship between the variables, and the constraints. Then, we measured and checked the consistency of the data in relation to the variables. Finally, we evaluated the data using statistical tools such as linear regression and Pearson correlation among the variables and the power noise levels.

The present paper examines possible noise sources correlated with noise levels in an attempt to help researchers who study how to reduce noise levels and improve sound quality in the vehicle interior.

In this paper, our key contributions are:

- Acoustic measurements were collected in several conditions (weather, car windows position, car speed, and traffic level).
- The data collected herein, including information on the conditions and location of each measurement, are freely available [44] and can help researchers in different purposes.
- Statistical evaluation of the different conditions in relation to noise levels was performed.

This paper is organized as follows. In Section 2, we describe all known and controlled variables and measurement conditions, presenting the process of select and organize the data. The measurement setup is described in Section 3. In Section 4, the main source contributions are evaluated and discussed quantitatively and qualitatively. In Section 5, we present our final remarks and further investigations.

2. Methods

2.1. Environment Variables

Natal is a city located in northeastern Brazil, and has a population of about 900,000 and an area of 167 km^2, considered the second smallest capital of Brazil. Natal has a typical tropical climate, with warm temperatures and high humidity throughout the year. The average low and high annual temperatures are 23 °C (73°F) and 29.7 °C (85.5°F), respectively, and the average annual precipitation in the year is 1721.4 mm (67.77 inches). The measurements were carried out in June and July, the coolest months with an average low temperature of 22 °C (71°F) and an average high temperature of 29 °C (84°F).

The sampling points were located in different streets and avenues spreading the uncontrolled conditions such as crowd and traffic, as illustrated in Figure 1. The traffic conditions were defined following the Google Maps traffic conditions policy [45]. Each sampling point was assigned to one of four possible traffic conditions in a specific location. Those locations may have different features, which is why the sampling point was spread for different regions of the city. For example, while the highway near the coast (with 6.2 m) has strong winds blowing from the ocean that may cause higher noise levels, the quiet streets usually present lower noise levels. All measurements were obtained on asphalt with smooth road surface conditions with no presence of potholes or unevenness.

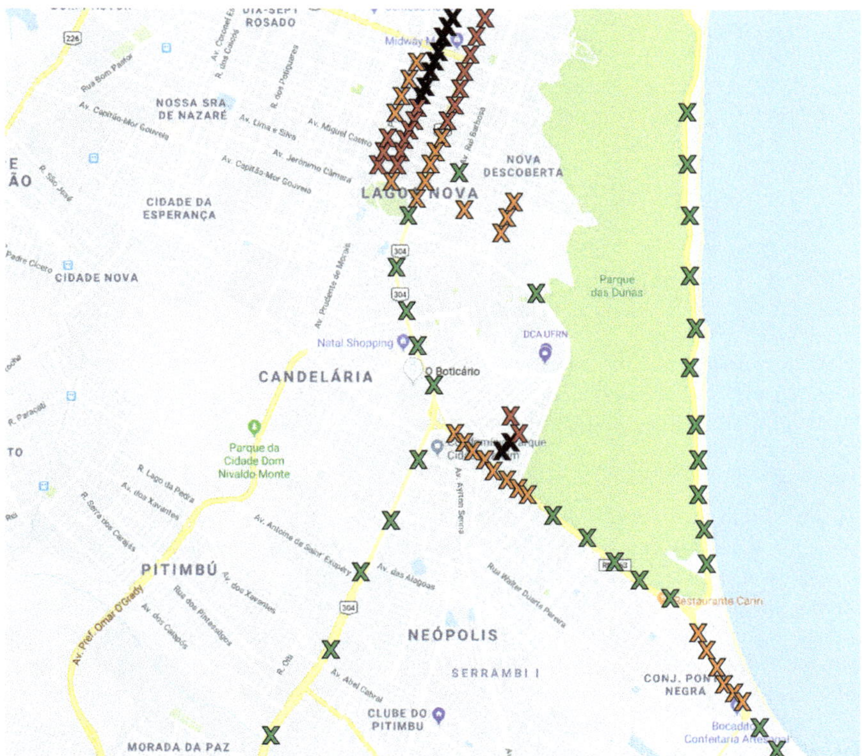

Figure 1. Map of Natal with traffic signs.

Table 1 presents all the possible conditions of the four environmental variables that were controlled during the measurements. Those are the position of the car windows, the presence of rain, the traffic condition, and the maximum speed of the car.

Table 1. Conditions of the controlled environment variables.

Variable	Possible Conditions	Constraints
Window Positions	Open; Closed	Open: only with no rain.
Rain	Yes; No	Yes: only with closed window.
Traffic	Black; Red; Orange; Green	Each traffic has a range of speed (see Table 2).
Speed	0–80 km/h	-

We acquired a large number of measurements. Care was taken to obtain data for the combination of all possible conditions of the controlled variables. During a measurement, the participants did not speak or make any noise. To identify outliers in the data, we reviewed the audio signals to check for highly impulsive events (such as sounds due to potholes in the road or a person's sudden shouting near the vehicle).

For measurements with no rain, all four windows were either fully open or closed. In the case of measuring during rainfall, the windows were kept closed.

We also measured noise levels in different traffic conditions, as shown in Table 2. To record this information, we utilized Google Maps' color codes for traffic and noted the color of the road displayed on the application during measurement. For example, when measuring in a high-speed highway with no traffic delays, the condition was recorded as Green.

Table 2. Conditions of the traffic.

Traffic Condition (Color)	Speed Interval	Description
Black	0	Indicates extremely slow or stopped traffic.
Red	<20 km/h	Highway traffic is moving slow and could indicate an accident or traffic jam on that route.
Orange	>20 km/h and <40 km/h	Indicate medium amount of traffic.
Green	>40 km/h	Indicate that traffic is fast.

Finally, we also recorded the maximum speed of the car during the interval of each measurement. The speed of the car was always compatible with the traffic condition displayed in Google Maps. Table 2 shows the speed intervals for each traffic category.

2.2. Statistical Methods

Our goal was to understand how each environmental factor affects the noise level inside a vehicle. To achieve this, we utilized visualization tools such as histograms and box plots to analyze the data. We also employed statistical modeling to highlight the relationship between the studied variables.

Initially, we quantified the signal power for each measurement. There are many different ways to calculate signal energy or power. One approach is to compute the energy from the cepstral coefficients [46]. The cepstral coefficients are a set of features obtained by first taking the natural logarithm of the magnitude of the Fourier transform of a signal, and then obtaining the inverse Fourier transform of the result. They are often applied in speech recognition and transcription tasks. Another approach is to use the Teager–Kaiser (TK) operator. The TK operator is a measure of energy that takes into account both the signal's amplitude and frequency. Despite their low complexity, the operators and their derivations are capable of estimating useful features of a signal such as instantaneous frequency and spatial envelope and phase [47]. They can be used, for example, in the instantaneous estimation of AM-FM signals and images. For our objective in this work, however, it was sufficient to compute the average power of the measurements in the following way:

$$P_{dB} = 20 \cdot \log \left(\frac{1}{N} \sum_{n=0}^{N-1} x^2(n) \right), \tag{1}$$

where N is the length of sampling and $x(n)$ is the voltage signal from the microphone.

The dataset contains 212 samples, with five features for each measurement: noise power, presence of rain, window position, traffic condition, and maximum speed. Power and speed are numeric, while the three other are categorical. We encoded the latter using natural numbers. The binary variables (window position and rain) were encoded with 0s and 1s. Traffic condition was encoded in descending order of severeness, i.e., "Green" corresponds to 3, and "Black" corresponds to 0.

We then performed an initial exploratory analysis. For the numeric features, we obtained a histogram plot to understand the distribution of power and speed data. We also obtained a histogram for noise power level by traffic condition to compare the distribution of noise level for each condition. For categorical data, we obtained the box plot of power levels for each category separately to highlight the difference of noise levels in them.

Next, we created a linear regression model for each feature. In the models, power is always the explanatory (dependent) variable and the other features are the response (independent) variable. The linear regression model, based on second-order, is the simplest feature extractor. It can be used to measure to what extent two or more variables have a linear relationship. Even if this relationship is only approximately linear, the model is a simple way to identify the influence of the inputs in the model output. To compute the models (estimate the coefficients of a linear regression model), we used the Ordinary Least Square (OLS) method [48]. It does so by minimizing the sum of the squared differences (residuals) between the observed dependent variable and the prediction line.

We computed three metrics of the goodness of fit to compare the influence of the environment features in the average noise power. First, we obtained the mean squared error (MSE) [49]. The MSE is the average of the square of the errors between the model and the actual values. A smaller MSE indicates a better fit, although the actual values of MSE depend on the scale of the data. It is mostly used to compare different models for the same response variable in the same scale. We also computed the coefficient of determination, R^2 [49]. This value is the ratio of the sum of squared residuals to the variance of the actual data values. It is always between 0 and 1 and represents the variation in the response variable that is accounted or explained by the model. In the context of acoustic noise, the R^2 is also related to the noise power, that is, how much of the noise power can be attributed to the explanatory variables.

The R^2 can highlight the correlation between variables. However, it is not a complete description of the goodness of fit of a model. The R^2 assumes that all independent variables in the model explain the variation in the response variable. It always increases when more variables are added to a model, even if they in reality do not affect the independent variable. Thus, it does not evaluate the significance of the relationships shown by the model [49].

A way to verify this significance is by computing the F-statistic [49]. Similar to the R^2, the F-statistic (or F-value) compares the explained and unexplained variation in the model, but weighted by the degrees of freedom of the model, that is, how many model coefficients are used in relation to the number of observations. Thus, it takes into account the complexity of the model.

The F-statistic is used in the F-test. In this test, the null hypothesis is that the model coefficients are zero, and the alternative hypothesis is that at least one coefficient is not zero. The F-test shows if the relationship between the variables is a result of chance or not. The higher is the F-value, the more significant are the results drawn from the model.

For categorical data, the linear regression model can be used to describe the relationship between two of more variables. However, it is not always adequate to represent this relationship as a linear function, as there is a limited, discrete range of values for the response variable. Thus, we also obtained a logistic regression model for the binary variables [49]. The logistic model was obtained by transforming the predicted values of the linear model to another scale that is bounded by 0 and 1. Thus, the output of the model can be interpreted as a probability that a data point belongs to a certain category, and the coefficients of the model are adjusted to find the best match of these probabilities to the data. For logistic models, the goodness of fit metrics described above are not used. To compare the models, we computed McFadden's Pseudo-R^2 [50]. While its calculation is different from the regular R^2, it has a similar interpretation.

We concluded our analysis by measuring the relationship between the environmental variables, highlighting how much correlation they present with each other. We also built a multiple variable regression model, using speed as the dependent variable and the reminder as explanatory variables. We compared the contribution of each variable, and how much better a model with multiple predictors is than the previous one variable model.

3. Measurement Setup

We selected a sedan C4 Lounge from Citroen with automatic transmission as the vehicle for our measurements. The measurement setup used is similar to the one presented in [51]. It consists of a ReSpeaker Core v1 (MT7688) board using the Analog-to-Digital Converter (ADC) AC108 with four ADC delta-sigma, with four microphones connected to a Raspberry Pi 3 (model B) processor to collect, compute the average power, and store the data. In our previous work [51], this setup was validated with a Data Acquisition (DAQ) NI-6361 from National Instruments. It was positioned on the panel, inside the cabin, similar to the position of the multimedia microphone, as illustrated in Figure 2.

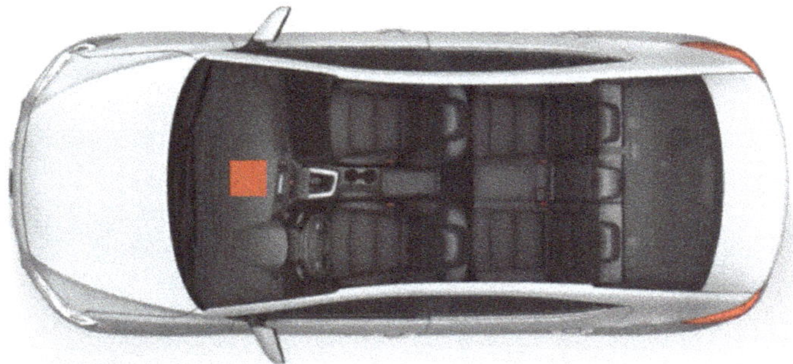

Figure 2. Top view of the car cabin showing the microphone as a red square [52].

Figure 3 shows the measurement setup suspended on the panel stuck on the windshield by a stand. The data were measured using the ReSpeaker at 48 kHz of sample rate. In total, 240,000 samples are collected in 5 s, from which the average power was computed.

Figure 3. The measurement setup.

The measurements were classified based on the known categorical parameters divided into 12 conditions and their combinations representing door window, rain, speed, and traffic conditions. For each measurement, all parameters and observations were manually recorded in a diary, and the power noise levels were computed in the Raspberry Pi. Moreover, we collected the spatial position, aiming to spread the observations regarding the unknown parameters making them independent.

4. Results and Discussions

4.1. Measurements Presentation

Table 3 presents the number of measurements for each condition of the controlled environmental variables. The number of samples is balanced between the conditions, except for the "Presence of Rain" variable, which contains a significantly higher number of measurements with no rain because of the weather conditions in northeastern Brazil. The table also shows the encoding information of each feature.

Table 3. Distribution of the 212 measurements by the possible states of each categorical variable.

	Position of Window		Presence of Rain		Traffic Condition			
Categories	**Open**	**Closed**	**Yes**	**No**	**Very Slow**	**Slow**	**Medium**	**Fast**
No. of samples	95	117	18	194	48	55	58	51
Encoding	1	0	1	0	0	1	2	3

Figure 4 shows the distribution of the noise power levels in dBV, along with some descriptive statistics. The histogram has a bell-like shape, with 50% of the measurements between −47.58 dBV and −30.35 dBV.

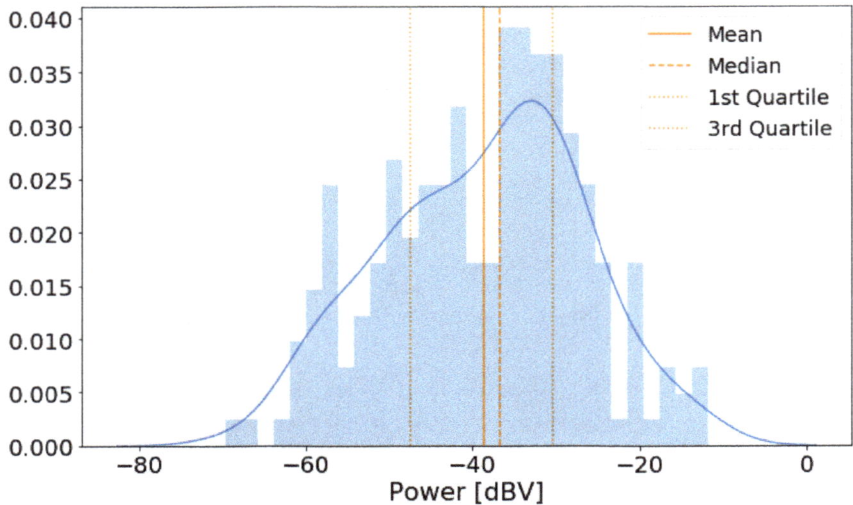

Figure 4. Histogram of measured noise power levels.

4.2. Traffic Analysis

Figure 5 presents the box plot of the power data grouped by traffic conditions. The box plots have an ascending order from "Black" to "Green", showing that as traffic becomes less severe, the noise power level in the car tends to increase. Figure 6 shows another visualization of the noise power level distribution. Analyzing the figure, the modes for each category are separate despite the significant overlap between the curves.

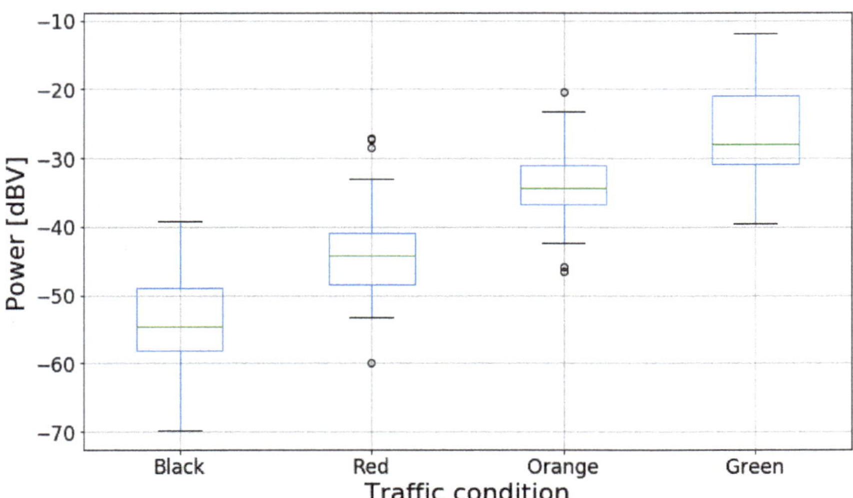

Figure 5. Box plot of noise power data divided by traffic categories.

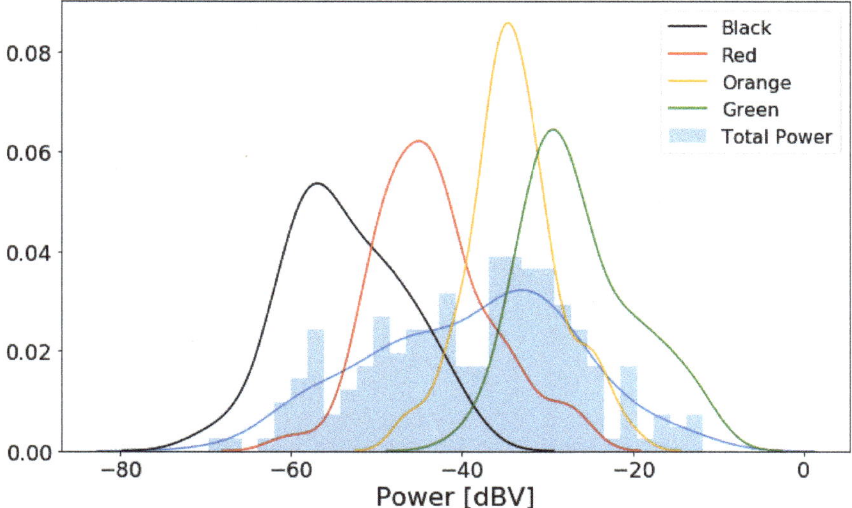

Figure 6. Histogram of noise power data with distribution curves for each traffic category.

These results suggest there is some correlation between noise power and traffic, and, by association, noise power and the speed of the car. To evaluate this relationship, we built a linear regression model in the form

$$Traffic \approx a_0 + a_1 \cdot power\,, \quad (2)$$

where a_0 is the intercept and a_1 is the coefficient of the explanatory variable (power level). Even though the response variable is categorical, we chose to fit a linear model due to the ordered nature of the traffic data, as well as due to the trend implied in Figures 5 and 6.

The model obtained is presented in Figure 7. The circles are the actual data points, and the diamonds are the predictions. The colors represent the actual traffic category of data points and predictions. The model shows that higher power level implies a better traffic condition, which agrees with the behavior displayed by the box plot. Visually, one can see that the predictions are centered around their actual values of traffic, although there is some variation that causes overlap between the categories. For example, the red diamonds are centered around $Traffic = 1$.

The goodness of fit metrics for the model are presented in Table 4. It also presents the coefficients of the model and their 95% confidence interval. The R^2 value indicates that 71.27% of the variance in Traffic is accounted for by the model. This implies a strong relation between the variables. The significance of this relationship is confirmed by the high F-value and its low probability. The MSE presents a low value; however, this is due to the categorical nature and small scale of the traffic data (from 0 to 3). Therefore, the MSE does not provide much information about the quality of the model in this case.

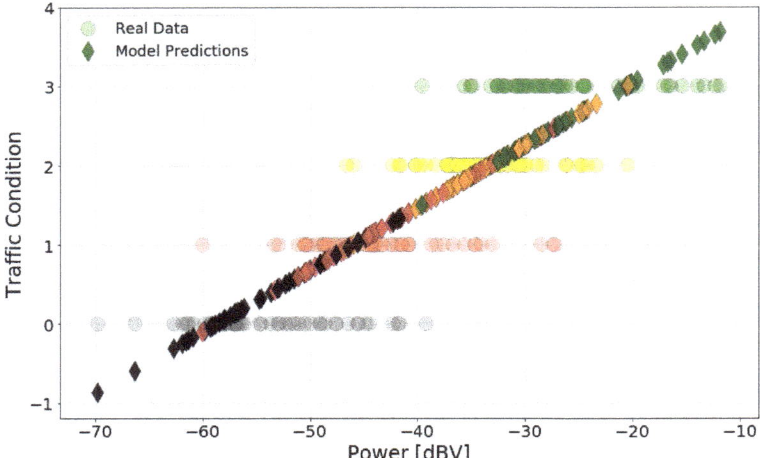

Figure 7. Traffic data and predictions using the linear model. Colors represent the real traffic category of the predictions.

Table 4. Model coefficients (with confidence interval in parenthesis) and goodness of fit metrics for traffic vs. power model.

Linear Regression Coefficients		Goodness of Fit			
a_0	a_1	**MSE**	**R-Squared**	**F-Statistic**	**Prob (F)**
4.6124 (4.342–4.883)	0.0787 (0.072–0.085)	0.3442	0.72	539.7	6.04×10^{-60}

4.3. Rain Analysis

Figure 8 presents the box plot of power data grouped by the presence of rain. Contrary to the previous case, the box plots for this variable have very distinct shapes. The "No Rain" category presents a much bigger variation in noise than the other case and its relation to noise levels are not intuitive. This behavior might be explained by the low number of measurements. Although we have only 18 points (Table 2), and those points may not be enough to represent this category statistically, these findings suggest a model where an external factor can contribute to the noise levels and do not have any relation to the position of the measured location, but represent an environmental parameter. This study, therefore, suggests that non-traditional factors can affect the noise level and they can even produce unexpected results. Most notably, this is the first study to our knowledge to investigate the rain contribution to the noise level measured in the setup located on the car panel.

Due to the distribution of data by category presented in Figure 8, it is expected that both the linear and the logistic models will fit poorly to the data. Nonetheless, we built those two models to evaluate the relationship between power and presence of rain, and also to provide a base of comparison with the next variable (window position). The linear model is in the form

$$Rain \approx b_0 + b_1 \cdot power\,, \tag{3}$$

where b_0 is the intercept and b_1 is the coefficient of the explanatory variable (power level). Tables 5 and 6 present the goodness of fit metrics for this model. Figure 9 presents the logistic model predictions. The circles are the actual data points, and the diamonds are the predictions. From the small R^2, F-Statistic, and Pseudo-R^2, the model shows no significant relationship between noise power and rainfall. Our data have more measurements for one category and, as shown in Figure 8, the noise power levels in the "Rain" scenario are completely contained in the range of values for the "No Rain" scenario. As presented in Figure 9, the model predicts all data points as belonging to the "No Rain" group. Thus, this model provides no information about whether it is raining or not based on noise power acquired inside the car. However, more data must be collected to create a more representative model of the raining scenario. This is a matter of our further studies, especially considering the shortage of rainfall at the measurement site.

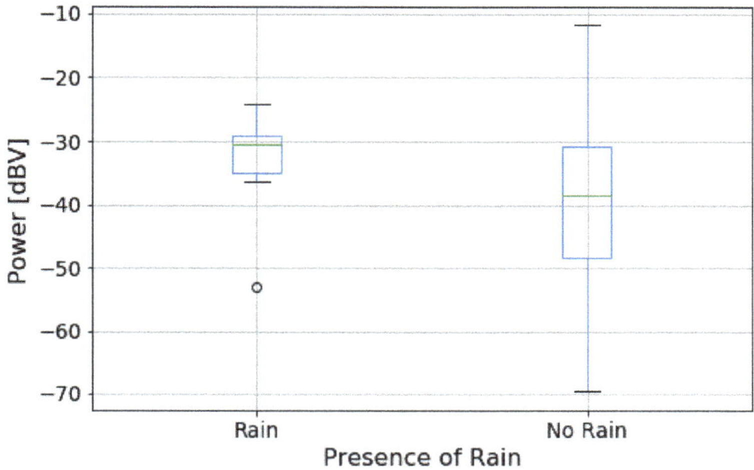

Figure 8. Box plot of data divided by presence of rain.

Table 5. Model coefficients (with confidence interval in parenthesis) and goodness of fit metrics for rain vs. power linear model.

Linear Regression Coefficients		Goodness of Fit			
b_0	b_1	MSE	R-Squared	F-Statistic	Prob (F)
0.2278 (0.100–0.355)	0.0037 (0.001–0.007)	0.07649	0.025	5.346	0.0217

Table 6. Model coefficients (with confidence interval in parenthesis) and goodness of fit metrics for rain vs. power logistic model.

Logistic Regression Coefficients		Goodness of Fit
b_0	b_1	Pseudo R-Squared
−0.5264 (−2.100 1.047)	0.0516 (0.007–0.097)	0.04482

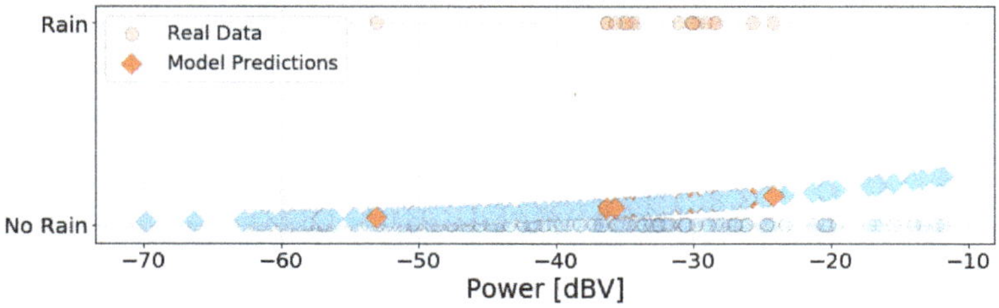

Figure 9. Data grouped by presence of rain and predictions using the logistic model.

4.4. Car Windows Analysis

Figure 10 presents the box plot of power data grouped by the position of the car windows. From the position of the plots, the noise power levels tends to be higher when the windows are open, as this allow for more external noise in the car. Unlike the previous variable, the shape of the box plots are similar, that is, the range of values is similar regardless of the condition. However, there is significant overlap between the two blox plots: only 8.45% of the measurements in the "Open" group have a power level above the maximum power level in the "Closed" group. This indicates that the linear and logistic models will not be able to represent the data, similar to the Rain variable, as there is not enough distinction in noise power between the two conditions. The linear model is in the form

$$Window \approx c_0 + c_1 \cdot power\,, \tag{4}$$

where c_0 is the intercept and c_1 is the coefficient of the explanatory variable (power level). Tables 7 and 8 present the goodness of fit metrics for the models, while Figure 11 shows the resulting logistic model. Visually, we see the poor distinction between the categories. Table 7 shows that, while the R^2 and F-value are slightly bigger for the window case (compared to the Rain analysis), both sets of models perform poorly in regards to the metrics and fail to represent the data. This implies a weak relationship between their response variables (rain and window position) and noise power.

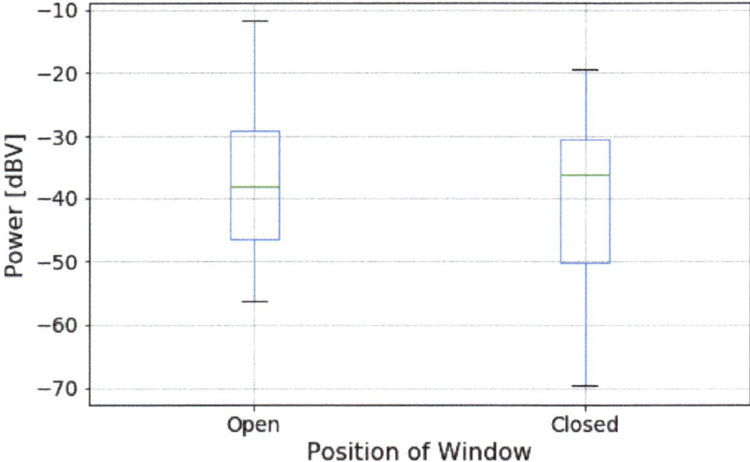

Figure 10. Box plot of data divided by the state of the car windows.

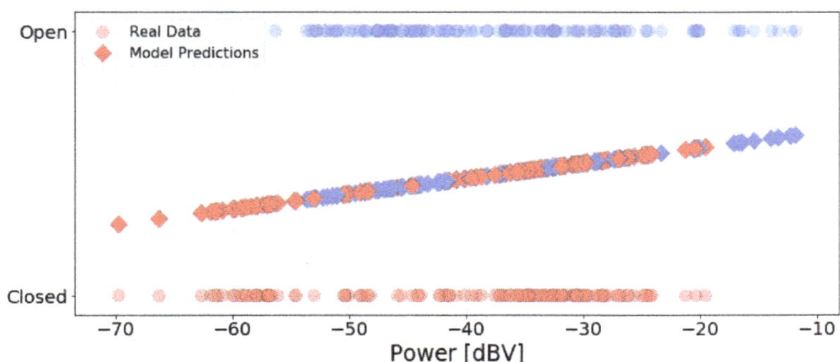

Figure 11. Data grouped by position of windows and predictions using the logistic model.

Table 7. Model coefficients (with confidence interval in parenthesis) and goodness of fit metrics for rain vs. power linear model.

Linear Regression Coefficients		Goodness of Fit			
c_0	c_1	MSE	R-Squared	F-Statistic	Prob (F)
0.6718 (0.444–0.900)	0.0058 (0.000–0.011)	0.2449	0.019	4.091	0.0444

Table 8. Model coefficients (with confidence interval in parenthesis) and goodness of fit metrics for rain vs. power logistic model.

Logistic Regression Coefficients		Goodness of Fit
c_0	c_1	Pseudo R-squared
0.7088 (−0.228 1.645)	0.0238 (0.000–0.047)	0.01403

4.5. Speed Analysis

Figure 12 presents the speed data of the measurements in a histogram. There is a higher number of measurements for the speed of zero. Those data points correspond to the "Black" traffic category, when the car is stationary due to a heavy traffic jam. As shown in Table 3, the numbers of measurements are balanced between traffic categories. Thus, the speed data are also balanced in accordance to the traffic categories.

Of all environmental variables, speed is the only one numeric in nature. Therefore, we obtained a linear regression model in the form:

$$Speed \approx d_0 + d_1 \cdot power\,, \tag{5}$$

where d_0 is the intercept and d_1 is the coefficient of the explanatory variable (power level). Figure 13 presents the resulting model predictions and the actual data. They show that a higher car speed implies a higher noise level inside the car, a result similar to that presented in Figure 7. As above, the circles are the actual data points, and the diamonds are the predictions.

Table 9, which lists the goodness of fit metrics, reaffirms the result that the model is a good representation of the data. The R^2 value indicates that roughly 67% of the variation in speed is explained by the model. The F-value is also high, indicating a significant relationship. These results, compared with the one for the window variable, indicates that there is more contribution to the interior noise levels from the vehicle itself than the wind [27].

The MSE value may seem high, specially compared to the three previous models. However, this comparison is not relevant as MSE is not an adequate metric for categorical data. Furthermore, the scale of speed data is bigger than that of the other variables, resulting in higher error values on average. Finally, the simple linear regression cannot model the fact that speed data cannot be negative. Since there is a high number of zero data points, a high MSE is expected.

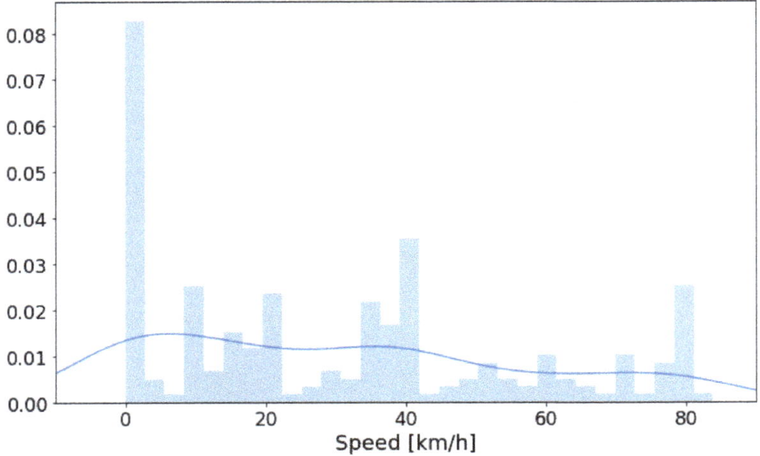

Figure 12. Histogram of the maximum speed of the car during the measurements.

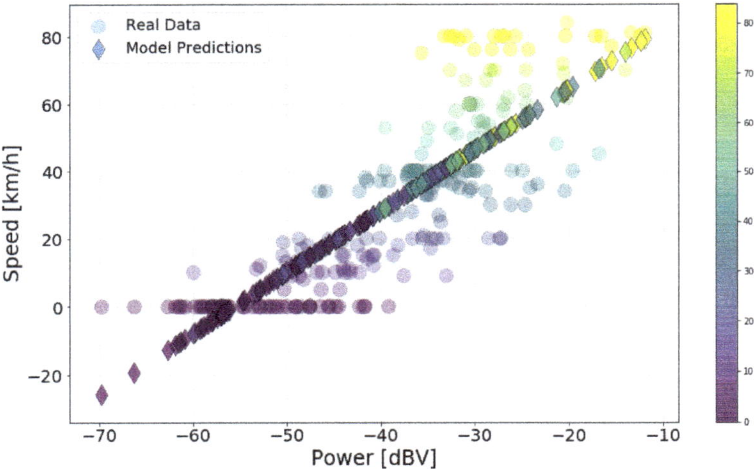

Figure 13. Speed data and predictions using the linear model.

Table 9. Model coefficients (with confidence interval in parenthesis) and goodness of fit metrics for speed vs. power model.

Linear Regression Coefficients		Goodness of Fit			
d_0	d_1	MSE	R-Squared	F-Statistic	Prob (F)
100.5496 (93.70–107.399)	1.8095 (1.640–1.979)	220.8105	0.68	445.2	8.60×10^{-54}

4.6. Multiple Variable Analysis

The previous analysis indicated the speed of the car and traffic conditions contribute the most to the noise power inside the car, while the position of the windows and rain presented a weak influence. Another way to verify how strong is the relationship between the variables and noise power is by computing their cross-correlation. Figure 14 presents a visualization of the correlation matrix of the data. Noise power has a high correlation with both speed and traffic, and a low correlation with the state of the windows. This confirms the behavior presented in Figures 7 and 13 that noise power tends to increase with the speed of the car. It also reaffirms the result presented in Figure 11 and Table 8 that the window variable has low explanatory value in the model.

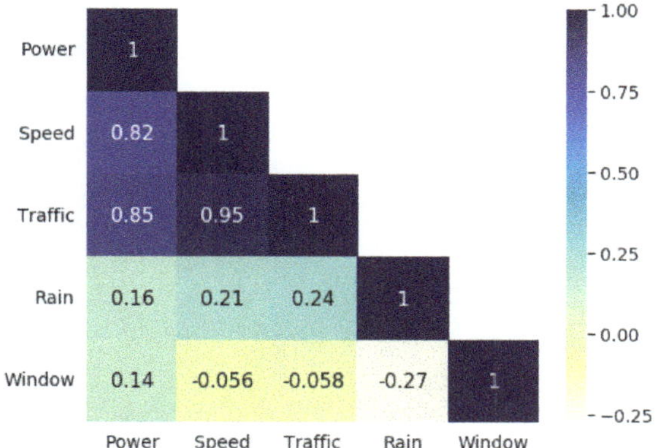

Figure 14. Correlation matrix of the dataset.

Figure 14 also shows a correlation of 0.94 between traffic and speed. This high correlation is expected. As stated in Section 2.1, the traffic categories were obtained in Google Maps by averaging the speed of the cars reported by the application's users. The correlation is not exactly 1 due to variations in driving speed during measurement for each traffic scenario. Nonetheless, in the context of statistical modelling, traffic and speed convey roughly the same information about the response variable and can be considered redundant.

To illustrate this redundancy between speed and traffic, we built a model to predict the speed of the car using all the other variables as independent.

$$speed \approx e_0 + e_1 \cdot power + e_2 \cdot traffic_{red} + e_3 \cdot traffic_{orange} + e_4 \cdot traffic_{green} + e_5 \cdot rain_{yes} + e_6 \cdot window_{open}\,, \quad (6)$$

where e_0 is the intercept and e_1 is the coefficient of the power level; e_2, e_3, and e_4 are the coefficient added when traffic conditions are red, orange and green, respectively; e_5 is the coefficient added when there is rain in the sample; and e_6 is the coefficient added when the windows are open. Although we do not expect a physical relationship between the rain and window variables and speed, we include these variables in this model to verify that they do not influence in the results. In order words, we want to verify that there is no bias in the measured speed data in relation to the absence or not of rain and the state of the car windows.

The model has prior-knowledge about the speed interval during measurement, which is conveyed by the traffic variable. The results shown in Figure 15 illustrate this. There are four groups of predictions divided by the traffic categories. No prediction is grouped incorrectly. Each traffic group has four lines, corresponding to the possible combinations of window and rain variables. These lines lie close to each other, indicating that the models for each pair of those conditions give similar predictions. This is expected, as speed has no relationship with rain and window, and the speed measurements were collected in a balanced quantity for all possible conditions of the variables. Effectively, the variable power, which determines the slope of all lines, is the one that determines the speed in each traffic group.

The goodness of fit metrics of the model are presented in Table 10. The high R^2 value indicates that most of the variation in speed is accounted for by the model. However, comparing Table 9 (noise power as the only explanatory variable) and Table 10, there is not much improvement in the F-value with the addition of the three other variables. Thus, the traffic variable does not contribute much to the model of speed, due to its redundancy. We conclude that either traffic or speed can be used as a good explanatory variable to noise power inside a vehicle, but not simultaneously.

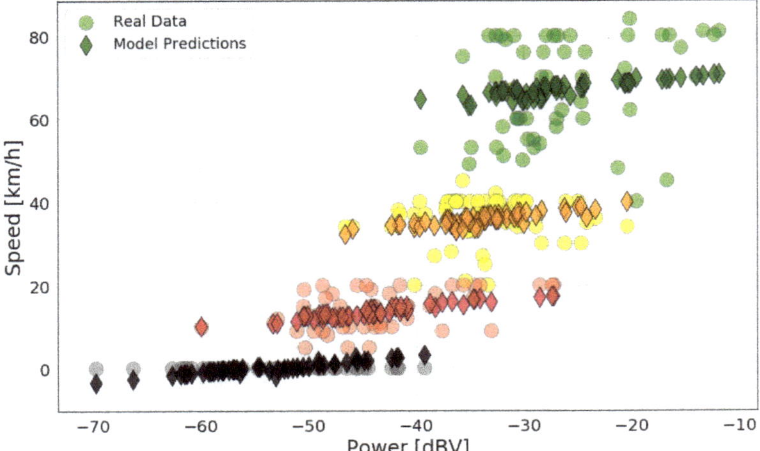

Figure 15. Speed data grouped by traffic conditions and predictions using the linear mode with all explanatory variables.

Table 10. Model coefficients and goodness of fit metrics for the speed vs. all other variables model.

Linear Regression Coefficients							Goodness of Fit			
e_0	e_1	e_2	e_3	e_4	e_5	e_6	MSE	R-Squared	F-Statistic	Prob (F)
14.19	0.2544	11.21	30.91	60.50	−2.70	−1.08	48.80	0.931	460.04	5.50×10^{-116}

5. Conclusions

Acoustic noise is a central issue in vehicle design. It is expected that will gain more attention as health concerns and multimedia, security, and autonomous vehicle applications become more prevalent. Prior work has shown different contributions to noise levels in the vehicle interior, investigating its subsystems and components. Those studies are mostly conducted in laboratory environments or using mathematical models. However, they may underestimate or ignore noise sources from specific conditions inside or outside the car. In this study, we presented an experimental evaluation of the contribution of different acoustic noise sources inside a car. The experiments were carried out by using a low-cost measurement setup inside a vehicle to acquire noise power levels in different traffic areas and different controlled conditions. Data visualization, statistical modeling, and goodness of fit metrics were used to assess the influence of speed, traffic, rain, and position of the car windows.

Our experiments in real traffic conditions showed a strong correlation between the speed of a car and its interior noise level, likely due to higher noise generation in the motor at higher speeds. Those results are correspondent with our general theoretical assumption. In contrast, the state of the car windows seems to not contribute significantly to the measured noise. The same, even with few collected data, can be speculated about the presence of rain. This could imply that most of the noise inside the vehicle can be attributed to its operation and movement, creating a higher variation on noise level and thus reducing the correlation of these less dominant factors. Thus, the results suggest that efforts to improve acoustic quality inside a vehicle should be focused on reducing the noise generated by the car itself.

To further our research on this topic, we plan to collect more noise data and study more variables, such as wind speed and different vehicles, as well as considering the noise in different car positions. We also plan to evaluate the vehicular scenario with the presence of human speech sources in the driver's seat, passenger's seat, and backseat. We also plan to investigate the spectral characteristics of the measured noise, which is especially important for noise suppression purposes. The data collected could then be explored in machine learning tasks and source location problems in the context of vehicle applications. In addition, more data may be acquired in rain conditions, enriching our study by providing more data points at this condition. Finally, an investigation of the impulsiveness of acoustic noise [51] in the vehicular scenario is warranted.

Author Contributions: Conceptualization, D.P.; Data curation, D.F., D.P., and L.P.; Formal Analysis, D.F and D.P.; Supervision, V.A.d.S. and A.M.; Visualization, D.F.; Writing—original draft, D.F., D.P., and L.P.; and Writing—reviewing and editing, V.A.d.S. and A.M. All authors have read and agreed to the published version of the manuscript.

Funding: This study was financed in part by the Coordenação de Aperfeiçoamento de Pessoal de Nível Superior-Brasil (CAPES)-Finance Code 001.

Conflicts of Interest: The authors declare that they have no competing interests.

Data Availability: The data collected and presented in this paper is avaiable in [44].

References

1. Almalkawi, I.T.; Guerrero Zapata, M.; Al-Karaki, J.N.; Morillo-Pozo, J. Wireless Multimedia Sensor Networks: Current Trends and Future Directions. *Sensors* **2010**, *10*, 6662–6717. [CrossRef] [PubMed]
2. Wang, X.; Ma, J.J.; Ding, L.; Bi, D.W. Robust Forecasting for Energy Efficiency of Wireless Multimedia Sensor Networks. *Sensors* **2007**, *7*, 2779–2807. [CrossRef]
3. Xue, W.; Bishop, D.W.; Ding, L.; Wang, S. Multi-agent Negotiation Mechanisms for Statistical Target Classification in Wireless Multimedia Sensor Networks. *Sensors* **2007**, *7*, 2201–2237. [CrossRef] [PubMed]
4. Jennehag, U.; Forsstrom, S.; Fiordigigli, F. Low Delay Video Streaming on the Internet of Things Using Raspberry Pi. *Electronics* **2016**, *5*, 60. [CrossRef]
5. Toma, C.; Alexandru, A.; Popa, M.; Zamfiroiu, A. IoT Solution for Smart Cities' Pollution Monitoring and the Security Challenges. *Sensors* **2019**, *19*, 3401. [CrossRef]
6. Kordov, K. A Novel Audio Encryption Algorithm with Permutation-Substitution Architecture. *Electronics* **2019**, *8*, 530. [CrossRef]
7. Garcia-Font, V.; Garrigues, C.; Rifà-Pous, H. Attack Classification Schema for Smart City WSNs. *Sensors* **2017**, *17*, 771. [CrossRef]
8. García-Hernández, A.; Galván-Tejada, C.; Galván-Tejada, J.; Celaya-Padilla, J.; Gamboa-Rosales, H.; Velasco-Elizondo, P.; Cárdenas-Vargas, R. A Similarity Analysis of Audio Signal to Develop a Human Activity Recognition Using Similarity Networks. *Sensors* **2017**, *17*, 2688. [CrossRef]
9. Liu, L.; Han, Z.; Fang, L.; Ma, Z. Tell the Device Password: Smart Device Wi-Fi Connection Based on Audio Waves. *Sensors* **2019**, *19*, 618. [CrossRef]
10. Kuhn, T.; Jameel, A.; Stumpfle, M.; Haddadi, A. Hybrid in-car speech recognition for mobile multimedia applications. In Proceedings of the 1999 IEEE 49th Vehicular Technology Conference (Cat. No.99CH36363), Houston, TX, USA, 16–20 May 1999; Volume 3, pp. 2009–2013. [CrossRef]
11. Higuchi, M.; Shinohara, S.; Nakamura, M.; Mitsuyoshi, S.; Tokuno, S.; Omiya, Y.; Hagiwara, N.; Takano, T. An effect of noise on mental health indicator using voice. In Proceedings of the 2017 International Conference on Intelligent Informatics and Biomedical Sciences (ICIIBMS), Okinawa, Japan, 24–26 November 2017; p. 202. [CrossRef]
12. Khan, S.; Akmal, H.; Ali, I.; Naeem, N. Efficient and unique learning of in-car voice control for engineering education. In Proceedings of the 2017 International Multi-Topic Conference (INMIC), Lahore, Pakistan, 24–26 November 2017; pp. 1–6. [CrossRef]
13. Tremoulet, P.D.; Seacrist, T.; Ward McIntosh, C.; Loeb, H.; DiPietro, A.; Tushak, S. Transporting Children in Autonomous Vehicles: An Exploratory Study. *Hum. Factors J. Hum. Factors Ergon. Soc.* **2019**. [CrossRef]
14. Chen, S.C. Multimedia for Autonomous Driving. *IEEE MultiMedia* **2019**, *26*, 5–8. [CrossRef]
15. Basner, M.; Babisch, W.; Davis, A.; Brink, M.; Clark, C.; Janssen, S.; Stansfeld, S. Auditory and non-auditory effects of noise on health. *Lancet* **2014**, *383*, 1325–1332. [CrossRef]
16. Rudolph, K.E.; Shev, A.; Paksarian, D.; Merikangas, K.R.; Mennitt, D.J.; James, P.; Casey, J.A. Environmental noise and sleep and mental health outcomes in a nationally representative sample of urban US adolescents. *Environ. Epidemiol.* **2019**, *3*, e056. [CrossRef] [PubMed]

17. Sygna, K.; Aasvang, G.M.; Aamodt, G.; Oftedal, B.; Krog, N.H. Road traffic noise, sleep and mental health. *Environ. Res.* **2014**, *131*, 17–24. [CrossRef] [PubMed]
18. Marco, C.; Elena, A.; Francesco, B.; Luca, F.; Gaetano, L.; Liliana, C. Chapter Global noise score indicator for classroom evaluation of acoustic performances in LIFE GIOCONDA project. *Noise Mapp.* **2016**, 157–171. [CrossRef]
19. Lercher, P.; Evans, G.W.; Meis, M. Ambient Noise and Cognitive Processes among Primary Schoolchildren. *Environ. Behav.* **2003**, *35*, 725–735. [CrossRef]
20. Vienneau, D.; Schindler, C.; Perez, L.; Probst-Hensch, N.; Röösli, M. The relationship between transportation noise exposure and ischemic heart disease: A meta-analysis. *Environ. Res.* **2015**, *138*, 372–380. [CrossRef]
21. Roswall, N.; Raaschou-Nielsen, O.; Ketzel, M.; Gammelmark, A.; Overvad, K.; Olsen, A.; Sørensen, M. Long-term residential road traffic noise and NO2 exposure in relation to risk of incident myocardial infarction—A Danish cohort study. *Environ. Res.* **2017**, *156*, 80–86. [CrossRef]
22. Roswall, N.; Raaschou-Nielsen, O.; Jensen, S.S.; Tjønneland, A.; Sørensen, M. Long-term exposure to residential railway and road traffic noise and risk for diabetes in a Danish cohort. *Environ. Res.* **2018**, *160*, 292–297. [CrossRef]
23. Wang, Y.; Guo, H.; Feng, T.; Ju, J.; Wang, X. Acoustic behavior prediction for low-frequency sound quality based on finite element method and artificial neural network. *Appl. Acoust.* **2017**, *122*, 62–71. [CrossRef]
24. Pietila, G.; Lim, T.C. Intelligent systems approaches to product sound quality evaluations—A review. *Appl. Acoust.* **2012**, *73*, 987–1002. [CrossRef]
25. Wang, Y.; Shen, G.; Xing, Y. A sound quality model for objective synthesis evaluation of vehicle interior noise based on artificial neural network. *Mech. Syst. Signal Process.* **2014**, *45*, 255–266. [CrossRef]
26. Huang, H.B.; Li, R.X.; Huang, X.R.; Yang, M.L.; Ding, W.P. Sound quality evaluation of vehicle suspension shock absorber rattling noise based on the Wigner–Ville distribution. *Appl. Acoust.* **2015**, *100*, 18–25. [CrossRef]
27. Jung, W.; Elliott, S.J.; Cheer, J. Local active control of road noise inside a vehicle. *Mech. Syst. Signal Process.* **2019**, *121*, 144–157. [CrossRef]
28. Praticò, F.G. On the dependence of acoustic performance on pavement characteristics. *Transp. Res. Part D Transp. Environ.* **2014**, *29*, 79–87. [CrossRef]
29. Pizzo, A.D.; Teti, L.; Moro, A.; Bianco, F.; Fredianelli, L.; Licitra, G. Influence of texture on tyre road noise spectra in rubberized pavements. *Appl. Acoust.* **2020**, *159*, 107080. [CrossRef]
30. Licitra, G.; Teti, L.; Cerchiai, M.; Bianco, F. The influence of tyres on the use of the CPX method for evaluating the effectiveness of a noise mitigation action based on low-noise road surfaces. *Transp. Res. Part D Transp. Environ.* **2017**, *55*, 217–226. [CrossRef]
31. Praticò, F.G.; Anfosso-Lédée, F. Trends and Issues in Mitigating Traffic Noise through Quiet Pavements. *Procedia Soc. Behav. Sci.* **2012**, *53*, 203–212, SIIV-5th International Congress—Sustainability of Road Infrastructures 2012. [CrossRef]
32. Licitra, G.; Cerchiai, M.; Teti, L.; Ascari, E.; Bianco, F.; Chetoni, M. Performance Assessment of Low-Noise Road Surfaces in the Leopoldo Project: Comparison and Validation of Different Measurement Methods. *Coatings* **2015**, *5*, 3–25. [CrossRef]
33. Shin, T.J.; Park, D.C.; Lee, S.K. Objective evaluation of door-closing sound quality based on physiological acoustics. *Int. J. Automot. Technol.* **2013**, *14*, 133–141. [CrossRef]
34. Parizet, E.; Guyader, E.; Nosulenko, V. Analysis of car door closing sound quality. *Appl. Acoust.* **2008**, *69*, 12–22. [CrossRef]
35. Li, Q.; Qiao, F.; Yu, L.; Shi, J. Modeling vehicle interior noise exposure dose on freeways: Considering weaving segment designs and engine operation. *J. Air Waste Manag. Assoc.* **2018**, *68*, 576–587. [CrossRef] [PubMed]
36. Huang, H.B.; Huang, X.R.; Yang, M.L.; Lim, T.C.; Ding, W.P. Identification of vehicle interior noise sources based on wavelet transform and partial coherence analysis. *Mech. Syst. Signal Process.* **2018**, *109*, 247–267. [CrossRef]
37. Yoshida, J.; Inoue, A. Road & wind noise contribution separation using only interior noise having multiple sound sources—Accuracy improvement and permutation solution-. *Mech. Eng. J.* **2017**, *4*, 17-00165. [CrossRef]

38. Talay, E.; Altinisik, A. The effect of door structural stiffness and flexural components to the interior wind noise at elevated vehicle speeds. *Appl. Acoust.* **2019**, *148*, 86–96. [CrossRef]
39. Cai, M.; Zhong, S.; Wang, H.; Chen, Y.; Zeng, W. Study of the traffic noise source intensity emission model and the frequency characteristics for a wet asphalt road. *Appl. Acoust.* **2017**, *123*, 55–63. [CrossRef]
40. Alonso, J.; López, J.; Pavón, I.; Recuero, M.; Asensio, C.; Arcas, G.; Bravo, A. On-board wet road surface identification using tyre/road noise and Support Vector Machines. *Appl. Acoust.* **2014**, *76*, 407–415. [CrossRef]
41. Kanarachos, S.; Kalliris, M.; Blundell, M.; Kotsakis, R. Speed-dependent wet road surface detection using acoustic measurements, octave-band frequency analysis and machine learning algorithms. In Proceedings of the ISMA Conference on Noise and Vibration Engineering, Leuven, Belgium, 17–19 September 2018. [CrossRef]
42. Soeta, Y.; Shimokura, R. Sound quality evaluation of air-conditioner noise based on factors of the autocorrelation function. *Appl. Acoust.* **2017**, *124*, 11–19. [CrossRef]
43. Volandri, G.; Di Puccio, F.; Forte, P.; Mattei, L. Psychoacoustic analysis of power windows sounds: Correlation between subjective and objective evaluations. *Appl. Acoust.* **2018**, *134*, 160–170. [CrossRef]
44. Pena, D.; Sousa, V.; Pena, L.; de Lucena Flor, D. Database of Acoustic Noise Power Levels in a Vehicle Interior under Different Conditions. 2019. Available online: https://figshare.com/articles/Noise_power_level_in_the_vehicle_interior_under_different_conditions/1370885 (accessed on 24 February 2020). doi:10.6084/m9.figshare.11370885.v1. [CrossRef]
45. Pokorný, P. Determining Traffic Levels in Cities Using Google Maps. In Proceedings of the 2017 Fourth International Conference on Mathematics and Computers in Sciences and in Industry (MCSI), Corfu, Greece, 24–27 August 2017; pp. 144–147. [CrossRef]
46. Přibil, J.; Přibilová, A.; Frollo, I. Analysis of the Influence of Different Settings of Scan Sequence Parameters on Vibration and Noise Generated in the Open-Air MRI Scanning Area. *Sensors* **2019**, *19*, 4198. [CrossRef]
47. Boudraa, A.O.; Salzenstein, F. Teager–Kaiser energy methods for signal and image analysis: A review. *Digit. Signal Process.* **2018**, *78*, 338–375. [CrossRef]
48. Stigler, S.M. Gauss and the Invention of Least Squares. *Ann. Stat.* **1981**, *9*, 465–474. [CrossRef]
49. Devore, J.L. *Probability and Statistics for Engineering and the Sciences*, 8th ed.; Cengage Learning: Boston, MA, USA, 2011.
50. Menard, S. Coefficients of Determination for Multiple Logistic Regression Analysis. *Am. Stat.* **2000**, *54*, 17–24. [CrossRef]
51. Pena, D.; Lima, C.; Dória, M.; Pena, L.; Martins, A.; Sousa, V., Jr. Acoustic Impulsive Noise Based on Non-Gaussian Models: An Experimental Evaluation. *Sensors* **2019**, *19*, 2827. [CrossRef] [PubMed]
52. Image of a Car Seen from Above. 2019. Available online: https://pngimage.net/carros-vistos-desde-arriba-png-4/ (accessed on 20 October 2019).

MDPI

Article

Aggregate Impact of Anomalous Noise Events on the WASN-Based Computation of Road Traffic Noise Levels in Urban and Suburban Environments

Francesc Alías *, Ferran Orga, Rosa Ma Alsina-Pagès and Joan Claudi Socoró

GTM—Grup de recerca en Tecnologies Mèdia, La Salle—Universitat Ramon Llull. c/Quatre Camins, 30, 08022 Barcelona, Spain; ferran.orga@salle.url.edu (F.O.); rosamaria.alsina@salle.url.edu (R.M.A.-P.); joanclaudi.socoro@salle.url.edu (J.C.S.)

* Correspondence: francesc.alias@salle.url.edu; Tel.: +34-932902440

Received: 10 December 2019; Accepted: 20 January 2020; Published: 22 January 2020

Abstract: Environmental noise can be defined as the accumulation of noise pollution caused by sounds generated by outdoor human activities, Road Traffic Noise (RTN) being the main source in urban and suburban areas. To address the negative effects of environmental noise on public health, the European Environmental Noise Directive requires EU member states to tailor noise maps and define the corresponding action plans every five years for major agglomerations and key infrastructures. Noise maps have been hitherto created from expert-based measurements, after cleaning the recorded acoustic data of undesired acoustic events, or Anomalous Noise Events (ANEs). In recent years, Wireless Acoustic Sensor Networks (WASNs) have become an alternative. However, most of the proposals focus on measuring global noise levels without taking into account the presence of ANEs. The LIFE DYNAMAP project has developed a WASN-based dynamic noise mapping system to analyze the acoustic impact of road infrastructures in real time based solely on RTN levels. After studying the bias caused by individual ANEs on the computation of the A-weighted equivalent noise levels through an expert-based dataset obtained before installing the sensor networks, this work evaluates the aggregate impact of the ANEs on the RTN measurements in a real-operation environment. To that effect, 304 h and 20 min of labeled acoustic data collected through the two WASNs deployed in both pilot areas have been analyzed, computing the individual and aggregate impacts of ANEs for each sensor location and impact range (low, medium and high) for a 5 min integration time. The study shows the regular occurrence of ANEs when monitoring RTN levels in both acoustic environments, which are especially common in the urban area. Moreover, the results reveal that the aggregate contribution of low- and medium-impact ANEs can become as critical as the presence of high-impact individual ANEs, thus highlighting the importance of their automatic removal to obtain reliable WASN-based RTN maps in real-operation environments.

Keywords: road traffic noise; noise monitoring; dynamic noise maps; anomalous noise events; individual impact; aggregate impact; WASN; sensor nodes; urban and suburban environments.

---

1. Introduction

Environmental noise can be defined as the accumulation of noise pollution caused by sounds generated by human activity outdoors, mainly produced by transport, road traffic, rail traffic, air traffic and industrial activities [1]. According to the World Health Organization, noise exposure produces a loss of around one million healthy life years in Western Europe every year due to different types of derived diseases [2,3]. Focusing on this public health problem, the European (EU) authorities published the Environmental Noise Directive (END) [1] in 2002, which requires the EU member states to tailor noise maps and to develop the subsequent action plans to mitigate noise every five years for major agglomerations and key infrastructures [4]. To address this issue in a harmonized manner, the Common Noise Assessment Methods in Europe (CNOSSOS-EU) was also developed, defining the measurement guidelines to allow comparable noise assessments across the EU [5]. However, as one of the first set of results obtained after the implementation of the END regulation showed [6], noise pollution continues to be one of the principal causes of health problems in Europe. This premise was further endorsed by [7,8], which led to the development of an updated version of the CNOSSOS-EU [9].

The aforementioned dramatic effects of noise pollution on citizens are mainly caused by traffic noise, as it is the main noise source in urban and suburban areas [10,11]. Road Traffic Noise (RTN) maps have been historically created from expert-based measurements using certified devices during specific time periods and locations, considering vehicle flows averaged over long periods of time [12]. During the recordings, the presence of acoustic events non-related to road traffic (e.g., sirens, horns, works, dogs' barks, airplanes flyovers, etc.) may occur [13]. As a consequence, the collected acoustic data should be cleaned of these undesired events before feeding the noise map creation software [13] to avoid biasing the computation of the A-weighted equivalent sound levels (L_{Aeq}) beyond 2 dB, as recommended by the European Commission Working Group Assessment of Exposure to Noise (WG-AEN) [14]. In this context, the Signal-to-Noise Ratio (SNR) of these acoustic events becomes a crucial parameter to evaluate and model [15,16]. Although some researchers have opted to control the SNR of the events by creating artificially mixed datasets (see e.g., [17–20]), their accurate characterization remains as an open research question as it is almost unfeasible to represent the wide diversity of acoustic data for real world [21].

The so-called Wireless Acoustic Sensor Networks (WASNs) have become an alternative to the creation of noise maps using real-life data, since they allow the ubiquitous monitoring of environmental noise [22–24]. During the last decade, several WASNs have been deployed in different smart cities such as Barcelona [25], Algemesí [26], Pisa [27], Monza [28], Halifax [29] and Milan and Rome [30] in Europe, or New York city [31], to name a few. In this WASN-based approach, the traditional manual cleaning of the Anomalous Noise Events (ANEs) on the noise pattern [32] becomes unfeasible due to the huge volume of data that have to be processed in real time [13]. As a consequence, the first generation of these WASN-based environmental noise monitoring systems have mainly been focused on measuring the global sound levels of the sensed locations, without considering the impact of the presence of specific acoustic events on the L_{Aeq} computation. To address this issue, some projects have started incorporating acoustic event detection techniques within the WASN-based noise monitoring pipeline. The Sounds of New York City (SONYC) project includes the real-time identification of 10 common classes of urban sound sources [31] through a machine listening system trained after artificially mixing the events with background noise in the UrbanSound dataset [16]. Moreover, the DYNAMAP project aims at developing a WASN-based dynamic noise mapping system to monitor the acoustic impact of road infrastructures through the creation of noise maps in real time [30]. The project includes two pilot areas: one in the District 9 of Milan as urban area [33], and another in the A90 highway surrounding Rome as a suburban area [34,35]. As the system focuses on measuring RTN levels solely, the ANEs present in the acoustic environments should be automatically removed. To that effect, a machine listening algorithm denoted as Anomalous Noise Events Detector [36] was designed and initially trained using a 9-h expert-based dataset collected from the two pilot areas before installing both sensor networks [21]. The analysis of that preliminary dataset highlighted the importance of the removal of individual ANEs based on their duration and SNR [37]. However, no evidence of a critical impact was yet observed in that dataset due to the presence of several ANEs within the same period of time, probably because the expert-based dataset missed several key aspects from real operation, such as different RTN patterns between day-night and weekday-weekends, or variable weather conditions, among others [38].

After the deployment of the two WASNs in the urban and suburban pilot areas, this paper evaluates the aggregate impact of ANEs on the L_{Aeq} computation of RTN in both environments in real operation. Besides analyzing the individual impact of ANEs on the measurements, the analysis methodology focuses on evaluating the bias caused by the presence of several ANEs within a given period of time, taking into account their impact range (low, medium or high) and sensor location. The study is conducted on 304 h and 20 min of WASN-based labeled acoustic data collected through both sensor networks, before proceeding to update the ANED algorithm with both WASN-based datasets (see [39,40] for a detailed description of the general characteristics of the urban and suburban datasets, respectively).

The paper is structured as follows. Section 2 reviews practices in acoustic environments where the salience and the impact of the events is a key issue. Section 3 presents the impact analysis methodology and impact-related measurements. Section 4 presents the conducted experiments and the results obtained from the analysis of the WASN-based urban and suburban acoustic datasets. Finally, after discussing several key aspects of this work in Section 5, the main conclusions and future work are described in Section 6.

2. Related Work

In this section, we review several works from the literature dealing with the identification of salient acoustic events regardless of the noise source; this issue together with the duration of the event sets the basis for the evaluation of the actual impact of these events on the L_{Aeq} computation.

One of the most challenging issues when working with environmental acoustic data recorded in real-life is their accurate characterization, which is supervised by experts. More precisely, this process deals with the parameterization of the data by means of several representative features, among which are the temporal limits of each sound event—i.e., its *actual* duration—by setting up its start and end boundaries [41,42], and its acoustic salience with respect to the background noise [15,16], i.e., the SNR of the event, which is a key parameter to consider. To properly address this issue, it should be taken into account that the events that need to be detected are usually independent one from each other, and typically present a variable duration and SNR. Furthermore, no temporal correlation can be found among them, which makes the challenge of parameterizing audio events particularly more complex compared to speech or music signal [43]. Consequently, the accurate characterization of environmental sound remains as an open research question in real world environments [21].

To work with a controlled environment, artificially-mixed datasets are usually built taking into account a predefined range of SNRs when mixing the events with the background noise during the dataset process generation. Some examples can be found in Foggia et al. [17], Stowell et al. [18] and Socoró et al. [19] (see [21] for further examples). The measurements of SNRs in audio fragments makes it possible to sort events by their degree of acoustic salience with respect to their environment. Moreover, datasets containing synthetic or artificially modified samples also respond to the need to generate more samples of a particular type of noise that is scarce , which is yet today one of the main limitations of acoustic event detection [44]. The explicit SNR measure can be evaluated by means of a closed set of saliency levels, such as −6 dB, 0 dB or +6 dB, as suggested by Stowell et al. in [18]; the authors also propose to record live scripted monophonic event sequences in acoustic environments under control. Foggia et al. [17] mixes several sounds related to surveillance (e.g., scream, glass breaking and gunshots) with both indoor and outdoor environments with six different levels of SNR (from 5 dB to 30 dB, with a step of 5 dB), after the observation of the occurrences of these events in a real-life environment. Socoró et al. [19] presents a dataset composed of a mixture of sound sources considering road traffic noise plus other type of sound events generated using two different SNRs (+6 dB and +12 dB) in order to assess the performance of an anomalous noise event detector. The original non-traffic-noise related audio fragments were extracted from Freesound (https://freesound.org/) while road traffic noise was recorded in a city ring road in real-life conditions. Nakajima et al. [45] works with a dataset recorded in real operation with several examples of noise sources of interest (e.g., cicadas, outside air conditioner, road traffic noise, and neighborhood noise). The work complements the dataset with artificial mixtures to increase the sound source diversity by means of varying the salience of the events using the SNR of three sound sources in the dataset, adapting the margins from −6 dB to +6 dB depending on the characteristics of the noise source. Finally, in Koizumi et al. [46], the authors conduct an objective evaluation on a synthetic dataset, using an open toy-car-running sound dataset; the dataset includes four types of factory noises, and it was generated by mixing synthetically those audio samples at a SNR = 0 dB, together with the audio files of less than 5-s duration from the Task-2 dataset of DCASE 2018 Challenge [47].

Following a different approach, several research works consider auditory attention when evaluating the impact of sound events on acoustic measurements through the evaluation of their SNR levels, whose focus may vary depending on the domain of application (e.g., noise monitoring or surveillance) or the signal of interest (see [48] and references therein for further details). These works analyze the perceptual relevance of audio events according to human response, as in [15], where De Coensel and Botteldooren design a salience-based map to simulate the capability of humans to switch the attention among several auditory stimuli along time, considering noise examples of means of transportation. This research approach is focused on the identification of the salient event. However, it ignores both its origin and its relative energy with respect to background noise. Following this approach, Salamon et al. [16] included a perceptually based binary descriptor in their dataset to discriminate whether the event was perceived as the main noise source or in the background of the recording. Afterwards, the dataset was used to evaluate the performance of a sound event classification algorithm, getting better accuracy results on foreground events rather than those perceived in the background. Annotating and evaluating a recorded set of audio files is a very time-consuming task. To address these limitations, Salamon et al. published Scaper [20], whose goal is to conduct soundscape synthesis together with data augmentation given a soundbank, controlling characteristics such as the number and type of events, their timing, duration and SNR with respect to a background sound. The final goal is to ease the dataset generation process but also to ensure that the sets of data evaluated present suitable statistical characteristics for training and test of acoustic event detection algorithms.

Finally, it is worth mentioning that a couple of WASN-based projects have recently incorporated the detection of acoustic events in urban and suburban environments in the environmental noise monitoring pipeline. To that effect, the SONYC project [31] has developed a representative dataset with diverse sounds of interest, using the data gathered from the 56 sensors deployed in different neighborhoods of New York, considering up to 10 different common urban sound sources from the urban soundscape (highly frequent in urban noise complaints). The UrbanSound dataset was created after artificially mixing the events coming from Freesound with the background noise collected in the project [16]. Our team, in the framework of the DYNAMAP project [30] made its first attempt to create an acoustic dataset of the urban and suburban pilot areas (District 9 in Milan and A90 highway surrounding Rome) before the sensors of the two WASN were deployed in those scenarios, by means of an expert-based recording campaign [21]. The analysis of those datasets showed the highly local and unpredictable nature of anomalous noise events, which were manually labeled and used to train the preliminary version of the ANED algorithm [36]. Recently, the deployment of the two WASNs in both pilot areas has led to the generation of a suburban acoustic dataset through the 19-nodes WASN in Rome [40], together with the completion of the first steps of the creation of an urban dataset through the 24-node WASN installed in Milan in real operation [39]. From these two experiences, it can be concluded that the evaluation of the acoustic salience of any environmental acoustic event is relevant in order to improve the accuracy of the derived machine listening approaches [43], an issue that was justified in [37] after evaluating the individual impact of the detected events on the overall equivalent noise level computation considering 9 h of real-life acoustic data collected through an expert-based recording campaign. However, as far as we know, no specific analysis has been conducted to assess to what extent the concentration of ANEs with low SNRs within a period of time may bias the WASN-based computation of the L_{Aeq} measurements.

3. Impact Analysis Methodology

This section describes the methodology followed to analyze the bias caused by ANEs on the L_{Aeq} computation for a given integration time T (hereafter denoted as $L_{Aeq,T}$), building on the analysis methodology presented in [37]. The impact analysis methodology permits the study of both individual and aggregate contributions of the anomalous noise events present within a specific period of time. To that effect, individual and aggregate impact histograms are obtained from the labeled data for each sensor of the network according to the considered impact ranges. As depicted in Figure 1, the analysis starts with the labeled acoustic data collected from a WASN of N_S sensors in real operation. After windowing the audio streams into frames of T seconds, the individual and aggregate impacts of the ANEs present in each period of time t are computed and stacked. Finally, both individual and aggregate impact histogram matrices are derived to account for the occurrences belonging to each impact range defined by a set of impact thresholds. The following paragraphs explain the key elements of the proposed analysis methodology in detail.

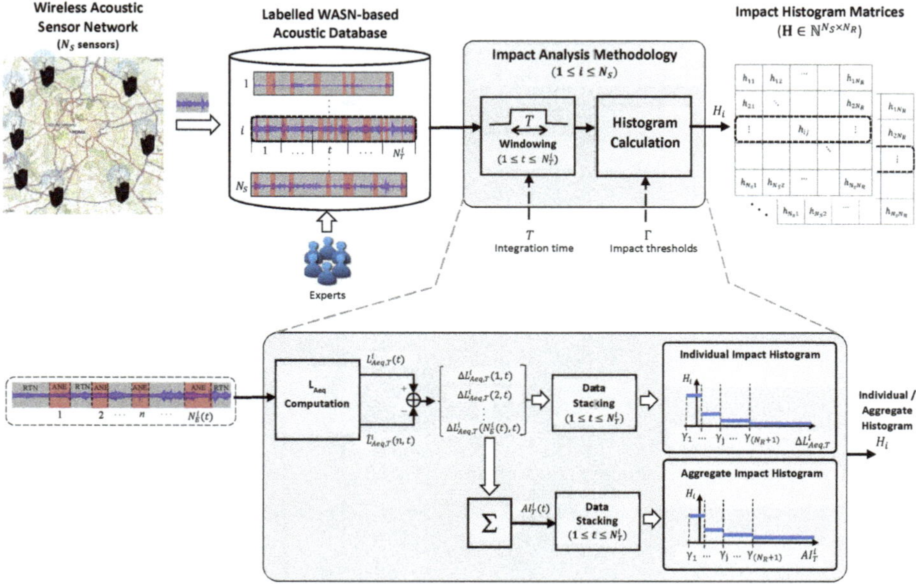

Figure 1. Block diagram of the impact analysis methodology on a labeled WASN-based acoustic dataset obtained from a N_S sensors network, where T is the integration time considered to compute $L^i{}_{Aeq,T}(t)$ and $\hat{L}^i_{Aeq,T}(n,t)$ for each sensor i and event n. Moreover, $\Delta L^i_{Aeq,T}(n,t)$ and $AI^i_T(t)$ denote the individual and aggregate impacts of the ANEs, respectively. Finally, h_{ij} represents the components of the histogram matrices $\mathbf{H}$ derived from the individual and aggregate impact histograms H_i, which account for the impact values according to N_R impact ranges defined by a set of impact thresholds $\Gamma = \{\gamma_1, \gamma_2, ..., \gamma_{(N_R+1)}\}$.

- **Aggregate impact computation per sensor**

  The *Aggregate Impact* (AI) of several acoustic events can be defined as the accumulated contribution of the individual impacts of all the ANEs present within a period of time and sensor node.

  It is denoted as $AI^i_T(t)$, where indexes i and t respectively represent the sensor number, for $i = \{1, 2, ..., N_S\}$, and the integration time period, for $t = \{1, 2, ..., N^i_T\}$, N^i_T being the total number of integration time periods of length T considered for its computation given a sensor i, and it is defined as

$$AI_T^i(t) = \sum_{n=1}^{N_E^i(t)} \Delta L^i_{Aeq,T}(n,t), \tag{1}$$

where $\Delta L^i_{Aeq,T}(n,t)$ is the individual impact of the n-th ANE on the $L_{Aeq,T}$ computation within the integration time period t, $N_E^i(t)$ being the total number of ANEs present in that time period for sensor i, and it is computed as

$$\Delta L^i_{Aeq,T}(n,t) = L^i_{Aeq,T}(t) - \hat{L}^i_{Aeq,T}(n,t), \tag{2}$$

$L^i_{Aeq,T}(t)$ being the total A-weighted equivalent sound level in the integration period of interest t for the i-th sensor (i.e., considering RTN and all ANEs found in that t), and $\hat{L}^i_{Aeq,T}(n,t)$ the corresponding noise level after removing the n-th ANE from the measurement through the linear interpolation of the $L_{Aeq,1s}$ values of the previous and subsequent RTN samples (the reader is referred to [37] for further details).

To that effect, first, the audio data collected from sensor i is divided into N_T^i windows of T seconds length (see Figure 1). Next, the A-weighted equivalent noise levels with and without ANEs are computed, whose difference gives the n-th individual ANE impact $\Delta L^i_{Aeq,T}(n,t)$. Then, the aggregate impact of window t is obtained by accumulating the individual impacts of all the ANEs it contains.

- **Range-based impact analysis per sensor**

The analysis methodology also aims at categorizing the relevance of both individual and aggregate impacts according to N_R impact ranges $\Theta = \{\theta_1, \theta_2, ..., \theta_{N_R}\}$ (in dB) delimited by a predefined set of impact thresholds $\Gamma = \{\gamma_1, \gamma_2, ..., \gamma_{(N_R+1)}\}$, and it is computed as

$$\Theta = \bigcup_{j=1}^{N_R} \theta_j = \bigcup_{j=1}^{N_R} [\gamma_j, \gamma_{j+1}), \tag{3}$$

where θ_j is defined as the impact range where $\gamma_j \leq \Delta L^i_{Aeq,T}(t) < \gamma_{j+1}$, for $j = \{1, 2, ..., N_R\}$.

This information is statistically analyzed through the histograms obtained for each sensor (see Figure 1) in the *impact histogram matrix* $\mathbf{H} = (h_{ij}) \in \mathbb{N}^{(N_S \times N_R)}$, h_{ij} being the number of occurrences of ANEs that account for an impact within θ_j observed in the i-th sensor as follows

$$\mathbf{H} = \begin{pmatrix} H_1 \\ H_2 \\ \vdots \\ H_i \\ \vdots \\ H_{N_S} \end{pmatrix} = \begin{pmatrix} h_{11} & h_{12} & \cdots & \cdots & \cdots & h_{1N_R} \\ h_{21} & h_{22} & \cdots & \cdots & \cdots & h_{2N_R} \\ \vdots & \vdots & \ddots & \ddots & \ddots & \vdots \\ \vdots & \vdots & \ddots & h_{ij} & \ddots & \vdots \\ \vdots & \vdots & \ddots & \ddots & \ddots & \vdots \\ h_{N_S1} & h_{N_S2} & \cdots & \cdots & \cdots & h_{N_SN_R} \end{pmatrix}, \tag{4}$$

where

$$h_{ij} = \begin{cases} \sum_{t=1}^{N_T^i} \sum_{n=1}^{N_E(t)} \mathbf{1}_{\theta_j}(\Delta L^i_{Aeq,T}(n,t)) & \text{for individual impact,} \\ \sum_{t=1}^{N_T^i} \mathbf{1}_{\theta_j}(AI_T^i(t)) & \text{for aggregate impact,} \end{cases} \tag{5}$$

with $\mathbf{1}_{\theta_j}(\cdot)$ being the indicator function defined for the interval range θ_j as

$$\mathbf{1}_{\theta_j}(x) = \begin{cases} 1 & \text{if } x \in \theta_j, \\ 0 & \text{if } x \notin \theta_j. \end{cases} \tag{6}$$

Notice that rows of $\mathbf{H}$ (denoted as H_i in Equation (4)) correspond to the impact histograms obtained from each i sensor.

- **Analysis of the critical aggregate impacts per impact range and sensor**

  To complement the previous analyses, it is also interesting to identify the origin of critical AIs for those cases that surpass the critical threshold γ_c. To that effect, the aggregate impact of ANEs for a given integration time period and sensor is computed considering only those individual ANEs which $\Delta L^i_{Aeq}(n,t)$ belongs to a particular impact range (i.e., $\Delta L^i_{Aeq}(n,t) \in \theta_j$) as follows

$$AI^i_T(\theta_j, t) = \sum_{n \in \Psi(\theta_j, t)} \Delta L^i_{Aeq,T}(n,t), \tag{7}$$

  where $\Psi(\theta_j, t)$ represents the subset of ANE indices within t which individual impact belongs to impact range θ_j.

  Finally, the *critical AI histogram matrix* $\mathbf{H_c} = (h^c_{ij}) \in \mathbb{N}^{N_S \times N_R}$ is defined as a particular case of $\mathbf{H}$ (see Equation (4)) considering the matrix components as

$$h^c_{ij} = \sum_{t=1}^{N^i_T} \mathbf{1}_{\theta_c}(AI^i_T(\theta_j, t)), \tag{8}$$

  the $\mathbf{1}_{\theta_c}(x)$ being a particular case of the indicator function defined by $\theta_j = \theta_c$ (see Equation (6)), where $\theta_c = [\gamma_c, +\infty)$ defines the range of critical impacts, as γ_c represents the threshold of a non-tolerable deviation of the A-weighted equivalent road traffic noise levels.

4. Experiments and Results

This section describes the results of the experiments from the impact analysis conducted on the two environmental WASN-based audio databases from the DYNAMAP's Milan and Rome pilot areas [39,40]. According to the project specifications, the considered integration time to update the $L_{Aeq,T}$ values of the RTN maps is 5 min [30], i.e., $T = 300$ s. To analyze to what extent the collected ANEs from each sensor location bias the $L_{Aeq,300s}$ measurement, the impacts are categorized within three impact ranges (i.e., $N_R = 3$) [37], accounting for those occurrences (from either individual or aggregate ANEs) causing a low-impact in $\theta_1 = (-\infty, 0.5)$ dB, a medium-impact in $\theta_2 = [0.5, 2)$ dB, and, finally, a high-impact in $\theta_3 = [2, +\infty)$ dB, $\theta_3 = \theta_c$ being as this last interval collects those cases that surpass the critical threshold $\gamma_c = 2$ dB according to the WG-AEN [14]. Regarding the two WASNs, the number of sensors N_S considered for the subsequent analyses is 19 for the suburban network, and 23 for the urban one, whereas the total number of evaluated segments of 5 min is 1812 in Milan and 1840 in Rome, respectively.

4.1. WASN-Based Environmental Databases

After the deployment of the sensor networks in the urban and suburban pilot areas of the DYNAMAP project, two WASN-based databases were obtained from environmental acoustic data in real-operation conditions. On the one hand, the nodes distribution across the urban area of Milan is based on the clustering of traffic noise profiles in order to place the best sensor locations for different road categories [33]. On the other hand, in the Rome suburban area, the sensor nodes have been spread along the A90 highway, considering several scenarios of different complexity (single road, crossings, nearby railways and multiple connections) [34,35]. Figure 2 depicts two examples of the sensor placements in both urban and suburban areas, and Appendix A details the sensors' Ids as well as the description of their locations within Tables A1 and A2 for the urban and suburban environments, respectively.

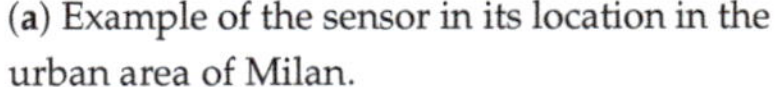

(**a**) Example of the sensor in its location in the urban area of Milan.

(**b**) Example of the sensor in its location in the suburban area of Rome (picture property of ANAS S.p.A.).

Figure 2. Examples of the location of the low-cost acoustic sensors in the DYNAMAP's urban and suburban pilot areas.

In both cases, the recorded databases include data from two days with different traffic conditions: one from a weekday (on Tuesday, the 28th of November 2017 for the urban area, and on Tuesday, the 2nd of November 2017 for the suburban environment), and another during the weekend (on Sunday, the 3rd of December 2017 on the urban area, and on Sunday, the 5th of November 2017 in the suburban environment). The audio recordings were collected in continuous raw audio clips from the first 20 min of each hour (considering a sampling frequency of 48 kHz) , as a trade-off between the storage capacity and communications resources of the nodes, and obtaining a representative sub-sampling of the L_{Aeq} measurements along the day [40]. The gathered acoustic data were manually labeled by experts in audio signal processing (see [39,40] for further details). As a result, up to 28 ANE subcategories were identified. Table 1 lists the 16 types of ANEs observed during the manual labeling process in the suburban environment (subcategories being *stru* and *trck* only specifically detected in this scenario), together with the 26 subcategories identified during the annotation of the urban dataset (being *bell*, *blin*, *dog*, *glas*, *peop*, *rubb*, *sqck*, *step*, *tram* and *wrks* those ANE subcategories typically found within this environment). Meteorological-related ANEs like *thun*, *rain* and *wind* cannot be attributed to any specific acoustic environment since they are highly dependent on the weather during the days of the WASN-based data collection. Finally, audio excerpts that contained a mixture of different sound sources (e.g., diverse ANEs together with RTN as background) were labeled as complex sound mixtures or CMPLX. Both CMPLX and ANEs are considered for the subsequent impact-related analyses as both contain undesired acoustic events, after windowing the audio streams into N_T^i frames of length T (see Figure 1).

Table 1. Description and % of occurrences of the 28 sound subcategories attributed to anomalous noise events found throughout the manual labeling process of the WASN-based urban and suburban acoustic databases.

Label	Suburban Counts (%)	Urban Counts (%)	Description
airp	0.1	1	Noise of airplanes and helicopters
alrm	0.2	0.3	Sound of an alarm or a vehicle beep moving backwards
bell	0	1.2	Church bells
bike	<0.1	3.6	Sound of bikes and bike chains
bird	15.1	14.7	Birdsong
blin	0	<0.1	Opening and closing of a blind
brak	23.1	12.7	Brakes and conveyor belts
busd	2.8	1.1	Opening bus door (or tramway), depressurized air
dog	0	2.5	Barking of dogs
door	2.6	14.7	Closing doors (vehicle or house)
glas	0	0.1	Sound of glass crashing
horn	6.7	3.7	Horns of vehicles (cars, motorbikes, trucks, etc.)
inte	0.3	0.2	Interfering signal from an industry or human machine
musi	<0.1	0.6	Music in car or in the street
peop	0	22.2	Sounds of people chatting, laughing, coughing, sneezing, etc.
rain	23.7	0.4	Sound of heavy rain
rubb	0	0.1	Rubbish service (engines and grabbing system)
sire	1.8	0.7	Sirens (ambulances, police, etc.)
sqck	0	0.8	Squeak sound of door hinges
step	0	13.7	Sounds of steps
thun	7.4	<0.1	Thunderstorm
trck	11.9	0	Noise when trucks or vehicles with heavy load passed over a bump.
tram	0	0.7	Stop, start and passby sounds of tramways
tran	2.7	<0.1	Sound of trains
trll	0	1	Sound of wheels of suitcases (trolley)
stru	1.4	0	Noise of highway portals structure caused by vibration of trucks passbys
wind	0	<0.1	Noise of wind (movement of the leaves of trees,...)
wrks	0	4.1	Works in the street (e.g., saws, hammer drills, etc.)

As a result, the subsequent analyses evaluate 153 h and 20 min of audio data obtained from the 19 sensors placed on the A90 highway portals along the Rome suburban environment, and 151 h obtained from 23 different sensors placed in the building façades of several public buildings across the District 9 of Milan, after discarding node hb114 due to technical problems during the data recording process, but keeping sensor hb119 despite missing some data from the Sunday recordings to $75\%N_T^i$.

Table 2 summarizes the general characteristics of both analyzed datasets. As can be observed, RTN is the majority class in both cases, as identified 83.7% of the time in the urban environments, while this value raised to 96.5% in the suburban scenario. Accordingly, ANEs were more frequently observed in the urban than in the suburban dataset, being more than four times detected in this environment compared to the suburban one (8.7% of ANE in urban while 1.9% of ANE in suburban). It should be also noticed that the increase of ANE occurrences in the urban environment also fostered the presence of highly complex audio passages.

Table 2. General characteristics of the WASN-based urban and suburban acoustic databases evaluated considering the impact analysis methodology.

Acoustic Environment	Total Duration	RTN (%)	ANE (%)	CMPLX (%)
Milan (Urban)	151 h	83.7%	8.7%	7.6%
Rome (Suburban)	153 h 20 min	96.5%	1.9%	1.6%

4.2. Individual Impact of ANEs

To understand the relevance of the events, first, a study of the individual ANE impact is conducted following the aforementioned impact analysis methodology. As an overall analysis, Table 3 details the number of occurrences and sensor activation ratios for each environment and recording day.

Table 3. Number of occurrences and sensor activation ratios per sensor for low, medium and high individual impact ranges.

Individual Impacts		**Low Impact** $(-\infty, 0.5)$ **dB**		**Medium Impact** $[0.5, 2)$ **dB**		**High Impact** $[2, +\infty)$ **dB**	
		Occurrences Count (%)	**Activation Count/**N_S	**Occurrences Count (%)**	**Activation Count/**N_S	**Occurrences Count (%)**	**Activation Count/**N_S
Milan	Tuesday	21,264 (99.5%)	23/23	76 (0.4%)	21/23	28 (0.1%)	16/23
	Sunday	15,215 (99.4%)	23/23	58 (0.4%)	20/23	29 (0.2%)	16/23
Rome	Tuesday	2105 (98.1%)	19/19	33 (1.6%)	13/19	7 (0.3%)	5/19
	Sunday	3415 (99.0%)	19/19	31 (0.9%)	11/19	5 (0.1%)	3/19

As can be observed, the presence of anomalous noise events is common in both environments, particularly in Milan which records 10 times more ANEs on Tuesday and 4 times more on Sunday than Rome. Specifically, all recording days have yielded a high percentage of low-impact ANEs, but in Milan, particularly, the presence of low-impact events in relation to the other impact ranges, is higher than in Rome, rising from 98.1 to 99.5% on Tuesday, and from 99.0 to 99.4% on Sunday. In Rome, however, the percentage of medium-impact events is higher than in Milan on both days, with a total of 134 ANEs in Milan and 64 in Rome, respectively. This implies that the sensors in Milan can detect this kind of event in almost all sensors, while only 60% of the sensors in Rome can detect these ANEs. Finally, concerning high-impact events, the percentage of occurrences is similar in both locations, despite Milan has 57 high-impact events detected in 16 sensors and Rome only 12, which activate few sensors.

In Figure 3, the corresponding impact histogram matrices for individual ANEs are detailed for each sensor location according to the three impact range intervals (low, medium and high). Notice that the number of occurrences in the low-impact intervals is depicted separately from the medium and high-impact intervals for illustration purposes, as it is more than two orders of magnitude larger.

It can be observed that the maximum number of low-impact ANEs has been found in sensor hb123 of Milan on Tuesday, with 2374 occurrences. In contrast, the maximum number of low-impact events in Rome is 379 for sensor hb143 on Sunday. Concerning the medium-impact events in Milan, the first day accounts for the highest number of events, coming from hb139, which obtains the maximum number of medium-impact ANEs, with 9 occurrences, also presents a significant number in Sunday, with 6 events. In the rest of the cases in Milan, no clear pattern is observed relating both recording days. In Rome, however, sensor hb104 attributes for the maximum number of medium-impact events, with 18 occurrences on Tuesday and 17 on Sunday. This is a particularly relevant case in the suburban area as the second closest sensor is hb134 with only 3 medium-impact events on Sunday. When looking at the column depicting high impact ANEs, it can be observed that a maximum of 5 events were captured on Sunday in sensor hb133 of Milan, while also a significant presence on Tuesday with 4 occurrences. In Rome, sensor hb104 accounts for the highest number of high-impact ANEs on Tuesday, with 3 events, which also recorded one of the highest number of occurrences on Sunday, with 2 events.

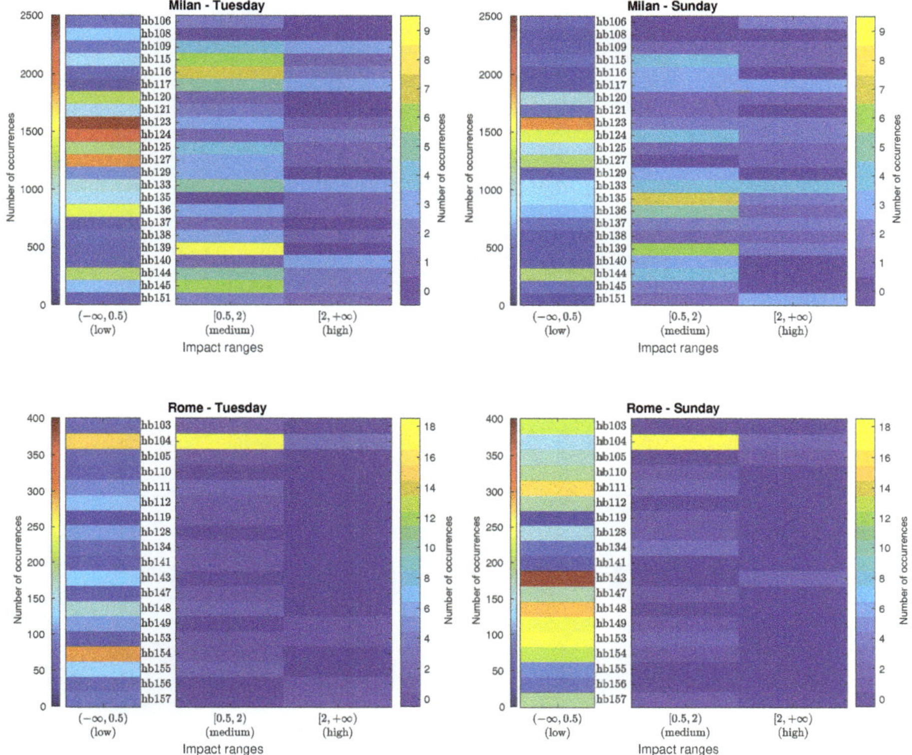

Figure 3. Individual impact histogram matrices (obtained using integration time $T = 300$ s) categorized in three impact ranges (low, medium and high) for the urban (Milan) and suburban (Rome) environments obtained from a weekday (Tuesday) and weekend day (Sunday).

4.3. Aggregate Impact of ANEs

This section details the results obtained from the analysis of the labeled data in order to find to what extent the presence of several ANEs with low and medium individual impacts within the same integration period can bias the $L_{Aeq,300s}$ computation.

First, Table 4 shows the number of occurrences and sensors activation ratios of the AI for environment and recording day. As it can be observed from the table, the overall presence of occurrences and activation ratios are similar for both days within each location. However, when comparing Milan with Rome, the distribution of the impact ranges differs. In the case of Milan, near 85% of the AIs entail a low impact on the $L_{Aeq300s}$. This percentage increases to almost 96% in Rome. For this reason, the presence, as well the sensor activation, of medium and high-level AIs in Rome is lower than in Milan. In Milan, only one sensor on Tuesday and two on Sunday fail to detect a medium-impact AI. However, in Rome, on Tuesday 7 sensors were not capable of detecting any event and on Sunday the number was 6. In the particular case of high-impact aggregates, their presence is reduced from near 4% in Milan to less than 1% in Rome. Most Milan sensors activate (18 on Tuesday and 17 on Sunday), but only 5 and 3 sensors detect ANEs of this category in Rome in the weekday and during the weekend, respectively.

Following the same analysis scheme described in the previous section, Figure 4 depicts the AI histogram matrices showing the number of occurrences of aggregate ANEs for each impact range and sensor location for both pilot areas. Again, the number of occurrences in the low-impact range is separated from the rest of occurrences for illustration purposes, due to the same reason indicated in the previous analysis. As can be observed, in Milan, low-impact AIs range from 28 to 43 on Tuesday, and from 24 to 36 on Sunday. A total of 107 intervals on the first day and 88 in the second day contain a medium-impact AI, highlighting sensor hb115 in Milan, with 11 occurrences on Tuesday and hb124 with 10 occurrences on Sunday. However, high-impact AIs record a lower presence of occurrences, with a highest value of 4 in sensors hb109 and hb140 on Tuesday, and in sensor hb133 on Sunday.

Table 4. Number of occurrences and sensor activation ratios per sensor for low, medium and high aggregate impact ranges.

Aggregate Impacts		**Low Impact** $(-\infty, 0.5)$ **dB**		**Medium Impact** $[0.5, 2)$ **dB**		**High Impact** $[2, +\infty)$ **dB**	
		Occurrences Count (%)	**Activation Count/** N_S	**Occurrences Count (%)**	**Activation Count/** N_S	**Occurrences Count (%)**	**Activation Count/** N_S
Milan	Tuesday	855 (85.5%)	23/23	107 (10.7%)	22/23	38 (3.8%)	18/23
	Sunday	693 (85.4%)	23/23	88 (10.8%)	21/23	31 (3.8%)	17/23
Rome	Tuesday	874 (95.8%)	19/19	29 (3.2%)	12/19	9 (1.0%)	5/19
	Sunday	887 (95.6%)	19/19	35 (3.8%)	13/19	6 (0.6%)	3/19

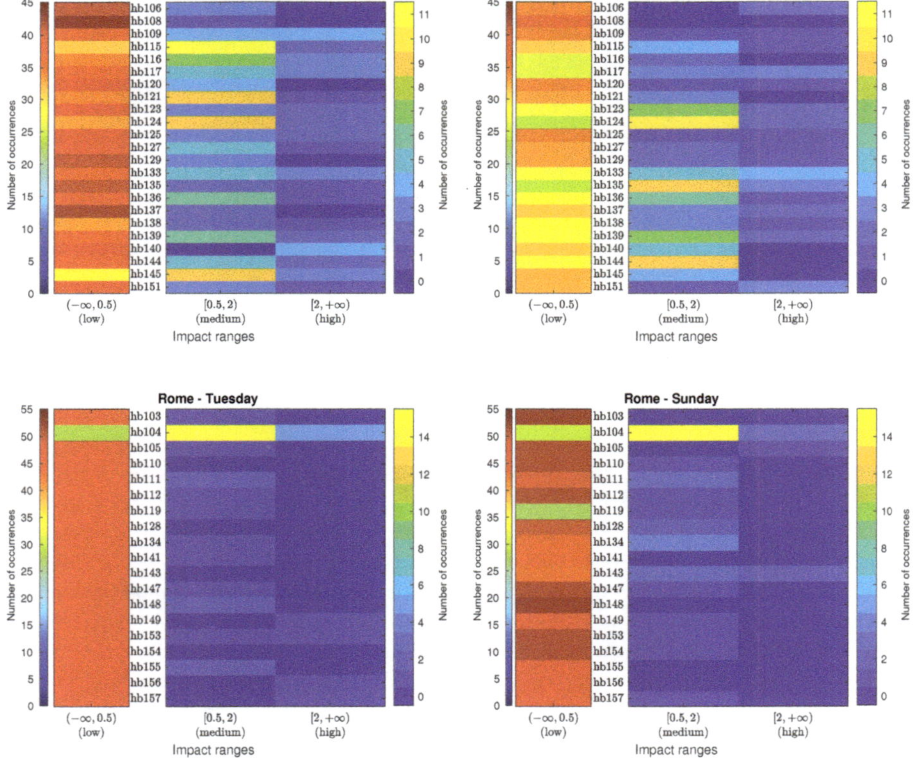

Figure 4. Aggregate impact histogram matrices (obtained using integration time $T = 300$ s) categorized in three impact ranges (low, medium and high) for the urban (Milan) and suburban (Rome) environments obtained from a weekday (Tuesday) and weekend day (Sunday).

Regarding the pilot area in Rome, the presence of low-impact AIs is clearly dominant. However, it is worth mentioning that sensor hb104 presents a completely different pattern, with 15 medium-impact AIs on Tuesday and Sunday. This reduces significantly the low-impact occurrences in that sensor in comparison to other nodes. Finally, as aforementioned, it is to note that sensor hb119 failed in recording several hours of Sunday.

4.4. Critical Aggregate Impacts Per Level

In this section, the occurrences that surpass the critical threshold $\gamma_c = 2$ dB, are analyzed in detail. First, the individual ANEs that bias the $L_{Aeq,300s}$ beyond threshold γ_c by themselves belong to the high-impact range. To analyze their distribution in detail, the critical individual ANEs observed in Section 4.2 (see Figure 3) are divided in 2-dB spans for each sensor in Figure 5. When analyzing this kind of anomalous noise events, Milan credits for most of the high-impact individual ANEs, most of them within the range of 2 to 4 dB, without belittling their presence in the other ranges for both days. Concerning Rome, sensor hb104 is the one that recorded the largest number of high-impact events, most of them belonging to the $[2, 4)$ dB range. Finally, it is to note that 10 events surpass the 10-dB impact range are *sirens*, being the event with the highest impact a 3-min siren with 29.4 dB of impact, recorded in sensor hb137 on Sunday. In contrast, no events surpassing the 10-dB threshold are present in Rome.

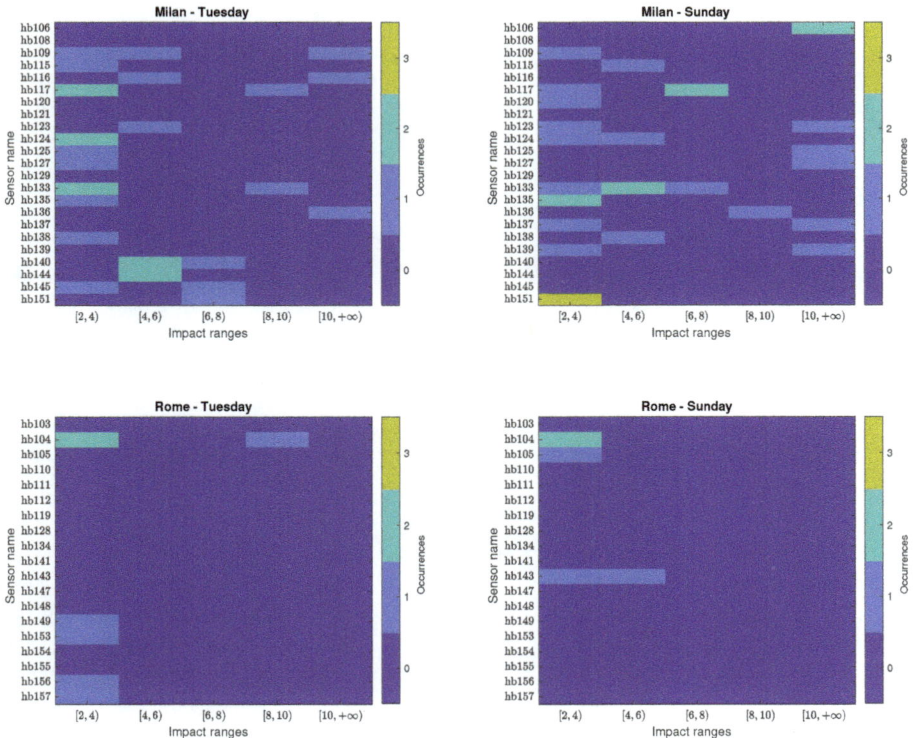

Figure 5. Critical AI histogram matrices ($\mathbf{H_c}$) of individual ANEs for the urban (Milan) and suburban (Rome) environments obtained from a weekday (Tuesday) and weekend day (Sunday).

On the other hand, in order to evaluate if the presence of several ANEs may contribute to the surpassing of the γ_c threshold, Figure 6 shows the critical AI histogram matrices $\mathbf{H_c}$ obtained for each network for different impact intervals. That is to say, it depicts the number of times the AI of

ANEs contribute to bias the $L_{Aeq,300s}$ of RTN critically for both pilot areas and recording day according to the type of impact range. To that effect, besides considering θ_1 (low), θ_2 (medium) and θ_3 (high) impact intervals to analyze the critical aggregate impacts, two more intervals are considered: $\theta_1 \bigcup \theta_2$ to account for co-occurring low and medium individual impact ANEs, and $\theta_1 \bigcup \theta_2 \bigcup \theta_3$ to quantify all the critical cases, disregarding the type of the ANE's individual impact.

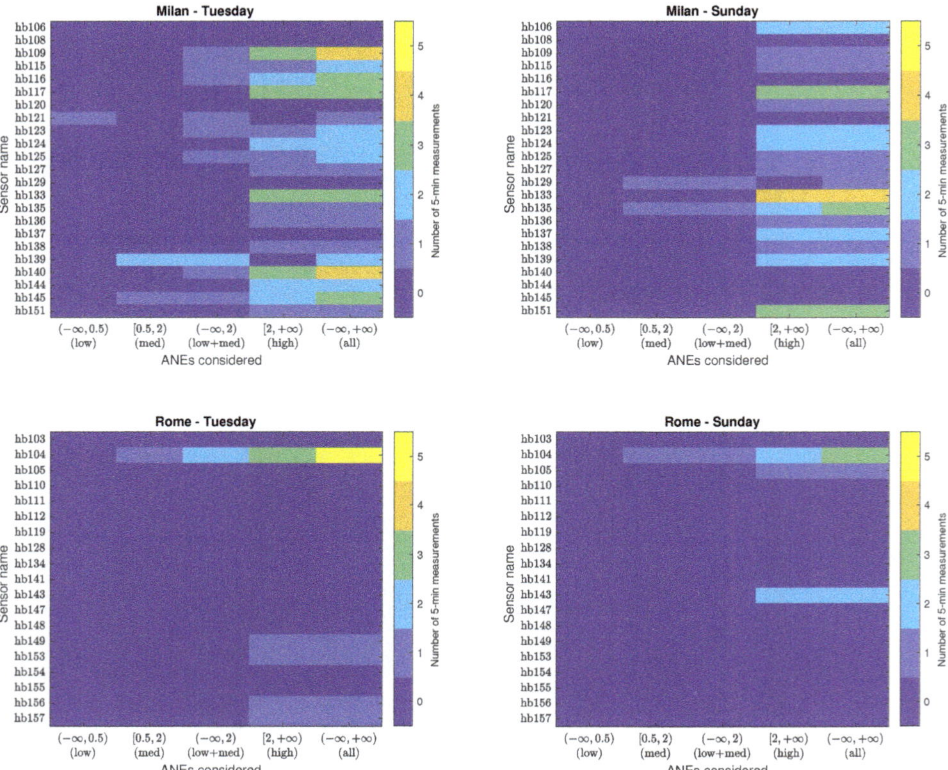

Figure 6. Critical AI histogram matrices ($\mathbf{H_c}$) categorized in the defined impact ranges for the urban (Milan) and suburban (Rome) environments obtained from a weekday (Tuesday) and weekend day (Sunday).

The first column of each $\mathbf{H_c}$ matrices depicted in Figure 6 shows those low-impact AIs causing a critical impact. It can be observed that there is one case accounting for a deviation of the AI higher than 2 dB for a particular period of time t of 5 minutes in sensor hb121 installed in Milan. It is due to 13 *wrks* sounds recorded on Tuesday ranging from 0.01 dB to 0.4 dB, i.e., all of them belong to the individual low-impact range θ_1, but due to their co-occurrence within the same period of time their AI becomes critical.

Likewise, the second column plots critical medium-impact AIs. In Milan, the threshold γ_c is surpassed three times on Tuesday and twice on Sunday, whereas in Rome, purely medium-impact occurrences cause a critical AI once each day. Specifically, sensor hb139 collected two of these pieces of evidence on Tuesday. In the first case, the two most significant ANEs are *horns*, with individual impacts of 0.8 and 1.2 dB, respectively (the third one is a *dog* bark with an impact of 0.03 dB). The second is composed of a *horn* of 1.3 dB and two CMPLX sounds, consisting on a mix of RTN and an undetermined beep noise of 0.8 and 0.5 dB. Moreover, sensor hb145 also recorded a period in which individual ANEs bias the $L_{Aeq,300s}$ critically on Tuesday, where the most important event is a *tram* passby of 1.5 dB and the second one is a 1.6-dB CMPLX event consisting of a mix of a *tram* passby and *birds* tweeting near the sensor. On Sunday, two of the periods recorded in the urban environment contain a combination of medium-impact ANEs that surpass the threshold: one in sensor hb129, composed of two distant *sirens* mixed with other sounds, and another due two CMPLX sounds in sensor hb135, containing unidentified mechanical sounds. In what concerns Rome, sensor hb104 presents critical impact evidence due the co-occurrence of purely medium-impact events for both week and weekend periods. On Tuesday two *train* passbys of 1.3 and 1.1 dB bias the $L_{Aeq,300s}$ more than 2 dB. On Sunday, the critical bias is caused by the presence of two *horns* of 1.2 and 1.9 dB, respectively.

The third column of the four AI critical matrices of Figure 6 show the number of times γ_c is surpassed for ANEs when considering low and medium-impact ANEs, i.e., it collects the occurrences of aggregate low-impact ANEs from θ_1 and the aggregate medium-impact ANEs from θ_2, as well as the the number of times that the critical threshold is surpassed as a result of the combination of the medium- and low-impact events. This last case is only observed during the weekday 6 times in Milan and once at sensor hb104 in Rome. The latter happens on Tuesday and it consists of the sum of several *train* passbys, with the most salient event an impact of 1.9 dB and the other ones of about 0.1 dB. The six cases in Milan have all been found on Tuesday in different sensors: in hb109, three CMPLX sounds have been found that consist of *train* passbys mixed with RTN of 1.8, 0.2 and 0.2 dB; in hb115, a sum of 13 *wrks* sounds with impacts from 0.01 dB to 0.9 dB; in hb116, a 1.9-dB siren co-occurring with a 0.4-dB CMPLX sound of *birds* mixed with RTN; in hb123, an *airp* of 1.9 dB and other *peop* and *brak*-related sound with impacts smaller than 0.02 dB; in hb125, all significant events are *dog* barks, with impacts of 0.9, 0.6, 0.4, 0.3 dB and decreasing; and in hb140, a *siren* of 1.9 dB has been found, jointly with *people*-related sounds of 0.2 dB.

The next column of critical AI matrices presents high-impact ANEs. For the data at hand, the aggregate high-impact occurrences coincide with the number of individual high-impact events depicted in Figure 3 (see also Table 3, where the number of occurrences in this level is quantified).

Finally, the last column of matrices $\mathbf{H_c}$ shows the critical AI histogram caused by the co-occurrence of ANEs of any individual impact range altogether. If we focus on the last three columns of Figure 6, it can be appreciated that in all cases, the sum of the low and medium-impact ANEs with the high-impact ANEs results in the total number of times the 2 dB threshold is surpassed. This result could have differed in the case that aggregate low and medium ANEs co-occurred with high-impact ANEs. Therefore, Figure 6 clarifies the fact that high-impact events have not co-occurred at the same 5-min interval for the datasets at hand, besides showing there is no situation in our datasets where low and medium impact aggregated surpass γ_c at the same 5-min slot t in which a high-impact ANE occurs.

To summarize, in Milan, the threshold has been surpassed due to low and medium aggregate impacts in 12 of the 69 critical cases, which correspond to 17% of cases. Likewise, in Rome, the ratio is 3 to 15, corresponding to 20% of the critical cases. Therefore, according to these results, it can be stated that the removal of low and medium-impact ANEs becomes as relevant as high-impact events in order to preserve the accuracy of the RTN level measurements in both urban and suburban environments.

5. Discussion

This section discusses several relevant aspects related to the results obtained after applying the impact analysis methodology to the two WASN-based datasets collected from the urban and suburban areas. First of all, it is to note that the individual analysis of the impact of each ANE of those co-occurring within the same integration period has been conducted as a baseline study, since the individual view of the impact of acoustic events unrelated to traffic noise is a straightforward but unrealistic approach to the problem at hand. However, this study has been useful to set the basis for the subsequent aggregate analyses. In this sense, it is worth noting that although the datasets have been collected during specific time periods, the analyzed data show the regular presence of anomalous events across all the days and locations in a real-operation context. Specifically, the number of ANEs found in the urban area is seven times greater than in the suburban environment on average (this ratio being ten times on the weekday). In the suburban environment, the weekday pattern is very similar to what is observed in during the weekend, although a larger number of events have been recorded during the weekend, which should be studied in the future with more detail.

In terms of the acoustic categories, it is worth mentioning that 7.7% of the urban WASN-based dataset and 1.6% of the suburban one has been annotated as CMPLX. As aforementioned, the CMPLX acoustic category can be either caused by a mix of RTN and ANEs or by unidentified ANEs by the experts. The conducted analyses have shown that these kinds of acoustic events can also have a significant impact on the $L_{Aeq,300s}$ computation, showing a similar presence in both datasets as the corresponding ANE acoustic category. Therefore, as well as ANEs, CMPLX audio passages should also be removed from the computation of road traffic noise levels to tailor reliable RTN maps.

When comparing the individual and aggregate impact occurrences for low, medium and high-impact ranges, the analyzed environments present a different distribution. In the case of the urban area, a larger number of low-impact events have been recorded than in the suburban environment. However, as far as AIs are concerned, the percentage of low-impact pieces of evidence are lower in the former than in the latter. In addition, medium and high-impact aggregate ANEs have a significant presence in the urban environment, being near the 15% of occurrences; however, in the suburban area, this value decreases to 5%, probably because also the high-impact ANEs present a lower number of instances. From these results it can be concluded that the detection and removal of ANEs will be more usual in a urban than in a suburban environment, since a significantly higher number of $L_{Aeq,300s}$ values can be biased critically. Furthermore, it is worth mentioning that the number of individual high-impact ANEs may not always coincide with the number of times these events bias the 2 dB threshold. This is because it could happen that two or more high-impact events co-occurred in the same evaluated period of time. However, as shown in the results of this work, this is not the case for the data at hand, thus, all high-impact ANEs occur in different integration times.

The impact patterns observed on both environments present different trends. From the analysis conducted in the suburban area, it was observed that sensor hb104 presents a clearly different pattern of the impact of ANEs compared to the rest of the nodes of that WASN for both week and weekend days. This sensor was installed on a major road with two lanes in each direction with a crossing highway under the bridge (see Table A2), which makes this location substantially different from the other sensors locations in Rome (as they do not correspond to major crossroads). For this sensor, the aggregate ANEs are more likely to bias the $L_{Aeq,300s}$, as a 40% of the analyzed measurements contain a medium or high aggregate impact considering both days. This result leads to the preliminary conclusion that in a suburban area, a crossroad is more susceptible to collect anomalous noise events that may distort the RTN level measurements critically. On the contrary, the data analyzed from the other sensor locations in Rome show that the AIs of the ANEs do not usually have a significant impact on the A-weighted equivalent RTN level measurements. In Milan, however, it becomes difficult to identify specific impact patterns according to the sensor locations due to the great variability of occurrences observed from the recordings of both week and weekend days. Nevertheless, note that all sensors have recorded ANEs with a significant impact—both evaluated individually and in an aggregate manner—being relevant enough to bias the RTN map representation in certain periods of time. Given the fact that the recordings were taken over two days, a relevant number of $L_{Aeq,300s}$ measurements could have been computed with an inaccuracy of more than 2 dB, we can conclude that is necessary to remove all kind of anomalous noise events from the final computation of the noise map.

Briefly, the results drawn from this work present a non-negligible number of anomalous noise events that occur randomly both in the DYNAMAP's pilot urban and suburban acoustic environments. This is a relevant issue, as we have to mention that the analyzed data correspond only to a recording campaign of two different days, which provide a relevant but limited scope of *all* the possible issues that may occur in all streets and ring road portals during any day of the year at any time. Nevertheless, although the amount of evidence observed in the gathered data may result statistically poor (i.e., only

84 critical pieces of evidence have been observed), their mere presence demonstrates the importance of their automatic removal to obtain reliable dynamic RTN maps through WASN-based approaches. That is, if the sub-sampling done in two days for several 20-min long audio files has led us to this conclusion, what will be the real impact on the measurements in a 24-h × 7-day WASN-based monitoring system? How many works around the city and the highway can occur throughout the year together some horns and sirens? How many sensors can be located close to a school (with the children in the playground) or next to a church with its bells?... This opens a much wider research goal, focused on the detailed analysis of the sensors location and the consequences it entails in terms of anomalous noise events detection and removal, as the election of the sensor's installation place is usually based on spatial coverage to draw the acoustic map, being also limited by the actual location of the portals and public buildings where the sensors are finally installed.

6. Conclusions

In this work, we have analyzed more than 300 h of labeled acoustic data collected through two WASNs after being deployed in the pilot urban and suburban areas of the DYNAMAP project. The study shows that ANEs can be widely found in acoustic environments when monitoring RTN levels in real-operation conditions, being particularly common in the data gathered from the urban area. Moreover, through the impact analysis methodology, it has been also concluded that the aggregate contribution of low and medium-impact ANEs can deviate the $L_{Aeq,300s}$ as critically as high-impact individual ANEs. Therefore, the obtained results highlight the importance of the automatic removal of low, medium and high-impact events to obtain reliable WASN-based RTN maps in real-operation environments.

Future work will be focused on the detailed analysis of the particularities of each acoustic environment and ANEs subcategories together with complex passages, not only to consider their global impact patterns in the urban and suburban, but also to study the spatio-temporal particularities of all the locations and periods of time. Finally, we plan to adapt the preliminary version of the ANED algorithm by using the two WASN-based datasets to improve its performance in both urban and suburban environments in real operations.

Author Contributions: F.A. has written most parts of the paper, besides participating in the design and formulation of the analysis methodology and in the study of the results. F.O. has worked and led the analysis of the data and participated in the writing of the results and discussion sections. R.M.A.-P. has written the related work, besides contributing to the discussion and reviewing the entire paper. J.C.S. has worked in the design and formulation of the analysis methodology, and its writing, besides summarizing the main characteristics of the WASN-based acoustic datasets. All authors have read and agreed to the published version of the manuscript.

Acknowledgments: The authors would like to thank ANAS S.p.A. for the picture of the sensor installed in the portal. The research presented in this work has been partially supported by the LIFE DYNAMAP project (LIFE13 ENV/IT/001254). Joan Claudi Socoró thanks the Obra Social La Caixa for grant ref. 2019-URL-IR1rQ-053. Ferran Orga thanks the support of the European Social Fund and the Secretaria d'Universitats i Recerca del Departament d'Economia i Coneixement of the Catalan Government for the pre-doctoral FI grant No. 2019_FI_B2_00168.

Conflicts of Interest: The authors declare no conflict of interest.

Abbreviations

The following abbreviations are used in this manuscript:

AI	Aggregate Impact
ANE	Anomalous Noise Event
ANED	Anomalous Noise Event Detection
CNOSSOS-EU	Common Noise Assessment Methods in Europe
DYNAMAP	Dynamic Noise Mapping
END	European Noise Directive
EU	European Union

RTN	Road Traffic Noise
SNR	Signal-to-Noise Ratio
SONYC	Sounds of New York City
WASN	Wireless Acoustic Sensor Network
WG-AEN	European Commission Working Group Assessment of Exposure to Noise

Appendix A

This section includes the description of the sensor locations for both urban and suburban environments by means of Tables A1 and A2, respectively.

Table A1. Sensor locations description for the urban environment. X-lane/Y-lane road stands for a two-way road that has X lanes in one sense and Y lanes in the opposite sense. X-lane road stands for a street with X lanes in the same sense.

Sensor Id	Sensor Location Description
hb106	1-lane/1-lane road with connection with 1 line road, area with parks nearby, no shops
hb108	1-lane/1-lane road, in front University exit, no shops
hb109	3-lane/3-lane road, near crossing with tramway and 1 line+2 line/2 line+1 line road, shopping and coffe/restaurant area
hb115	1-lane road with shopping in front
hb116	1-lane/1-lane road with connection with 1-lane road, residential area
hb117	3-lane/3-lane road, near school, area with parks nearby, no shops
hb120	1-lane/1-lane road, residential area, no shops
hb121	2-lane/2-lane road, connection with 1-lane road, University area, no shops
hb123	2-lane/2-lane road with hotel and traffic light nearby
hb124	1-lane road, no shops
hb125	1-lane road with connection with 1-lane/1-lane road, mix of residential with some shops
hb127	1-lane road near bifurcation with 1 line road, some shop nearby
hb129	1-lane/1-lane road, bike line, connection with 1-lane road, some shop
hb133	1-lane road, residential area, no shops, little park area in front
hb135	1-lane road with connection with 1-lane road (low speed), near University campus (students), no shops, in front of park area
hb136	1-lane/1-lane road with connection with 1-lane road, area with parks nearby, no shops
hb137	1-lane road with connection with 1 line road, in front of park, residential area, no shops
hb138	1-lane road near connection with other 1-lane road, no shops
hb139	1-lane road, residential area, some shop/enterprise
hb140	2-lane/2-lane road with parking area and traffic light with crossing nearby, no shops near and high traffic
hb144	1-lane road in residential area, one shop far away
hb145	1-lane road, in front of park
hb151	1-lane/1-lane road, bike line, some shop and restaurant

Table A2. Sensor locations description for suburban environment. X-lane/Y-lane road stands for a two-way road that has X lanes in one sense and Y lanes in the opposite sense. X-lane road stands for a street with X lanes in the same sense.

Sensor Id	Sensor Location Description
hb103	Highway with 3-lane/3-lane
hb104	Major road with 2-lane each direction crossing a highway under bridge (out of major ring)
hb105	Highway with 4-lane (only 1 direction, and near exits/crossings)
hb110	Highway with 3-lane/3-lane

Table A2. *Cont.*

Sensor Id	Sensor Location Description
hb111	Highway with 3-lane/3-lane
hb112	Highway with 3-lane/3-lane (near exit and near crossings)
hb119	Highway with 3-lane/3-lane
hb128	Highway with 3-lane/3-lane
hb134	Highway with 4-lane/4-lane (near bridge and crossings)
hb141	Highway with 5-lane/5-lane (near crossings)
hb143	Highway with 2-lane/2-lane (out of major ring)
hb147	Highway with 3-lane/3-lane
hb148	Highway with 3-lane/3-lane
hb149	Highway with 3-lane (near tunnel)
hb153	Major road with 2-lane each direction crossing a highway under bridge (out of major ring)
hb154	Highway with 4-lane/4-lane
hb155	Highway with 2-lane (near connection but out major ring) plus 1 road same sense next to
hb156	Highway with 3-lane/3-lane
hb157	Highway with 5-lane/5-lane

References

1. Directive, E.U. Directive 2002/49/EC of the European Parliament and the Council of 25 June 2002 relating to the assessment and management of environmental noise. *Off. J. Eur. Communities* **2002**, *L 189/12*.
2. World Health Organization Regional Office for Europe. Burden of Disease from Environmental Noise: Quantification of Healthy Life Years Lost in Europe 2011. Available online: https://www.who.int/quantifying_ehimpacts/publications/e94888/en (accessed on 21 January 2020).
3. World Health Organization Regional Office for Europe. Environmental Noise Guidelines for the European Region 2018. Available online: https://apps.who.int/iris/handle/10665/279952 (accessed on 21 January 2020).
4. Licitra, G.; Ascari, E.; Fredianelli, L. Prioritizing Process in Action Plans: a Review of Approaches. *Curr. Pollut. Rep.* **2017**, *3*, 151–161. [CrossRef]
5. Kephalopoulos, S.; Paviotti, M.; Anfosso-Lédée, F. *Common Noise Assessment Methods in Europe (CNOSSOS-EU)*; Report EUR 25379 EN; Publications Office of the European Union: Luxembourg, 2002; pp. 1–180.
6. King, E.A.; Murphy, E.; Rice, H.J. Implementation of the EU environmental noise directive: Lessons from the first phase of strategic noise mapping and action planning in Ireland. *J. Environ. Manag.* **2011**, *92*, 756–764. [CrossRef] [PubMed]
7. Carrier, M.; Apparicio, P.; Séguin, A.M.; Crouse, D. School locations and road transportation nuisances in Montreal: An environmental equity diagnosis. *Transp. Policy* **2019**, *81*, 302–310. [CrossRef]
8. Tsai, K.T.; Lin, M.D.; Lin, Y.H. Noise exposure assessment and prevention around high-speed rail. *Int. J. Environ. Sci. Technol.* **2019**, *16*, 4833–4842. [CrossRef]
9. Morley, D.; de Hoogh, K.; Fecht, D.; Fabbri, F.; Bell, M.; Goodman, P.; Elliott, P.; Hodgson, S.; Hansell, A.; Gulliver, J. International scale implementation of the CNOSSOS-EU road traffic noise prediction model for epidemiological studies. *Environ. Pollut.* **2015**, *206*, 332–341. [CrossRef] [PubMed]
10. Botteldooren, D.; Dekoninck, L.; Gillis, D. The influence of traffic noise on appreciation of the living quality of a neighborhood. *Int. J. Environ. Res. Public Health* **2011**, *8*, 777–798. [CrossRef] [PubMed]
11. Alberts, W.; Roebben, M. Road Traffic Noise Exposure in Europe in 2012 based on END data. In Proceedings of the 45th International Congress and Exposition on Noise Control Engineering (INTER-NOISE 2016), Hamburg, Germany, 21–24 August 2016; pp. 1236–1247.
12. Garcia, A.; Faus, L. Statistical analysis of noise levels in urban areas. *Appl. Acoust.* **1991**, *34*, 227–247. [CrossRef]
13. Alsina-Pagès, R.M.; Alías, F.; Socoró, J.C.; Orga, F.; Benocci, R.; Zambon, G. Anomalous events removal for automated traffic noise maps generation. *Appl. Acoust.* **2019**, *151*, 183–192. [CrossRef]
14. European Commission Working Group Assessment of Exposure to Noise (WG-AEN). *Good Practice Guide for Strategic Noise Mapping and the Production of Associated Data on Noise Exposure Ver.2, 13 August*; Technical Report; 2007. Available online: https://www.lfu.bayern.de/laerm/eg_umgebungslaermrichtlinie/doc/good_practice_guide_2007.pdf (accessed on 21 January 2020).

15. De Coensel, B.; Botteldooren, D. A model of saliency-based auditory attention to environmental sound. In Proceedings of the 20th International Congress on Acoustics (ICA-2010), Sydney, Australia, 23–27 August 2010; pp. 1–8.
16. Salamon, J.; Jacoby, C.; Bello, J.P. A dataset and taxonomy for urban sound research. In Proceedings of the 22nd ACM International Conference on Multimedia, Orlando, FL, USA, 3–7 November 2014; pp. 1041–1044.
17. Foggia, P.; Petkov, N.; Saggese, A.; Strisciuglio, N.; Vento, M. Reliable detection of audio events in highly noisy environments. *Pattern Recognit. Lett.* **2015**, *65*, 22–28. [CrossRef]
18. Stowell, D.; Giannoulis, D.; Benetos, E.; Lagrange, M.; Plumbley, M.D. Detection and Classification of Acoustic Scenes and Events. *IEEE Trans. Multimedia* **2015**, *17*, 1733–1746. [CrossRef]
19. Socoró, J.C.; Ribera, G.; Sevillano, X.; Alías, F. Development of an Anomalous Noise Event Detection Algorithm for dynamic road traffic noise mapping. In Proceedings of the 22nd International Congress on Sound and Vibration, Florence, Italy, 12–16 July 2015; pp. 1–8.
20. Salamon, J.; MacConnell, D.; Cartwright, M.; Li, P.; Bello, J.P. Scaper: A library for soundscape synthesis and augmentation. In Proceedings of the 2017 IEEE Workshop on Applications of Signal Processing to Audio and Acoustics (WASPAA), New Paltz, NY, USA, 15–18 October 2017; pp. 344–348.
21. Alías, F.; Socoró, J.C. Description of anomalous noise events for reliable dynamic traffic noise mapping in real-life urban and suburban soundscapes. *Appl. Sci.* **2017**, *7*, 146. [CrossRef]
22. Botteldooren, D.; De Coensel, B.; Oldoni, D.; Van Renterghem, T.; Dauwe, S. Sound monitoring networks new style. In Proceedings of Acoustics 2011: Breaking New Ground – Proceedings of the Annual Conference of the Australian Acoustical Society, Gold Coast, Australia, 2–4 November 2011.
23. Manvell, D. Utilising the Strengths of Different Sound Sensor Networks in Smart City Noise Management. In Proceedings of the EuroNoise 2015, EAA-NAG-ABAV, Maastrich, Netherlands, 31 May–3 June 2015; pp. 2305–2308.
24. Alías, F.; Alsina-Pagès, R.M. Review of Wireless Acoustic Sensor Networks for Environmental Noise Monitoring in Smart Cities. *J. Sens.* **2019**, *2019*, 13. [CrossRef]
25. Camps, J. Barcelona noise monitoring network. In Proceedings of the EuroNoise 2015, EAA-NAG-ABAV, Maastrich, The Netherlands, 1 May–3 June 2015; pp. 2315–2320.
26. Segura-Garcia, J.; Pérez Solano, J.J.; Cobos Serrano, M.; Navarro Camba, E.A.; Felici-Castell, S.; Soriano Asensi, A.; Montes Suay, F. Spatial Statistical Analysis of Urban Noise Data from a WASN Gathered by an IoT System: Application to a Small City. *Appl. Sci.* **2016**, *6*, 380. [CrossRef]
27. Nencini, L.; De Rosa, P.; Ascari, E.; Vinci, B.; Alexeeva, N. SENSEable Pisa: A wireless sensor network for real-time noise mapping. In Proceedings of the EuroNoise 2012, Czech Acoustical Society, Prague, Czech Republic, 10–13 June 2012.
28. Bartalucci, C.; Borchi, F.; Carfagni, M.; Furferi, R.; Governi, L. Design of a prototype of a smart noise monitoring system. In Proceedings of the 24th International Congress on Sound and Vibration (ICSV24); The International Institute of Acoustics and Vibration, London, UK, 23–27 July 2017.
29. Rainham, D. A wireless sensor network for urban environmental health monitoring: UrbanSense. *IOP Conf. Ser. Earth Environ. Sci.* **2016**, *34*, 012028. [CrossRef]
30. Sevillano, X.; Socoró, J.C.; Alías, F.; Bellucci, P.; Peruzzi, L.; Radaelli, S.; Coppi, P.; Nencini, L.; Cerniglia, A.; Bisceglie, A.; et al. DYNAMAP—Development of low cost sensors networks for real time noise mapping. *Noise Mapp.* **2016**, *3*, 172–189. [CrossRef]
31. Bello, J.P.; Silva, C.; Nov, O.; Dubois, R.L.; Arora, A.; Salamon, J.; Mydlarz, C.; Doraiswamy, H. SONYC: A System for Monitoring, Analyzing, and Mitigating Urban Noise Pollution. *Commun. ACM* **2019**, *62*, 68–77. [CrossRef]
32. Zambon, G.; Benocci, R.; Bisceglie, A.; Roman, H.E.; Bellucci, P. The LIFE DYNAMAP project: Towards a procedure for dynamic noise mapping in urban areas. *Appl. Acoust.* **2017**, *124*, 52–60. [CrossRef]
33. Zambon, G.; Benocci, R.; Bisceglie, A. Development of optimized algorithms for the classification of networks of road stretches into homogeneous clusters in urban areas. In Proceedings of the 22nd International Congress on Sound and Vibration (ICSV22). International Institute of Acoustics and Vibrations, Florence, Italy, 12–16 July 2015; pp. 1–8.
34. Bellucci, P.; Peruzzi, L.; Zambon, G. LIFE DYNAMAP project: The case study of Rome. *Appl. Acoust.* **2017**, *117*, 193–206. [CrossRef]

35. Benocci, R.; Bellucci, P.; Peruzzi, L.; Bisceglie, A.; Angelini, F.; Confalonieri, C.; Zambon, G. Dynamic Noise Mapping in the Suburban Area of Rome (Italy). *Environments* **2019**, *6*, 79. [CrossRef]
36. Socoró, J.C.; Alías, F.; Alsina-Pagès, R.M. An Anomalous Noise Events Detector for Dynamic Road Traffic Noise Mapping in Real-Life Urban and Suburban Environments. *Sensors* **2017**, *17*, 2323. [CrossRef] [PubMed]
37. Orga, F.; Alías, F.; Alsina-Pagès, R.M. On the Impact of Anomalous Noise Events on Road Traffic Noise Mapping in Urban and Suburban Environments. *Int. J. Environ. Res. Public Health* **2017**, *15*, 13. [CrossRef] [PubMed]
38. Socoró, J.C.; Alsina-Pagès, R.M.; Alías, F.; Orga, F. Adapting an Anomalous Noise Events Detector for Real-Life Operation in the Rome Suburban Pilot Area of the DYNAMAP's Project. In Proceedings of the EuroNoise 2018, EAA—HELINA, Heraklion, Crete, Greece, 27–31 May 2018; pp. 693–698.
39. Alías, F.; Socoró, J.C.; Orga, F.; Alsina-Pagès, R.M. Characterization of a WASN-Based Urban Acoustic Dataset for the Dynamic Mapping of Road Traffic Noise. Available online: https://www.researchgate.net/publication/338103848_Characterization_of_A_WASN-Based_Urban_Acoustic_Dataset_for_the_Dynamic_Mapping_of_Road_Traffic_Noise (accessed on 22 January 2020). [CrossRef]
40. Alsina-Pagès, R.M.; Orga, F.; Alías, F.; Socoró, J.C. A WASN-Based Suburban Dataset for Anomalous Noise Event Detection on Dynamic Road-Traffic Noise Mapping. *Sensors* **2019**, *19*, 2480. [CrossRef] [PubMed]
41. Giannoulis, D.; Stowell, D.; Benetos, E.; Rossignol, M.; Lagrange, M.; Plumbley, M.D. A database and challenge for acoustic scene classification and event detection. In Proceedings of the 21st European Signal Processing Conference (EUSIPCO 2013), Marrakech, Morocco, 9–13 September 2013.
42. Mesaros, A.; Heittola, T.; Virtanen, T. Metrics for Polyphonic Sound Event Detection. *Appl. Sci.* **2016**, *6*, 162. [CrossRef]
43. Alías, F.; Socoró, J.C.; Sevillano, X. A Review of Physical and Perceptual Feature Extraction Techniques for Speech, Music and Environmental Sounds. *Appl. Sci.* **2016**, *6*, 143. [CrossRef]
44. Zhang, J.; Ding, W.; He, L. Data augmentation and prior knowledge-based regularization for sound event localization and detection. Tech. Report of Detection and Classification of Acoustic Scenes and Events (DCASE2019 Challenge), Online, 4 March–30 June 2019. Available online: http://dcase.community/documents/challenge2019/technical_reports/DCASE2019_He_97.pdf (accessed on 21 January 2012).
45. Nakajima, Y.; Sunohara, M.; Naito, T.; Sunago, N.; Ohshima, T.; Ono, N. DNN-based Environmental Sound Recognition with Real-recorded and Artificially-mixed Training Data. In Proceedings of the 45th International Congress and Exposition on Noise Control Engineering (INTER-NOISE 2016), Hamburg, Germany, 21–24 August 2016; pp. 3164–3173.
46. Koizumi, Y.; Saito, S.; Yamaguchi, M.; Murata, S.; Harada, N. Batch Uniformization for Minimizing Maximum Anomaly Score of DNN-Based Anomaly Detection in Sounds. In Proceedings of the 2019 IEEE Workshop on Applications of Signal Processing to Audio and Acoustics (WASPAA), New Paltz, NY, USA, 20–23 October 2019.
47. Fonseca, E.; Plakal, M.; Font, F.; Ellis, D.P.; Favory, X.; Pons, J.; Serra, X. General-purpose Tagging of Freesound Audio with AudioSet Labels: Task Description, Dataset, and Baseline. In Proceedings of the Detection and Classification of Acoustic Scenes and Events 2018 Workshop (DCASE2018), Surrey, UK, 19–20 November 2018; pp. 69–73.
48. Schauerte, B.; Stiefelhagen, R. Wow! Bayesian surprise for salient acoustic event detection. In Proceedings of the 38th IEEE International Conference on Acoustics, Speech and Signal Processing (ICASSP2013), Vancouver, BC, Canada, 26–31 May 2013; pp. 6402–6406.

Article

Accuracy of the Dynamic Acoustic Map in a Large City Generated by Fixed Monitoring Units

Roberto Benocci [1,*], Chiara Confalonieri [1], Hector Eduardo Roman [2], Fabio Angelini [1] and Giovanni Zambon [1]

1 Department of Earth and Environmental Sciences (DISAT), University of Milano-Bicocca, Piazza della Scienza 1, 20126 Milano, Italy; c.confalonieri12@campus.unimib.it (C.C.); fabio.angelini@unimib.it (F.A.); giovanni.zambon@unimib.it (G.Z.)

2 Department of Physics "G. Occhialini", University of Milano-Bicocca, Piazza della Scienza 3, 20126 Milano, Italy; eduardo.roman@mib.infn.it

* Correspondence: roberto.benocci@unimib.it

Received: 18 November 2019; Accepted: 9 January 2020; Published: 11 January 2020

Abstract: DYNAMAP, a European Life project, aims at giving a real image of the noise generated by vehicular traffic in urban areas developing a dynamic acoustic map based on a limited number of low-cost permanent noise monitoring stations. The system has been implemented in two pilot areas located in the agglomeration of Milan (Italy) and along the Motorway A90 (Rome-Italy). The paper reports the final assessment of the system installed in the pilot area of Milan. Traffic noise data collected by the monitoring stations, each one representative of a number of roads (groups) sharing similar characteristics (e.g., daily traffic flow), are used to build-up a "real-time" noise map. In particular, we focused on the results of the testing campaign (21 sites distributed over the pilot area and 24 h duration of each recording). It allowed evaluating the accuracy and reliability of the system by comparing the predicted noise level of DYNAMAP with field measurements in randomly selected sites. To this end, a statistical analysis has been implemented to determine the error associated with such prediction, and to optimize the system by developing a correction procedure aimed at keeping the error below some acceptable threshold. The steps and the results of this procedure are given in detail. It is shown that it is possible to describe a complex road network on the basis of a statistical approach, complemented by empirical data, within a threshold of 3 dB provided that the traffic flow model achieves a comparable accuracy within each single groups of roads in the network.

Keywords: noise mapping; noise mitigation; DYNAMAP project

1. Introduction

Road traffic noise is one of the foremost problems in Europe, with more than 100 million people exposed to Lden (day-evening-night) levels higher than 55 dB (A) [1]. Consequently, scientific communities and authorities started observing the surge of noise-related health problems such as sleep disorders and tiredness associated with a long-term road traffic noise exposure [2,3], relationships between annoyance and exposure to transportation noise [4], increased cardiovascular risk and hypertension [4–6], mental performance [7], and students cognitive disorders [8,9].

The increasing awareness on these issues, promoted by EU policies through the Environmental Noise Directive (END) of 2002, its revision [10,11] and integrated approaches (CNOSSOS-EU) [12,13], encouraged the use of distributed monitoring systems and noise mapping in the control of noise exposure.

Mitigation measures in urban and near-urban contexts need to be identified according to a realistic picture of noise distribution over extended areas. This requirement demands for real-time measurements and processing to assess the acoustic impact of noise sources. In this framework, noise maps might represent an important tool. They are based on collecting and processing information on the traffic flow averaged over long periods of time [14] using acoustic models [15] rather than unattended phono-metric measurements, which, on the other hand, are typically used to validate results from computational models [16,17].

Recently, the development of dynamic noise maps is gaining interest because of the realistic soundscape picture they can provide in complex traffic network. Different approaches have been pursued, motivated by the fact that noise fluctuations might be important to evaluate sleep disturbance and noise annoyance [18,19]. Other recent approaches regard participatory sensing, which enables any person to take measurements using either specific measurement equipment or mobile phones [20,21]. In addition, mobile sampling could, in principle, increase temporal and spatial resolution, even with short length samples [22], in a more controlled environment than participatory sensing, since the measurement is carried out by trained people. A usual practice, which integrates traditional noise mapping and participatory sensing, is to take on-site measurements to calibrate the noise map based on computational models or to use them to dynamically update noise maps based on interpolation schemes [23,24].

Noise mapping recently moved towards a multi-source approach [25,26]. Specifically, advanced probabilistic noise modelling based on source-oriented sound maps within an open-source Geographic Information System (GIS) environment allows the production of traffic, fountains, voices and birds sound maps and to investigate the competition between sound sources [27].

In this continuously evolving scenario, DYNAMAP, a co-financed project by the European Commission through the Life+ 2013 program, started its activities in 2014 [28]. It aimed at developing a dynamical acoustic map in two pilot areas: a large portion of the urban area of the city of Milan (District 9) and the motorway surrounding Rome. In both cases, we developed a method for predicting the traffic noise in an extended area using a limited number of monitoring sensors and the knowledge of traffic flows.

The development of automatic noise mapping systems delivering short-term noise maps (dynamic noise maps) are not explicitly required by the END. However, their automatic generation is estimated to reduce the cost of long-term noise assessment by 50%, adding significant benefits for noise managers and the public through updated information and dedicated web tools with the opportunity to control noise with alternative measures based on traffic control and management. While this approach seems quite promising in purely suburban areas, where noise sources are well identified, in complex urban scenarios further considerations are needed.

Regarding a suburban area, a detailed study has been performed for the motorway zone around the city of Rome. The pilot area of Rome is located along a six-lane ring road (A90) surrounding the city, going through many suburban areas where the presence of single or multiple noise sources, such as railways, crossing, and parallel roads, impact the residents. Pre-calculated basic noise maps, prepared for different sources, traffic, and weather conditions, are updated from the information retrieved from 19 distributed noise sensors. Difficulties lied in the contribution of multiple noise sources and in the influence of meteorological conditions when receptors are located at a distance from the road greater than 80 m. The final assessment on DYNAMAP reliability and accuracy in the suburban area of Rome can be found in [29].

For an urban environment, we review in this paper the case of Milan, where DYNAMAP has been implemented in a pilot area, namely District 9, consisting of about 2000 road arches in the north-east part of the city. Due to the high number of potential noise sources needed to be monitored, we decided to adopt a statistically based approach. This is the outcome of previous investigations [30], proving that the noise emission from a street generally depends on its use and activity in the urban context, therefore suggesting a stratified sampling aimed at optimizing the number of monitoring sites.

In both scenarios, being urban or suburban, the presence of anomalous noise events (ANEs), that is events that are extraneous to the actual vehicle noise, may alter the noise levels represented by DYNAMAP. For this reason, dedicated algorithms have been implemented in ARM-based (see

Acronyms Sect. for definition) acoustic sensors showing the feasibility of the method both in terms of computational cost and classification performance [31] with the purpose of identifying and removing ANEs from the time series, thus restricting the acoustic data to the traffic source only [32]. In particular, different typologies of anomalous noise events have been described statistically and associated with the identified street clusters of the city of Milan [33]. Similar approaches based on permanent monitoring network and street categorization are now adopted in other cities [34].

In this paper, we provide a review of DYNAMAP project in the pilot area of the city of Milan. It represents the final assessment on its accuracy and reliability obtained from the comparison between field measurements and map predictions.

2. Materials and Methods

In this section, we provide an overview of the general scheme of DYNAMAP implemented in Milan, including the initial sampling campaign, the statistical analysis, the calculation and map updating procedure, and the methodology for the system validation (calibration of sensors, their reliability, and field measurements). Figure 1 shows a general block diagram of the following processes for illustrative purposes.

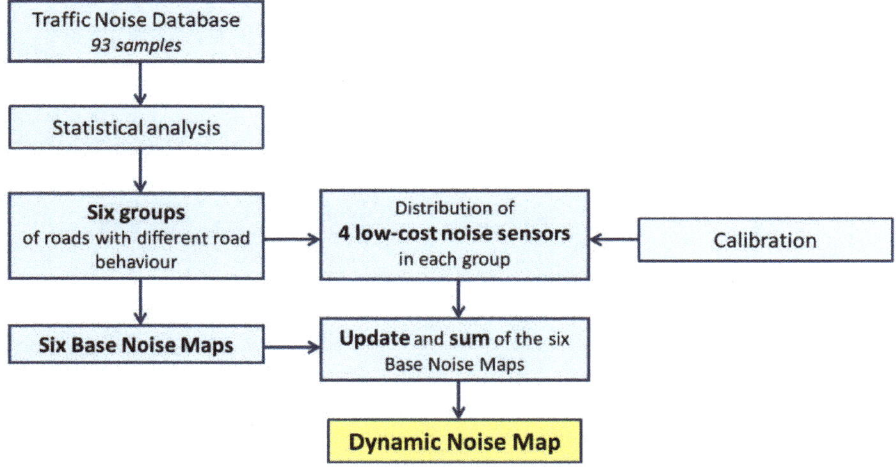

Figure 1. General block diagram of the followed processes in DYNAMAP.

2.1. Initial Sampling Campaign and Statistical Analysis

A database of noise time series belonging to the road network of Milan was necessary to characterize the traffic noise of the city. For this reason, 93 traffic noise recordings of 24 h represented our initial large-scale noise monitoring investigation [35]. Given the large number of roads in the city of Milan, in order to determine the acoustic behavior of different roads we applied a statistical approach based on a cluster analysis. The open source software "R" [36] was applied for clustering and the package "clValid" [37,38] was used for validating the results of the different cluster algorithms. The ranking provided by the "clValid" R-package showed the best performance of the hierarchical clustering with Ward algorithm [39], as detailed in [40].

The results showed two main noise behaviors correlated to vehicle flow patterns [41,42]. Its extension to non-monitored roads needed an available non-acoustic road-related parameter [43] and we found the logarithm of the total daily traffic flow, $\mathrm{Log}(T_T)$ to be a convenient quantity. The number of events and the cumulative probability for the resulting two clusters, as a function of the non-acoustic parameter $x = \mathrm{Log}(T_T)$, are illustrated in Figure 2.

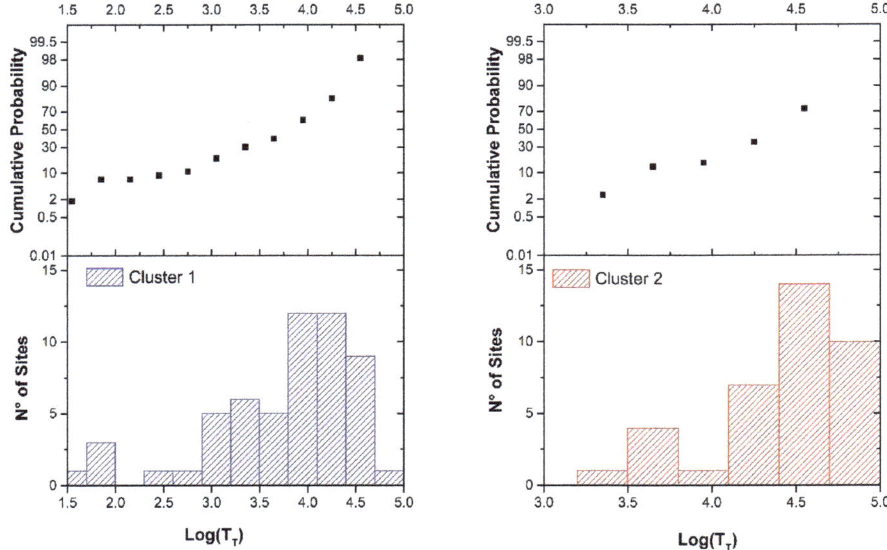

Figure 2. Number of sites and cumulative probability for Cluster 1 (left side) and Cluster 2 (right side) as a function of the non-acoustic parameter $x = \text{Log}(T_T)$. Bin size is 0.3.

As we are interested in finding an analytical representation for the distribution functions, $P(x)$, in each cluster, we studied the corresponding cumulative distributions of x, $I(x)$, which have been fitted using an analytical expression:

$$I(x) = 10^{f(x)} = \int_0^x dy\, P(y);\ \text{with}\ \lim_{x\to\infty} I(x) = 1. \tag{1}$$

Deriving $I(x)$, we get

$$\frac{dI(x)}{dx} = P(x) - P(0). \tag{2}$$

The probability distribution $P(x)$ can be obtained from the analytical fit of the cumulative distribution $I(x)$ according to the relation:

$$P(x) = \ln(10) f'(x) I(x), \tag{3}$$

where $f(x)$ is a polynomial of third degree and $f'(x)$ is the derivative of $f(x)$. The results of $I(x)$ for Clusters 1 and 2 are reported in Figures 3 and 4.

Analytical fit functions f_1 and f_2 for $P(x)$ for the two clusters are:

$$f_1(x) = -1.55545 - 0.24459\, x + 0.28834\, x^2 - 0.03526\, x^3 \tag{4}$$

$$f_2(x) = -15.21817 + 7.01263\, x - 1.02922\, x^2 + 0.04708\, x^3 \tag{5}$$

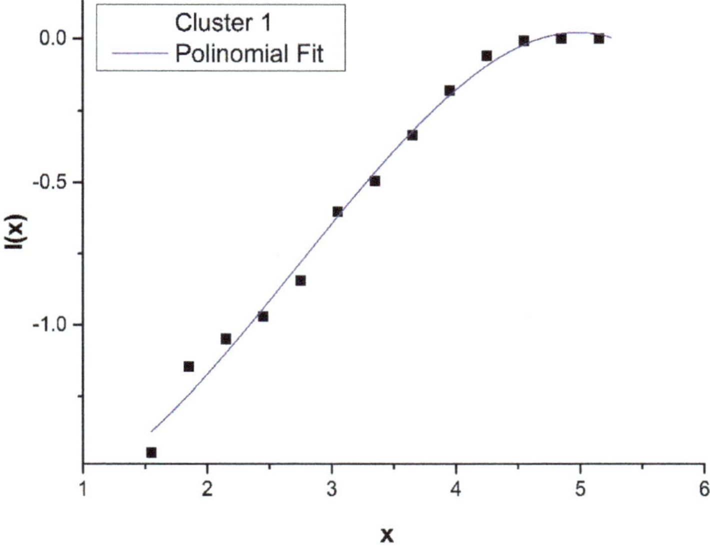

Figure 3. Cumulative distribution $I(x)$ for Cluster 1 fitted using the analytical expression, Equation (1), $I(x) = 10^{f(x)}$, where $f(x)$ is a polynomial of third degree and $x = \text{Log}(T_T)$.

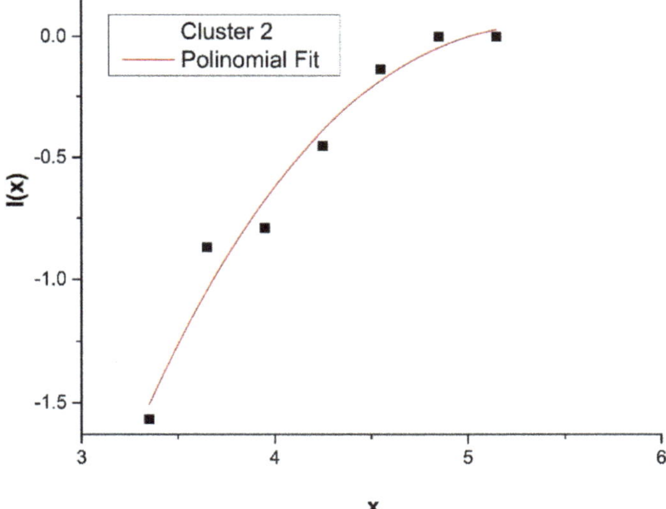

Figure 4. Cumulative distribution $I(x)$ for Cluster 2 fitted using the analytical expression, Equation (1), $I(x) = 10^{f(x)}$, where $f(x)$ is a polynomial of third degree and $x = \text{Log}(T_T)$.

In Figure 5, the histograms and density function, $P_1(x)$ and $P_2(x)$ for Cluster 1 and 2 (for the initial 93 sample measurements) are illustrated as a function of the non-acoustic parameter, x. Here, $P_1(x)$ and $P_2(x)$ represent the "probability" that a road with a given x belongs to Clusters 1 and 2, respectively.

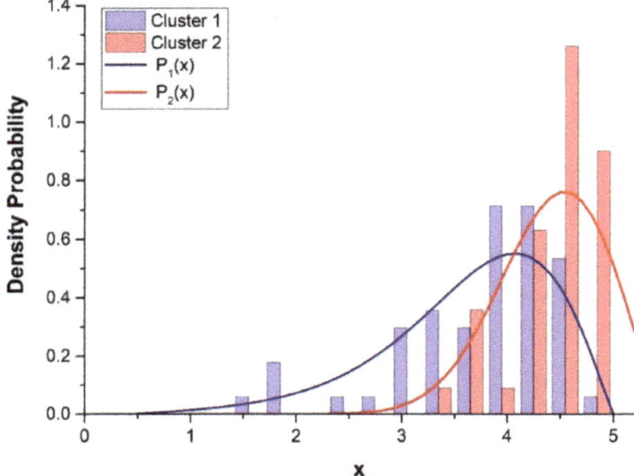

Figure 5. Distribution functions $P_1(x)$ and $P_2(x)$ (Equations (3)–(5)) for Clusters 1 and 2; x = Log (T_T).

In general, owing to the large superposition of the two cluster distributions $P_1(x)$ and $P_2(x)$, we might consider a linear combination between the two mean normalized cluster profiles to describe the noise behavior of a road with a given value of x.

The weights (α_1, α_2) of the linear combination can be obtained, for each value of x, using the relations: $\alpha_1 = P_1(x)$ and $\alpha_2 = P_2(x)$. Therefore, the values of $\alpha_{1,2}$ represent the probability that a given road characterized by its own value of x belongs to the corresponding Clusters, 1 and 2. By denoting as β, the normalized values of $\alpha_{1,2}$, we obtain:

$$\begin{aligned} \beta_1 &= \frac{\alpha_1}{\alpha_1+\alpha_2} \\ \beta_2 &= \frac{\alpha_2}{\alpha_1+\alpha_2} \end{aligned} \tag{6}$$

For practical use, we cannot describe the behavior of each single road in the network, therefore, the entire range of variability of the non-acoustic parameter has been divided into six intervals in such a way that each group contains approximately the same number of roads. In this way, all road stretches within a group are represented by the same acoustic map, while six groups are found to be suitable for our purposes. The noise in a given location will be predicted by a combination of the six acoustic base noise maps whose variation (dynamic feature) is provided by field stations. The process for updating the pre-calculated six base noise maps is based on the average of noise level variations recorded by the monitoring stations, according to two different procedures described below.

2.2. Dynamic Map

For the actual implementation of DYNAMAP, we relied on 24 monitoring stations that have been installed homogeneously in the six groups g (four in each group), in such a way to reproduce the empirical distribution of the non-acoustic parameter in District 9 (to be noted that the 24 fixed monitoring units have been installed in sites belonging to the pilot area and not corresponding to the locations where the 93 sample measurements have been recorded).

The noise signal from each station j is filtered from any anomalous events not belonging to road traffic noise prior to its integration to obtain $Leq^{\tau}{}_j$ over a predefined temporal interval τ (τ = 5, 15, 60 min) [32–34]. Thus, we get 24 $Leq^{\tau}{}_{g,j}$ values every τ min, each one corresponding to a recording station j and belonging to a group g. To update the acoustic maps, we deal with variations, $\delta^{\tau}_{g,j}(t)$, where the time t is discretized as $t = n\tau$ and n is an integer, defined according to:

$$\delta^{\tau}_{g,j}(t) = Leq^{\tau}_{g,j}(t) - Leq_{ref\ g,j}\left(T_{ref}\right) \tag{7}$$

where $Leq_{ref\,g,j}\left(T_{ref}\right)$ is a reference value calculated from the acoustic map of group g (using CADNA model) at the time interval T_{ref} = (08:00–09:00) at the point corresponding to the position of the (g, j)-th station. The CADNA software provides mean hourly Leq values over the entire city of Milan at a resolution of 10 m given a set of input traffic flow data, thus representing a reference static acoustic map, $Leq_{ref\,g,j}\left(T_{ref}\right)$. Here, we have chosen the reference time T_{ref} = (08:00–09:00) for convenience, since it displays rush-hour type of behavior. The predefined temporal ranges within the day are:

τ = 5 min for (07:00–21:00); τ = 15 min for (21:00–01:00); τ = 60 min for (01:00–07:00).

This choice has been motivated by the need to provide the shortest time interval for the update of the acoustic maps keeping the associated error approximately constant over the entire day [44].

2.3. Average Over the Monitoring Stations in Each Group: 1st Method

In this section, we discuss how to use the 24 $\delta^{\tau}_{g,j}(t)$ defined in Equation (7) in such a way to bring DYNAMAP to operation. We used two methods: the first described in this section and the second in Section 2.4. The first method is quite straightforward and implies that once all the $\delta^{\tau}_{g,j}(t)$ values are provided, the six acoustic maps corresponding to each group g can be updated by averaging the variations in Equation (7) over the four monitoring station values j in each group, according to [43,45]:

$$\delta^{\tau}_{g}(t) = \frac{1}{4}\Sigma^{4}_{j=1}\delta^{\tau}_{g,j}(t) \tag{8}$$

2.4. Clustering of the 24 Monitoring Stations: 2nd Method

The second procedure for updating the acoustic maps is based on a two-cluster expansion scheme, which uses all the 24 stations to determine $\delta^{\tau}_{g}(t)$ simultaneously (see Section 2.6 for details on the stations network). The clustering method, as described in Section 2.1, is applied here to determine the two corresponding clusters. For this purpose, we used the 24 h noise profiles recorded by each monitoring sensor over the period from 13th November 2018 to 5th February 2019. From this ANE-free dataset, we excluded all festivities, weekends, rainy, and windy days. In order to get robust noise profiles, we manually calculated, for each sensor, its median. For this analysis, we chose two time resolutions, τ, constant for all the day: $\tau =$ 60 and 5 min. The results of the analysis, performed on the 24 median profiles, are reported in Figures 6 and 7.

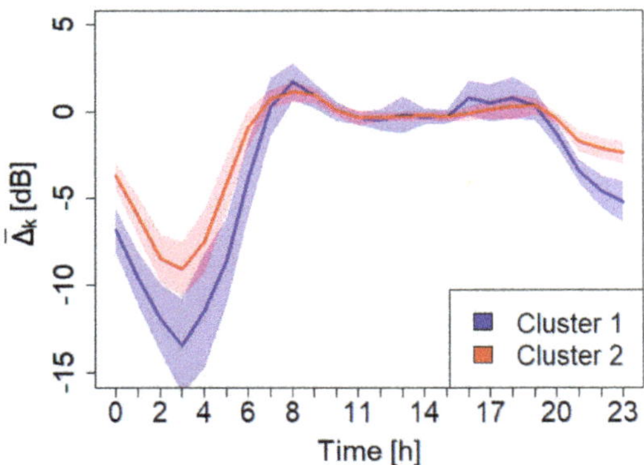

Figure 6. Mean normalized cluster profiles, $\overline{\Delta}_k$, and the corresponding error band, k indicates the cluster index. Time resolution $\tau =$ 60 min. The colored band represents the 1σ confidence level. In these calculations, the normalized noise level is obtained following the procedure described in [40].

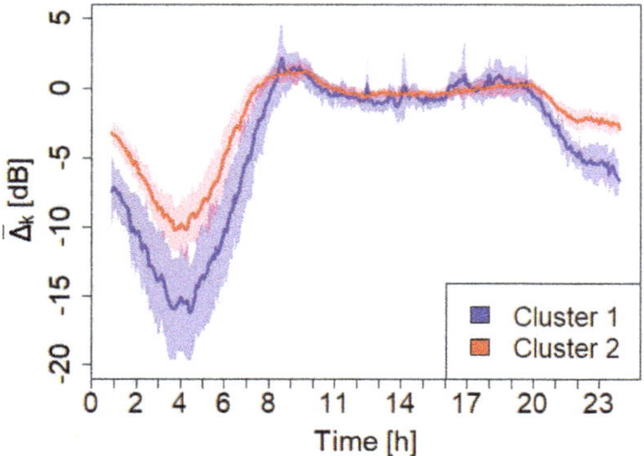

Figure 7. Mean normalized cluster profiles, $\overline{\Delta}_k$, and the corresponding error band, k indicates the cluster index. Time resolution $\tau = 5$ min. The colored band represents the 1σ confidence level. The normalized level used here is the same as the one determined in Figure 6.

From this analysis, it appears very clearly the robustness of the clustering method of the 24 monitoring sensors (for both $\tau = 60$ and 5 min). In fact, the 24 sensors result perfectly distributed in the two clusters mimicking the trend obtained with the original sampling measurements taken over the entire city. In Table 1, the information regarding the monitoring sensors together with their cluster membership are reported.

Table 1. Monitoring sensor information: code, group membership, non-acoustic parameter, $x = \text{Log}(T_T)$, and cluster membership according to the performed analysis (12 sensors belong to Cluster 1 and 12 to Cluster 2).

Sensor Code	Group g_i	$x = \text{Log}(T_T)$	Cluster
135	1	2.89	2
137	1	1.90	2
139	1	1.13	2
144	1	2.94	2
108	2	3.06	1
124	2	3.50	2
125	2	2.69	2
145	2	3.42	2
115	3	3.58	2
116	3	3.60	2
120	3	3.74	1
133	3	3.75	2
121	4	4.06	1
127	4	3.90	2
129	4	3.94	1
138	4	4.19	2
106	5	3.90	1
123	5	4.30	1
136	5	4.21	1
151	5	4.40	1
109	6	4.75	1
114	6	4.58	1
117	6	4.85	1
140	6	4.70	1

Updating Procedure for the 2nd Method

Once the compositions of Clusters 1 and 2 have been found (meaning that there are N_1 stations in Cluster 1, $k_1 = (1, \ldots, N_1)$, and N_2 stations in Cluster 2, $k_2 = (1, \ldots, N_2)$, such that $N_1 + N_2 = 24$), we need to rearrange the variations obtained from Equations (7) and (8) according to the indices $C_{1,k1}$ and $C_{2,k2}$, which we denote as $\delta^{\tau}_{C1,k1}(t)$ and $\delta^{\tau}_{C2,k2}(t)$ within each cluster, C_1 and C_2. Then, we calculate the mean variations, $\delta^{\tau}_{C1}(t)$ and $\delta^{\tau}_{C2}(t)$, for each cluster according to,

$$\begin{aligned} \delta^{\tau}_{C1}(t) &= \tfrac{1}{N_1}\Sigma^{N_1}_{k1=1}\, \delta^{\tau}_{C1,k1}(t) \\ \delta^{\tau}_{C2}(t) &= \tfrac{1}{N_2}\Sigma^{N_2}_{k2=1}\, \delta^{\tau}_{C2,k2}(t), \end{aligned} \qquad (9)$$

where $C_{1,k1}$ and $C_{2,k2}$ are indices of stations belonging to Cluster 1 and Cluster 2, respectively. In Figure 8, the histograms of the non-acoustic parameter, $x = \mathrm{Log}(\mathrm{T_T})$, for Clusters 1 and 2 of the 24 sensors (shown in Figures 6 and 7) are illustrated. For comparison, the density function $P_1(x)$ and $P_2(x)$ obtained for the initial 93 sample noise time series (shown in Figure 5) are also included. The rather good agreement allows using such distribution functions to express the mean variation $\delta^{\tau}_{g}(t)$ associated with each group g using the formula:

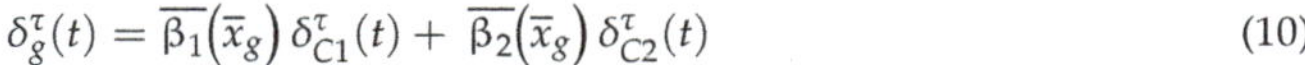

$$\delta^{\tau}_{g}(t) = \overline{\beta_1}(\overline{x}_g)\, \delta^{\tau}_{C1}(t) + \overline{\beta_2}(\overline{x}_g)\, \delta^{\tau}_{C2}(t) \qquad (10)$$

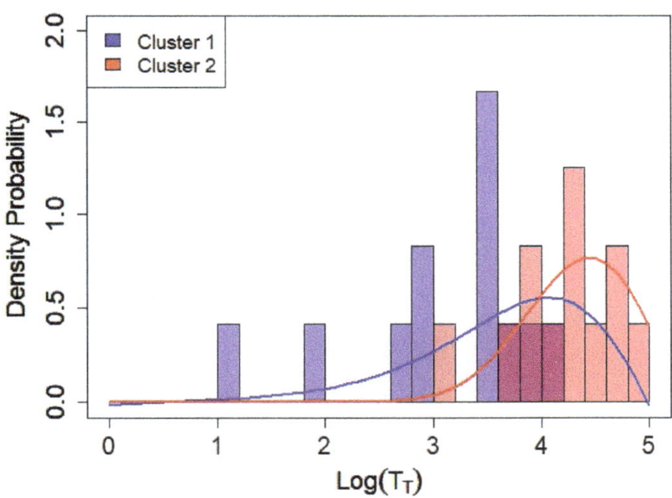

Figure 8. Histograms (from the 24 monitoring stations) and probability distributions, $P_1(x)$ and $P_2(x)$, as a function of the non-acoustic parameter, $x = \mathrm{Log}(\mathrm{T_T})$, for Clusters 1 and 2. Bin size is 0.2. $P_1(x)$ and $P_2(x)$ are the same functions shown in Figure 5.

Here, the value $\overline{x}_g$ represents the mean non-acoustic parameter associated with group g, and $\overline{\beta}_1(\overline{x}_g)$, $\overline{\beta}_2(\overline{x}_g)$ the corresponding probabilities to belong to Clusters 1 and 2, respectively (see Table 2 for the mean values of $\overline{\beta}_1$ and $\overline{\beta}_2$ for the six groups and Equation (6) for their definition).

Table 2. Mean values of $\overline{\beta}_1$ and $\overline{\beta}_2$ for the six groups of $x = \mathrm{Log}(\mathrm{T_T})$ within District 9.

Range of x	**0.0–3.0**	**3.0–3.5**	**3.5–3.9**	**3.9–4.2**	**4.2–4.5**	**4.5–5.2**
$\overline{\beta}_1$	0.99	0.81	0.63	0.50	0.41	0.16
$\overline{\beta}_2$	0.01	0.19	0.37	0.50	0.59	0.84

2.5. Dynamic Noise Level at an Arbitrary Location

The absolute level $Leq^{\tau}{}_{s}(t)$ at an arbitrary site s at time t can be obtained from the measured values of $\delta_g^{\tau}(t)$ using either Equation (8) or Equation (10). The first quantity we need to know is the value of $Leq_{ref\,g,s}$ that is the reference Leq calculated in the point s at the reference time (8:00–9:00) due to group g, which is provided by CADNA model (acoustic base map). The absolute level $Leq^{\tau}{}_{s}(t)$ at location s at time $t = n\tau$ can then be obtained by combining the level contribution of each base map with its variation $\delta_g^{\tau}(t)$:

$$Leq_s^{\tau}(t) = 10\,Log\sum_{g=1}^{6} 10^{\frac{Leq_{ref_{g,s}} + \delta_g^{\tau}(t)}{10}} \tag{11}$$

This operation provides what we called the "scaled map" (dynamic map).

2.6. Measurement Campaign

A measurement campaign, completed in 2019, aimed at testing the results of DYNAMAP predictions. This has been justified by the updated release of anomalous noise events detection (ANED) algorithm which acts directly on the recorded noise time series from the 24 monitoring stations prior to their use in the DYNAMAP calculation process (see below). It presented a higher recognition efficiency of anomalous events (less false positives) than the previous release, therefore, allowing a more reliable comparison between field measurements and DYNAMAP predictions [32].

The test measurements were performed in 21 locations within District 9 (purple stars in Figure 9 and Table 3 for detailed addresses) equally distributed in the six groups of roads. The measurement sites were located at arbitrary points distributed within the pilot area of Milan and with different noise propagation conditions. In particular, sites were selected in order to test the system in complex scenarios where the noise from roads belonging to different groups may contribute. Special attention was given to avoid non-traffic noise sources such as technical systems (thermal power stations or ventilation systems), construction sites, railway, and tram lines, interfering with the measurements. Figure 9 also contains the position of the 24 monitoring stations together with the indication of the six groups of roads represented by different colors.

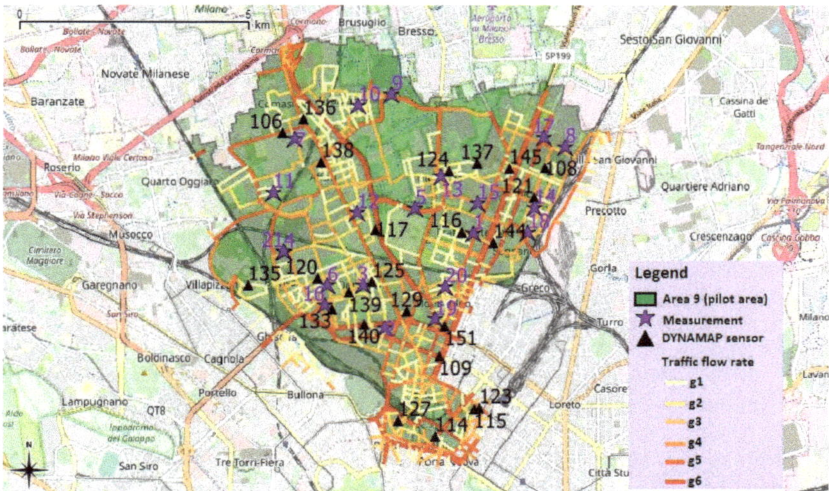

Figure 9. District 9 of the city of Milan city. Streets color corresponds to the different groups of streets according to range of non-acoustic parameter x: (0.0–3.0) (g_1), (3.0–3.5) (g_2), (3.5–3.9) (g_3), (3.9–4.2) (g_4), (4.2–4.5) (g_5), (4.5–5.20) (g_6). Black triangles and purple stars represent the sites where the monitoring stations are installed and the position of test measurements, respectively.

Table 3. Location of noise monitoring stations and measurement sites (cfr. Figure 9). The group index of each sensor and site are indicated within parenthesis.

Station (Group g_i)	Address	Station (Group g_i)	Address
106 (5)	Via Modigliani	127 (5)	Via Quadrio
108 (2)	Via Pirelli	129 (4)	Via Crespi
109 (6)	Viale Stelvio	133 (3)	Via Maffucci
114 (6)	Via Melchiorre Gioia	135 (1)	Via Lambruschini
115 (3)	Via Fara	136 (5)	Via Comasina
116 (3)	Via Moncallieri	137 (1)	Via Maestri del Lavoro
117 (6)	Viale Fermi	138 (4)	Via Novaro
120 (3)	Via Baldinucci	139 (1)	Via Bruni
121 (4)	Via Pirelli	140 (6)	Viale Jenner
123 (5)	Via Galvani	144 (1)	Via D'intignano
124 (2)	Via Grivola	145 (2)	Via F.lli Grimm
125 (2)	Via Abba	151 (5)	Via Veglia
Site (Group g_i)	**Address**	**Site (Group g_i)**	**Address**
1 (5)	Via Suzzani	12 (2)	Via Pastro
2 (2)	Via Bernina	13 (4)	Via Bauer
3 (3)	Via Ciaia	14 (2)	Via Polvani
4 (3)	Via Cosenz	15 (4)	Via Gregorovius
5 (5)	Via Majorana	16 (4)	Via Catone
6 (3)	Via Maffucci	17 (6)	V.le Sarca
7 (2)	Via Ippocrate	18 (1)	Via Boschi Di Stefano
8 (3)	Via Chiese	19 (6)	Via Murat
9 (5)	Via Moro	20 (1)	Via Sarzana
10 (1)	Via Marchionni	21 (3)	Via Cosenz
11 (1)	Via Gabbro		

2.7. DYNAMAP Sensors Calibration

The correct assessment of DYNAMAP operation needs a careful evaluation of noise sensor network. The first evaluation activity involved DYNAMAP sensors calibration. The sensors have a characteristic accuracy which needed to be verified prior to their use. To this end, a field calibration procedure has been implemented with the help of a Class 1 calibrator (emission level 94 dB at 1 kHz, see Figure 10). The deviations of DYNAMAP sensors with respect to the calibrator are reported in Table 4. This value has been employed to correct the noise levels recorded by the corresponding noise sensor. In Table 4, the label N.C. (Not Calibrated), referred to three monitoring stations and means that these sensors could not be on-site calibrated by the operator because of safety reasons.

Figure 10. Operation of calibration on DYNAMAP sensor.

Table 4. Calibration deviations of DYNAMAP sensors (values are in dB; N.C.: Not Calibrated).

Sensor	Site	Deviation [dB]
145	Via F.lli Grimm	−0.1
136	Via Comasina	−0.5
138	Via Novaro	+0.2
125	Via Abba	−0.6
123	Via Galvani	−0.2
115	Via Fara	−0.1
114	Via Melchiorre Gioia	−0.8
127	Via Quadrio	N.C.
140	Viale Jenner	−0.9
133	Via Maffucci	−0.3
120	Via Baldinucci	−0.5
129	Via Crespi	−0.7
151	Via Veglia	−0.4
116	Via Moncalieri	−0.2
124	Via Grivola	−0.6
137	Via Maestri del Lavoro	−0.5
144	Via d'Intignano	−0.5
121	Via Pirelli	0.0
108	Via Pirelli	−0.2
135	Via Lambruschini	−0.1
109	Viale Stelvio	N.C.
106	Via Litta Modignani	+0.2
117	Viale Fermi	0.0
139	Via Bruni	N.C.

2.8. DYNAMAP Sensors Reliability

The second evaluation activity aimed at verifying the reliability of DYNAMAP sensors by comparing their readouts with a Class 1 sound level meter. The sound levels measurements (10 short duration measurements (≈1 h) and 2 measurements of 24 h) were performed on 12 monitoring sites (two sites for each group of roads), placing the microphone in the same position of the DYNAMAP sensor. The results of the tests expressed in $Leq^{\tau}{}_{S}$ with τ = 5 min, are summarized in Figure 11, showing the correlation between Class 1 sound level meter and DYNAMAP sensor. We obtained a high correlation (R^2 = 0.99) with a mean deviation between the two sets of measurements of 1.0 ± 0.9 dB.

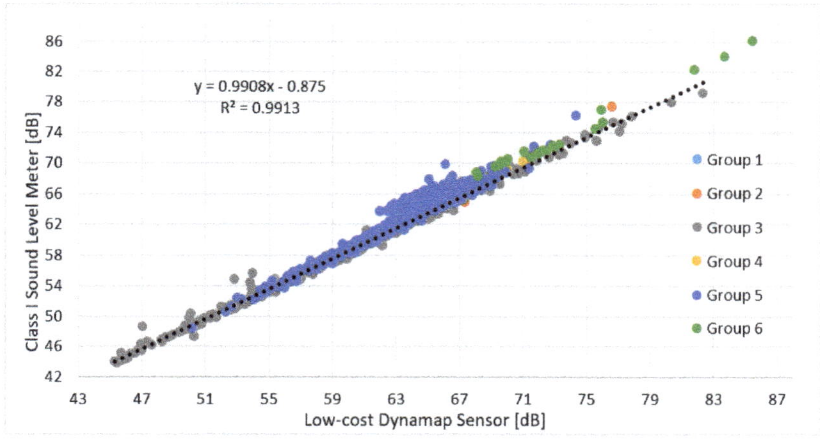

Figure 11. Correlation between the Class 1 Sound Level Meter and DYNAMAP Sensors. Different colors refer to sensors in each group of streets.

3. Results

In this section, we will describe the major steps to obtain an overall assessment of the project in terms of accuracy and reliability. A preliminary investigation [45] showed that the system is affected by different sources of error whose origin must be taken into account to minimize and eventually correct them. In the following, we provide a description of the measurement campaign and of the accuracy of both traffic model and DYNAMAP prediction.

3.1. Traffic Flow Data

In order to assess the validity of the traffic flow model, used to describe the non-acoustic parameter x, we performed a series of measurements of both traffic flow and noise at randomly selected sites and in correspondence of the noise monitoring stations, and compared them with the traffic model database. This test is important because the parameter x determines the group membership and therefore its dynamic behavior. In case the traffic model prediction is not accurate enough, DYNAMAP prediction could be sensibly affected.

As one can see from Figure 12, there are significant differences between the traffic flow model predictions and measurements. Possible causes can be found in changes of traffic conditions (the model refers to a 2012 road network) and the incapability of the model to manage traffic conditions characterized by low flows (it has been designed and calibrated to deal with critical traffic situations). Consequently, in some cases the "real" total daily vehicle flow can significantly differ from the one attributed to a specific road using the flow model. This may result in jumps of group membership and, therefore, inaccurate predictions.

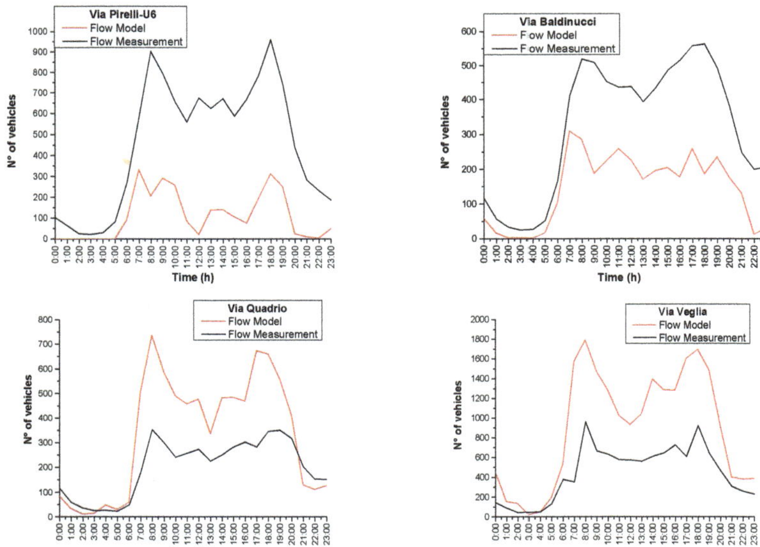

Figure 12. Comparison between hourly traffic flow (number of vehicles per hour) and traffic flow model calculations. Here, Via Pirelli-U6 (g_2), Via Baldinucci (g_2), Via Quadrio (g_4), Via Veglia (g_5) are some selected locations corresponding to the position of monitoring stations.

In Table 5, we report the comparison between the results of total traffic flows, in the form of the non-acoustic parameter x, obtained from the model calculations and the recent measurements in the same sites. Differences, or group jumps, occurred in particular for the case of Via Pirelli which became a congested road in recent years (from g_2 to g_4). As is apparent from Table 5, deviations of the predicted values x are within about 10% for groups g_3–g_5, and much higher for other groups.

Table 5. Group assignment according to model calculations and flow measurements in correspondence to 10 noise monitoring stations.

Site (Street Name)	Group g_i		Values of x = Log(T_T)	
	Model	Meas.	Model	Meas.
Via Lambruschini	1	2	2.95	3.50
Via Maestri del Lavoro	1	2	2.90	3.41
Via Grivola	2	1	3.29	2.92
Via Pirelli	2	4	3.42	4.04
Via Fara	3	3	3.75	3.66
Via Baldinucci	3	3	3.54	3.89
Via Quadrio	4	3	3.90	3.68
Via Crespi	4	4	4.15	4.08
Via Comasina	5	5	4.33	4.25
Via Veglia	5	4	4.33	4.03

3.2. DYNAMAP Predictions

In the following, we report the comparison between traffic noise measurements with the corresponding DYNAMAP predictions, $Leq_s^\tau(t)$, with t = (5, 15, 60) min. The different updating time intervals correspond to the three time-periods within the 24 h of a day: t = 5 min (07:00–21:00), t = 15 min (21:00–01:00), and t = 60 min (01:00–07:00). The DYNAMAP prediction of $Leq_s^\tau(t)$ at a site s within the network can be obtained from the relation reported in Equation (11). The reference values for the 21 selected sites $Leq_{ref\,g,s}$ that is the "static" level contribution from different groups are reported in Table 6. They illustrate how different groups contribute to the local noise level. The major contribution to the local site level, $Leq_{ref\,g,s}$, in general, comes from the group g the site belongs to (see bold figures in Table 6). For example, for Site 1, which belongs to group g_5, the most significant contribution comes from $Leq(s)_{ref(g5)}$. However, each local site level is subject to the influence of nearby streets through other groups, as is apparent from Table 6. In particular, roads characterized by low traffic flow generally are mostly influenced by neighboring higher flow roads (see as an example Site 10, 18, and 20 of group g_1).

Table 6. Level contributions of each group, $Leq_{refg,s}$ at 21 arbitrary chosen sites of District 9 (Figure 9). The group indices of each sites are shown in the second column. Bold figures represent the major contribution to the local site level.

Site	Group g_i	$Leq_{ref\,g1,s}$	$Leq_{ref\,g2,s}$	$Leq_{ref\,g3,s}$	$Leq_{ref\,g4,s}$	$Leq_{ref\,g5,s}$	$Leq_{ref\,g6,s}$
1	**5**	21.1	47.8	56.8	28.3	**64.9**	37.7
2	**2**	12.0	**64.6**	15.0	15.0	15.0	59.6
3	**3**	0.0	56.1	**62.7**	0.0	0.0	0.0
4	**3**	17.5	25.3	**59.4**	48.8	51.9	0.0
5	**5**	29.7	25.9	32.4	29.4	**67.8**	33.6
6	**3**	41.3	45.9	**66.4**	34.5	27.0	28.0
7	**2**	24.1	**58.1**	51.4	17.8	42.2	45.6
8	**3**	21.1	21.9	**53.9**	49.4	26.9	29.9
9	**5**	8.1	32.5	35.2	43.6	**62.3**	0.0
10	**1**	38.2	**43.0**	27.2	25.2	32.7	28.4
11	**1**	**55.8**	20.6	32.0	37.7	42.9	0.0
12	**2**	41.1	**62.4**	24.5	20.8	48.8	40.2
13	**4**	42.1	56.0	38.3	**69.2**	41.9	38.8
14	**2**	44.3	**61.1**	51.9	45.6	36.0	34.4
15	**4**	12.5	29.7	29.8	**70.2**	50.5	33.2
16	**4**	33.2	30.6	47.9	**68.6**	54.3	37.1
17	**6**	25.4	24.0	34.4	51.3	50.6	**69.7**
18	**1**	49.0	45.0	**59.1**	56.9	57.0	53.3
19	**6**	24.2	32.6	39.7	38.3	37.3	**71.7**
20	**1**	51.0	38.7	**52.1**	30.6	35.6	36.1
21	**3**	17.0	15.8	**56.8**	48.2	51.7	0.0

The comparison between traffic noise measurements, $Leq_s(t)_m$ and DYNAMAP predictions, $Leq_s(t)$ at site s is based on the evaluation of the mean deviation:

$$< \varepsilon_{Leq_s} > = \frac{1}{N} \sum_{k=1}^{N} \left| Leq_s(k) - Leq_s(k)_m \right| \tag{12}$$

where the summation index k extends over three time periods (24 h-N_{Tot} = 190; day 07:00–21:00 h-N_{5min} = 168; evening 21:00–01:00 h-N_{15min} = 16; night 01:00–07:00 h-N_{1h} = 6). The results of the comparison between measurements and predictions (cfr. Equation (11)) according to the two calculation methods (cfr. Equation (8) or Equation (10)) are reported in Figure 13 for a representative number of sites (Sites 6, 16, 19, 20).

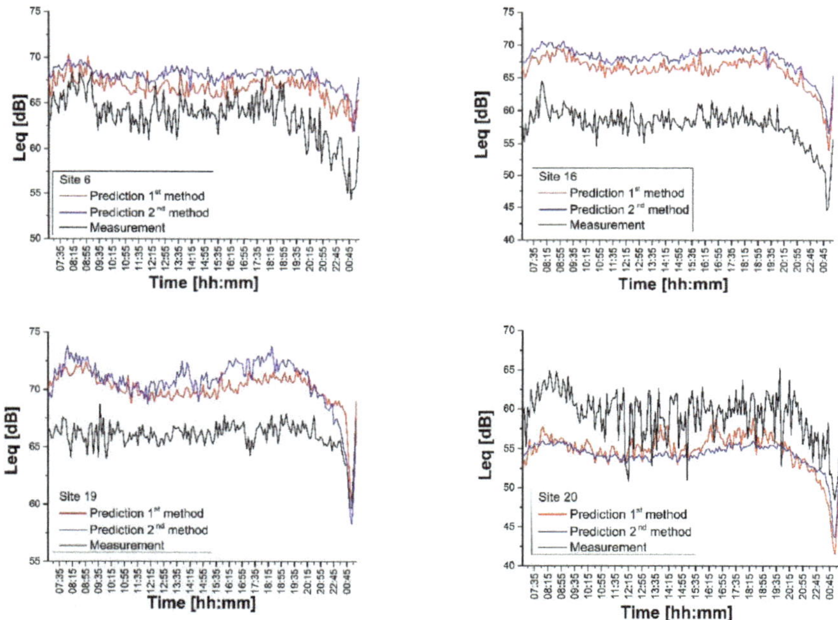

Figure 13. Comparison between traffic noise measurements at Sites 6, 16, 19, 20 and the corresponding DYNAMAP predictions according to two calculation methods (cfr. Equation (8) or Equation (10)).

Figure 13 shows how both methods provide predictions with similar trends and deviations. Both methods are affected by a systematic, almost constant, error, most likely introduced by the traffic flow model (see discussion below). The latter should have a higher influence on the second prediction method as it takes on the contribution of all noise monitoring stations (see Equation (10)). However, the second method should be more robust in case one or more noise monitoring are offline. Site 20 presents higher discrepancies with high intermittency patterns especially during the day-time due to both the small integration time (5 min) and the irregular traffic flows in local roads.

In Table 7, we report the total daily mean deviation (24 h) for the two prediction methods in all the 21 test measurements.

Table 7. Summary of the total daily mean deviation (24 h) for the two prediction methods in all the 21 test measurements.

Site	Group g_i	Mean Deviation–1st Method (24 h) [dB]	Mean Deviation–2nd Method (24 h) [dB]
10	1	6.4 ± 2.5	11.0 ± 2.6
11	1	3.2 ± 2.3	3.8 ± 2.1
18	1	6.5 ± 1.5	7.5 ± 1.4
20	1	4.7 ± 2.3	5.1 ± 2.3
7	2	1.7 ± 1.5	3.6 ± 1.4
12	2	7.5 ± 2.333	2.7 ± 2.0
14	2	3.4 ± 1.6	1.4 ± 0.9
3	3	2.8 ± 1.7	5.2 ± 2.0
4	3	3.0 ± 2.7	2.0 ± 1.9
6	3	3.2 ± 1.9	4.5 ± 2.0
8	3	2.0 ± 1.3	1.2 ± 1.1
21	3	1.7 ± 1.1	0.9 ± 0.7
13	4	4.0 ± 1.3	5.5 ± 1.4
15	4	3.0 ± 1.2	4.7 ± 1.0
16	4	8.4 ± 1.5	10.0 ± 1.3
1	5	4.9 ± 1.3	6.5 ± 1.5
5	5	2.0 ± 1.3	1.9 ± 1.4
9	5	1.3 ± 0.9	2.1 ± 1.2
17	6	1.4 ± 0.8	2.8 ± 1.1
19	6	4.0 ± 1.2	4.7 ± 1.6

4. Discussion

In the following, we will discuss a possible solution to improve DYNAMAP prediction within a reasonable range of error. For simplicity, we will consider 1h as updating time scale and the first prediction method based on Equations (8) and (11) for the calculation of the mean variation of each group, $\delta_g^{\tau}(t)$ and presented in Section 2.3.

Prediction Corrections

A number of selected sites have been chosen to compare the results of field measurements with the corresponding DYNAMAP predictions.

Figure 14 (left part) presents a relevant discrepancy between predictions and measurements, which can be higher during the daytime. Each figure shows the error bands obtained from the propagation error associated with the variability of $\delta_g^{\tau}(t)$ within each group g. During the day time (07:00–21:00) the mean group discrepancy remains within 1 dB, whereas in the evening-night time (21:00–07:00) the high "volatility" of traffic noise pushes it to about (2–4) dB.

The almost constant gap between measurements and predictions in different period of the day suggested us to search for a systematic error inherent the DYNAMAP calculation method; systematic error which is most likely correlated to the vehicular flow employed in the prediction model. In fact, $\delta_g^{\tau}(t)$ is calculated with respect to $Leq_{ref\,g}$, obtained from CADNA software using as input information on the number of vehicles/hour at the reference hour (8:00–9:00).

During the measurement campaign, we simultaneously recorded the traffic flows. This allowed us to compare the logarithm of traffic flow measurements with the traffic flow model calculations for Sites 6 (g_3), 16 (g_4), 19 (g_6), and 20 (g_1) as illustrated in Figure 14 (right part), respectively. The traffic flow data have been provided by Agenzia Mobilità Ambiente Territorio (AMAT), the agency in charge of the traffic mobility at the City Hall [46]. In the described examples, the model yields more reliable results for highly traffic roads belonging to groups g_3, g_4, and g_6, than for lower flow roads as in g_1, as already reported in a previous preliminary work [45].

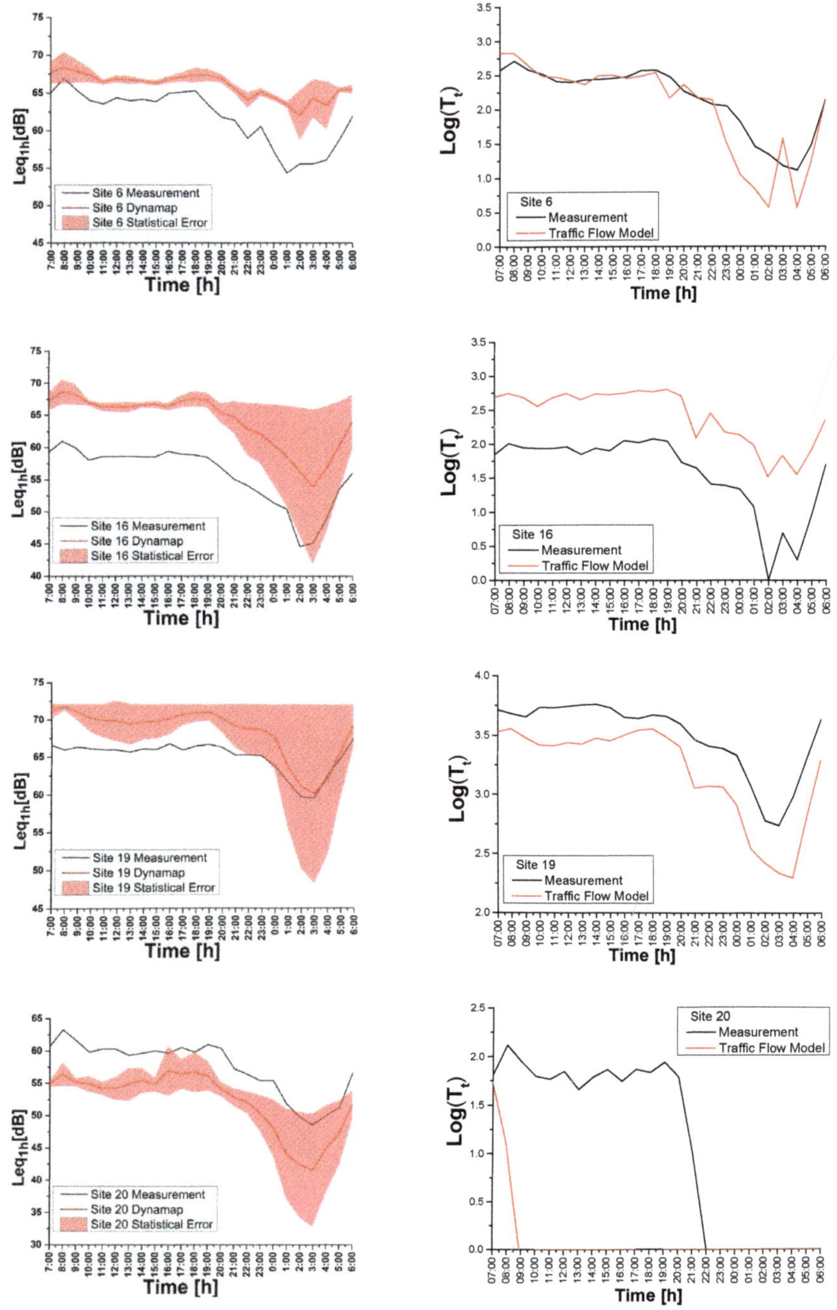

Figure 14. (Left side) Comparison between traffic noise measurements and DYNAMAP predictions at: Site 6 (group $g = 3$), Site 16 (group $g = 4$), Site 19 (group $g = 6$), Site 20 (group $g = 1$). The colored band represents the 1σ confidence level. (Right side) Comparison between traffic flow measurements and AMAT traffic model at the same sites.

As it is apparent from Figure 14, there is a gap between the prediction and the measurements of $Leq_s^T(t)$. The observed constant shift might be the result of inaccuracies of the traffic model in describing the traffic flow, especially for low traffic roads. Such shift is regarded as a systematic error.

To quantify this discrepancy and try to correct it, we calculate for each site the relative mean deviation (ε_L) between hourly traffic noise measurement level, $Leq_s(1h)_m$, and the corresponding hourly DYNAMAP prediction level, $Leq_s(1h)$, over the day and night period, defined as

$$\varepsilon_L = \frac{1}{N}\sum_{k=1}^{N} \frac{(Leq_s(k)_m - Leq_s(k))}{Leq_s(k)_m} \tag{13}$$

where the summation index k extends over two time zones (day 07:00–21:00 h $\rightarrow N_{1h} = 14$; evening-night 21:00–07:00 h $\rightarrow N_{1h} = 10$). The relative error is then averaged over all roads belonging to the same group, in order to represent the average hourly values of the road group ($\overline{\varepsilon_L}$). Furthermore, we consider the relative deviation (ε_F) between measurement and model for the logarithm of the traffic flow at the reference time, $Log\ F_{(8:00-9:00)}$,

$$\varepsilon_F = \frac{Log\left(F_{(8:00-9:00)Meas.}\right) - Log\left(F_{(8:00-9:00)Model}\right)}{Log\left(F_{(8:00-9:00)Meas.}\right)} \tag{14}$$

where $Log(F_{(8:00-9:00)Model})$ is the logarithm of the flows from 8:00 to 9:00 of the 2012 traffic model. Then we calculate the mean deviation of all sites belonging to the same group, $\overline{\varepsilon_F}$.

These values for $\overline{\varepsilon_L}$ and $\overline{\varepsilon_F}$ are plotted in Figure 15, illustrating, to some degree, a relationship between traffic flow deviations and noise level errors. This relationship will be treated as a systematic error and taken into account within the DYNAMAP scheme.

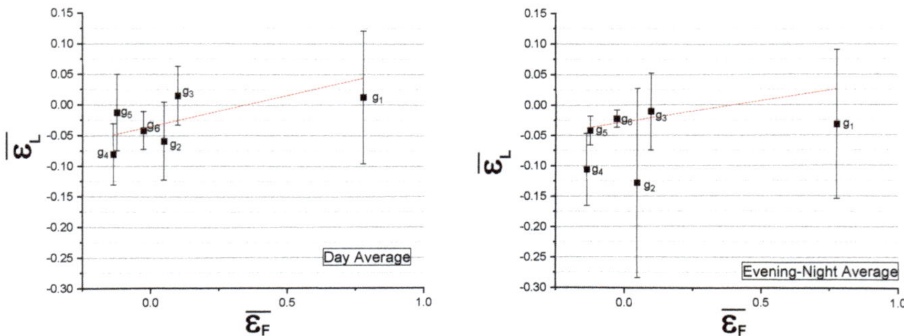

Figure 15. Relative mean hourly deviation between traffic noise measurements and the corresponding DYNAMAP predictions $\overline{\varepsilon_F}$ vs. the relative deviation between the logarithm of traffic flow measurements and the corresponding model calculations at the reference hour (8:00–9:00) $\overline{\varepsilon_F}$ for each group separately. The results refer to: Day time (07:00–21:00) (Left panel), and Evening-Night time (21:00–07:00) (Right panel) periods. The dashed line is just a guide for the eye.

We thus obtain the corrected hourly value for the predicted noise level ($Leq(1h)$), by multiplying the different hourly values of the predicted noise level times the relative mean group deviation, expressed in percentage terms [$1 + \overline{\varepsilon_L}\ (g)$]. The results of this operation are shown in Figure 16 (Right part, red line). We observe a general improvement of the prediction for these sites. In the graphics, the uncertainty bands include both the statistical and systematic errors (total error).

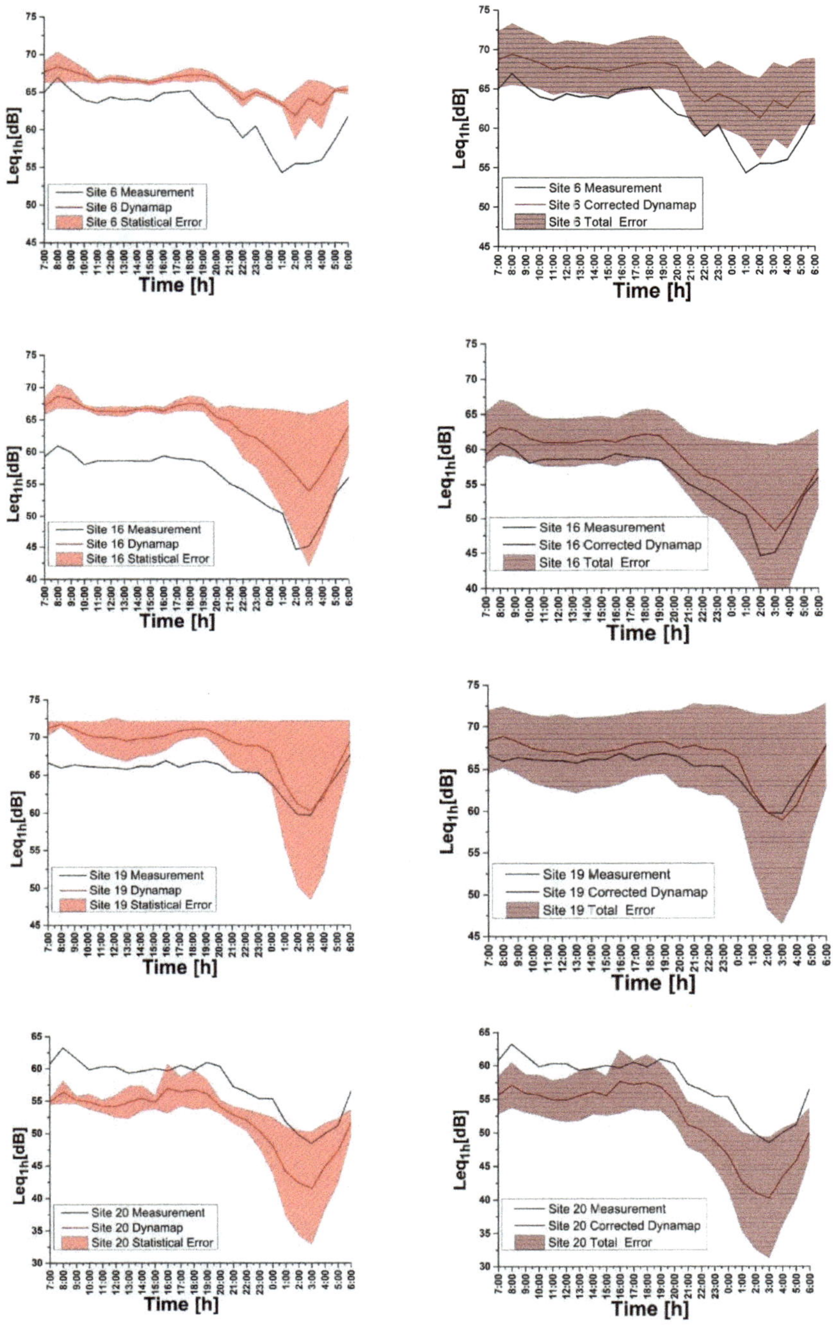

Figure 16. (Left part) Comparison of traffic noise measurements and DYNAMAP (non-corrected) prediction for: Site 16 (Upper panel), Site 19 (Middle panel), Site 20 (Lower panel). (Right part) Comparison of traffic noise measurements and DYNAMAP (corrected) prediction for the same sites. In the figure, the total error is displayed.

In Table 8, we report both the site mean hourly non-corrected, $<\varepsilon_{Leq}>_N$, and corrected prediction errors, $<\varepsilon_{Leq}>_C$, for all measurement sites, obtained through the comparison between the hourly non-corrected or corrected prediction levels and the hourly measurement levels, as shown in Equation (12).

The correction yields better predictions in many cases, but in others it remains poor. A median-based correction, $<\varepsilon_{Leq}>_M$, is also reported in Table 8. This quantity is less sensitive to outliers and, consequently, it provides more realistic estimates of the corrections. Finally, the right column of Table 8 shows the group mean errors calculated by averaging over the roads belonging to each group. The highest discrepancies are found for group g_1 as a consequence of the poor descriptive capabilities of the traffic flow model. Except for this, the results obtained for the group median-average error, $<\varepsilon_{Leq}>_M$, is below 3 dB.

Table 8. (Left part) Mean site prediction error without systematic error correction, $<\varepsilon_{Leq}>_N$, with systematic error correction, $<\varepsilon_{Leq}>_C$, and median average of the corrected prediction, $<\varepsilon_{Leq}>_M$. (Right part) Mean group non-corrected prediction error, $<\varepsilon_{Leq(g)}>_N$, mean group corrected error, $<\varepsilon_{Leq(g)}>_C$, and group median average, $<\varepsilon_{Leq(g)}>_M$. All values are in dB.

Site	Group g_i	$<\varepsilon_{Leq}>_N$	$<\varepsilon_{Leq}>_C$	$<\varepsilon_{Leq}>_M$	Group g_i	$<\varepsilon_{Leq(g)}>_N$	$<\varepsilon_{Leq(g)}>_C$	$<\varepsilon_{Leq(g)}>_M$
10	1	5.0	5.2	5.2	1	5.3	5.1	5.2
11	1	4.5	4.0	4.1	2	4.2	3.2	2.8
18	1	6.4	6.1	6.1	3	2.5	2.5	2.8
20	1	5.3	5.5	5.6	4	5.3	2.4	2.1
7	2	1.9	4.0	2.9	5	2.6	2.4	2.2
12	2	7.8	2.5	3.8	6	3.4	1.3	1.3
14	2	2.8	3.1	1.6				
3	3	1.8	2.3	2.5				
6	3	4.2	4.5	5.9				
4	3	2.1	1.5	2.0				
21	3	1.8	1.5	0.7				
13	4	4.1	2.0	0.8				
15	4	3.3	2.6	1.3				
16	4	8.4	2.6	4.2				
1	5	4.5	3.0	4.4				
5	5	1.9	2.4	1.2				
9	5	1.4	1.9	1.0				
19	6	3.4	1.3	1.3				

Therefore, excluding group g_1, for which a specific analysis needs to be developed, the prediction error of roads belonging to other groups, upon a systematic error correction $<\varepsilon_{Leq}>_C$, remains below 3 dB for each site, with the exception of Sites 6 (g_3) and 7 (g_2). The latter must be treated differently if we require that the 3 dB constrains must apply to all sites belonging to a group. We took 3 dB as a reference accuracy value as retrieved from the Good Practice Guide for strategic noise mapping [47]. As an example, consider site 6 (g_3). Correcting the predicted noise level using its own relative traffic flow deviation (not the group mean), we obtain the results reported in Figure 17, that correspond to $<\varepsilon_{Leq}>_C = 1.1$ dB.

This result suggests that in order to get an effective correction, the relative error between the measured and the model traffic flow (8:00–9:00) in a given road stretch has to be bound within an interval that depends on the group it belongs to. In Figure 18, for example, we report the relative mean hourly deviation between traffic noise measurements and the corresponding DYNAMAP predictions, ε_L, against the relative deviation between the logarithm of traffic flow measurements and the corresponding model calculations at the reference hour (8:00–9:00), ε_F, for each site of group g_3.

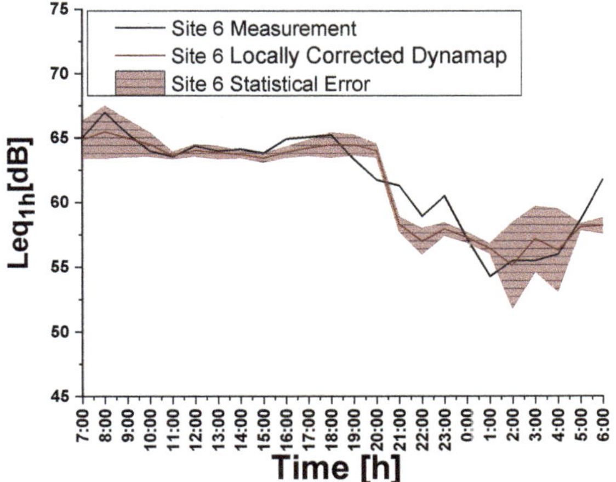

Figure 17. Comparison between traffic noise measurement and the "locally corrected" DYNAMAP prediction for Site 6.

Figure 18 has been obtained assuming for simplicity that the relation between ε_L and ε_F is linear within group g_3. In this case, in order to get a prediction error <3 dB for each site, the relative error on the traffic flow can vary by about ±0.10 with respect to the minimum found for the single site, as it can be observed in Figure 19 for Sites 3, 6, and 21.

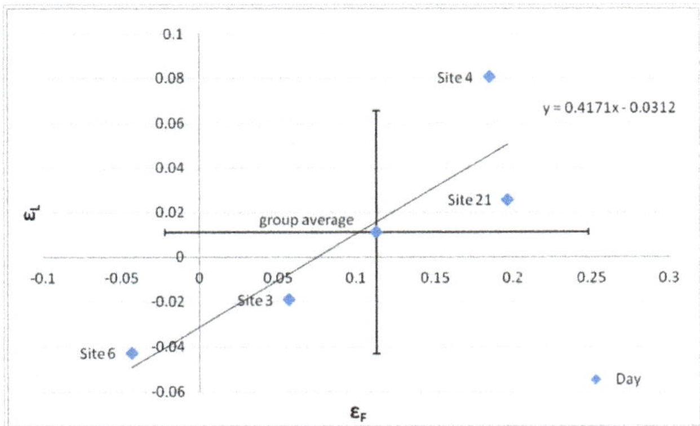

Figure 18. Relative mean hourly deviation between traffic noise measurements and the corresponding DYNAMAP predictions, ε_L, versus the relative deviation between the logarithm of traffic flow measurements and the corresponding model calculations at the reference hour (8:00–9:00), ε_F, for each site of group 3.

In Figure 19, the minimum prediction error is obtained near the corresponding site-specific flow error. It does not match exactly the value reported in Figure 18 because we are using a linear dependence between ε_L and ε_F (see Figure 18). In other words, the mean value of the relative error on the traffic flow of a given group g, $\overline{\varepsilon_F}$, (the one that has been used in the correction procedure of DYNAMAP prediction) must be bound within an interval that can be determined as follows: if we take $\overline{\varepsilon_F}$ centered at the minimum of the relative error of the site-specific traffic flow, $\varepsilon_{F,S6m}$ for the case of Site 6, it means that $\overline{\varepsilon_F} = \varepsilon_{F,S6m}$ can have a maximum standard deviation $\sigma = \pm(0.10)$ to satisfy the condition about the mean prediction error, $<\varepsilon_{Leq}> < 3$ dB. Therefore, $\overline{\varepsilon_F}$ must belong to an interval $(\varepsilon_{F,S6m} - 0.10, \varepsilon_{F,S6m} + 0.10)$. This procedure has to be repeated for each site of the group. If these conditions are met, all sites will have $<\varepsilon_{Leq}> < 3$ dB. This means that the traffic model must provide flow values for the streets belonging to each group with comparable accuracy in order that the error remains within the same threshold for all sites of the group.

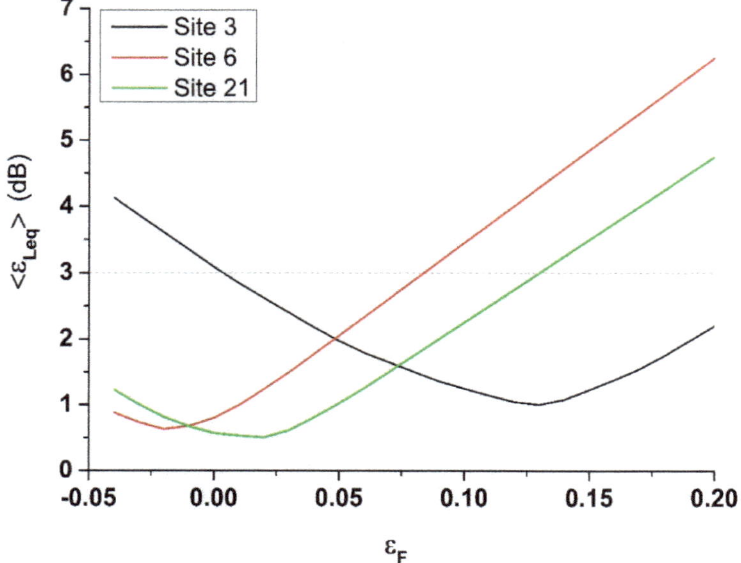

Figure 19. Mean prediction error $<\varepsilon_{Leq}>$ as function of the relative traffic flow error (8:00–9:00) ε_F for Sites 3, 6, and 21. The graphs have been obtained assuming for simplicity that the relation between ε_L and ε_F is linear within group g_3 (see Figure 18). The dashed line represents the 3 dB threshold.

As for roads characterized by low traffic flows, such as those belonging to group g_1, the application of the correction based only on the relative deviation of the local traffic flow is not effective, because in these cases, the noise level is not mainly determined by the local traffic, but by that of busier nearby roads. In these cases, we may think to reassign them to other groups, applying the correction of the group whose contribution in the prediction of the noise level is predominant.

5. Conclusions

DYNAMAP is an automatic monitoring system, based on customized low-cost sensors and a software tool implemented in a general purpose GIS platform. It has been developed and built in two pilot areas located along the A90 motorway that surrounds the city of Rome (Italy) and inside the agglomeration of Milan (Italy). This paper describes the final assessment of DYNAMAP system implemented in the pilot area of Milan. The statistical-based nature of the project relies on the high degree of correlation between what we called as non-acoustic parameter (total traffic flow) and traffic noise levels. This correlation allowed an accurate description of the traffic noise due to clusters of roads (described as a single noise map) from the information recorded from a few monitoring stations distributed all over the pilot area.

The paper includes the description of two procedures for updating the acoustic maps: one based on the average of the noise recorded by the monitoring stations in each group (1st method) and the other based on a two-cluster expansion scheme performed directly over the noise recorded by the 24 monitoring sensors distributed over the six groups of roads (2nd method). Both methods provided similar results though the second one was more robust in the case where one or more noise monitoring stations went offline. This is because the lack of information from one sensor (or more than one) is not as disruptive as for the first method. Indeed, we will have a 25% of missing information (1st method) against 4% (2nd method) in case of missing data from one sensor. In order to validate the system, each monitoring station was calibrated and cross-checked with Class I sound meters. A field measurement campaign was performed in order to compare the results of noise measurements and traffic flow with the corresponding estimated values of the noise map and of the traffic model.

In terms of accuracy, the predictive capability of DYNAMAP was mainly associated with the related accuracy of the chosen non-acoustic parameter (traffic flow). For this reason, a poor accuracy of the non-acoustic parameter is directly reflected on the noise prediction error. A method to correct the predicted noise levels in an arbitrary location and, therefore, limit the overall mean error within 3 dB for all groups of roads was illustrated. However, the requirements to keep the prediction error within 3 dB for each site established a serious constraint on the traffic flow model accuracy. This means that a significant improvement would be obtained by implementing a more realistic traffic flow model. This would reduce the systematic error and, therefore, enhance the overall reliability of DYNAMAP prediction. Hopefully, the implementation of mobile sampling and, more generally, of participatory sensing both for noise and traffic data would help reduce the uncertainty of noise maps. Conversely, this result may cause either an incorrect evaluation of the exposed population or improper noise action plans. Therefore, the uncertainty analysis in the creation of noise maps is a fundamental key tool to design noise action plans on extended areas.

Author Contributions: Methodology R.B., H.E.R., validation F.A. and C.C.; formal analysis, R.B.; data curation, C.C. and F.A.; writing—original draft preparation, R.B.; writing—review and editing, R.B., H.E.R.; supervision, R.B. and G.Z.; project administration, G.Z.; funding acquisition, G.Z. All authors have read and agreed to the published version of the manuscript.

Funding: This research has been co-funded by the European Commission under project LIFE13 ENV/IT/001254 DYNAMAP.

Conflicts of Interest: The authors declare no conflicts of interest.

Acronyms

In this section, we provide the list of acronyms employed throughout the manuscript:

AMAT	Agenzia Mobilità Ambiente Territorio
ANEs	Anomalous Noise Events
ANED	Anomalous Noise Events Detection
ARM	Advanced RISC Machines
CADNA	Computer Aided Noise Abatement
cfr	confer
CNOSSOS-EU	Common Noise Assessment Methods in Europe
DYNAMAP	DYNamic Acoustic MAPping
END	Environmental Noise Directive
e.g.,	exempli gratia
GIS	Geographic Information System
N.C.	Not Calibrated
RISC	Reduced Instruction Set Computer

References

1. Licitra, G.; Ascari, E.; Fredianelli, L. Prioritizing process in action plans: A review of approaches. *Curr. Pollut. Rep.* **2017**, *3*, 151–161. [CrossRef]
2. Muzet, A. Environmental noise, sleep and health. *Sleep Med. Rev.* **2007**, *11*, 135–142. [CrossRef]
3. de Kluizenaar, Y.; Janssen, S.A.; van Lenthe, F.J.; Miedema, H.M.; Mackenbach, J.P. Long-term road traffic noise exposure is associated with an increase in morning tiredness. *J. Acoust. Soc. Am.* **2009**, *126*, 626–633. [CrossRef]
4. Babisch, W.; Beule, B.; Schust, M.; Kersten, N.; Ising, H. Traffic noise and risk of myocardial infarction. *Epidemiology* **2005**, *16*, 33–40. [CrossRef]

5. Babisch, W. Road traffic noise and cardiovascular risk. *Noise Health* **2008**, *10*, 27–33. [CrossRef]
6. Belojevic, G.; Jakovljevic, B.; Slepcevic, V. Noise and mental performance: Personality attributes and noise sensitivity. *Noise Health* **2003**, *6*, 77–89. [PubMed]
7. Van Kempen, E.; Babisch, W. The quantitative relationship between road traffic noise and hypertension: A meta-analysis. *J. Hypertens.* **2012**, *30*, 1075–1086. [CrossRef] [PubMed]
8. Lercher, P.; Evans, G.W.; Meis, M. Ambient noise and cognitive processes among primary schoolchildren. *Environ. Behav.* **2003**, *35*, 725–735. [CrossRef]
9. Chetoni, M.; Ascari, E.; Bianco, F.; Fredianelli, L.; Licitra, G.; Cori, L. Global noise score indicator for classroom evaluation of acoustic performances in LIFE GIOCONDA project. *Noise Mapp.* **2016**, *3*, 157–171. [CrossRef]
10. European Union. Directive 2002/49/EC of the European Parliament and the Council of 25 June 2002 relating to the assessment and management of environmental noise. *Off. J. Eur. Commun.* **2002**, *L189*, 12.
11. European Commission. *Report From The Commission To The European Parliament And The Council On the Implementation of the Environmental Noise Directive in accordance with Article 11 of Directive 2002/49/EC*; European Commission: Brussels, Belgium, 2011.
12. Kephalopoulos, S.; Paviotti, M.; Anfosso Lédée, F. *Common Noise Assessment Methods in Europe (CNOSSOS-EU)*; Publications Office of the European Union: Luxembourg, 2002; pp. 1–180.
13. Morley, D.; de Hoogh, K.; Fecht, D.; Fabbri, F.; Bell, M.; Goodman, P.; Elliott, P.; Hodgson, S.; Hansell, A.; Gulliver, J. International scale implementation of the CNOSSOS-EU road traffic noise prediction model for epidemiological studies. *Environ. Pollut.* **2015**, *206*, 332–341. [CrossRef] [PubMed]
14. Garcia, A.; Faus, L. Statistical analysis of noise levels in urban areas. *Appl. Acoust.* **1991**, *34*, 227–247. [CrossRef]
15. Licitra, G.; Ascari, E.; Brambilla, G. Comparative analysis of methods to estimate urban noise exposure of inhabitants. *Acta Acust. United Acust.* **2012**, *98*, 659–666. [CrossRef]
16. Ausejo, M.; Recuero, M.; Asensio, C.; Pavon, I.; Pagan, R. Study of uncertainty in noise mapping. In Proceedings of the 39th International Congress on 25 Noise Control Engineering, INTERNOISE 2010, Lisbon, Portugal, 13–16 June 2010; pp. 6210–6219.
17. Licitra, G.; Gallo, P.; Rossi, E.; Brambilla, G. A novel method to determine multiexposure priority indices tested for Pisa action plan. *Appl. Acoust.* **2011**, *72*, 505–510. [CrossRef]
18. Wei, W.; van Renterghem, T.; De Coensel, B.; Botteldooren, D. Dynamic noise mapping: A map-based interpolation between noise measurements with high temporal resolution. *Appl. Acoust.* **2016**, *101*, 127–140. [CrossRef]
19. Brambilla, G.; Confalonieri, C.; Benocci, R. Application of the intermittency ratio metric for the classification of urban sites based on road traffic noise events. *Sensors* **2019**, *19*, 5136. [CrossRef]
20. Ventura, R.; Mallet, V.; Issarny, V. Assimilation of mobile phone measurements for noise mapping of a neighbourhood. *J. Acoust. Soc. Am.* **2018**, *144*, 1279. [CrossRef]
21. Quintero, G.; Aumond, P.; Can, A.; Balastegu, A.; Romeu, J. Statistical requirements for noise mapping based on mobile measurements using bikes. *Appl. Acoust.* **2019**, *156*, 271–278. [CrossRef]
22. Cana, A.; Dekoninckb, L.; Botteldooren, D. Measurement network for urban noise assessment: Comparisonof mobile measurements and spatial interpolation approaches. *Appl. Acoust.* **2014**, *83*, 32–39. [CrossRef]
23. King, E.A.; Rice, H.J. The development of a practical framework for strategic noise mapping. *Appl. Acoust.* **2009**, *70*, 1116–1127. [CrossRef]
24. Aumond, P.; Can, A.; Mallet, V.; De Coensel, B.; Ribeiro, C.; Botteldooren, D.; Lavandier, C. Kriging-based spatial interpolation from measurements for sound level mapping in urban areas. *J. Acoust. Soc. Am.* **2018**, *143*, 2847–2857. [CrossRef]
25. Licitra, G. *Noise Mapping in the EU: Models and Procedures*; CRC Press: Boca Raton, FL, USA, 2012.
26. Kang, J.; Aletta, F.; Gjestland, T.T.; Brown, L.A.; Botteldooren, D.; Schulte-Fortkamp, B.; Lercher, P.; van Kamp, I.; Genuit, K.; Fiebig, A.; et al. Ten questions on the soundscapes of the built environment. *Build Environ.* **2016**, *108*, 284–294. [CrossRef]
27. Aumond, P.; Jacquesson, L.; Can, A. Probabilistic modelling framework for multisource sound mapping. *Appl. Acoust.* **2018**, *139*, 34–43. [CrossRef]
28. DYNAMAP. 2014. Available online: http://www.life-dynamap.eu/ (accessed on 13 December 2019).
29. Benocci, R.; Bellucci, P.; Peruzzi, L.; Bisceglie, A.; Angelini, F.; Confalonieri, C.; Zambon, G. Dynamic noise mapping in the suburban area of Rome (Italy). *Environments* **2019**, *6*, 79. [CrossRef]

30. Romeu, J.; Jimenez, S.; Genescà, M.; Pamies, T.; Capdevila, R. Spatial sampling for night levels estimation in urban environments. *J. Acoust. Soc. Am.* **2006**, *120*, 791–800. [CrossRef]
31. Socoró, J.C.; Alías, F.; Alsina-Pagés, R.M. An anomalous noise events detector for dynamic road traffic noise mapping in real-life urban and suburban environments. *Sensors* **2017**, *17*, 2323. [CrossRef]
32. Alsina-Pagès, R.M.; Alías, F.; Socoró, J.C.; Orga, F.; Benocci, R.; Zambon, G. Anomalous events removal for automated traffic noise maps generation. *Appl. Acoust.* **2019**, *151*, 183–192. [CrossRef]
33. Orga, F.C.; Socoró, J.; Alías, F.; Alsina-Pagés, R.M.; Zambon, G.; Benocci, R.; Bisceglie, A. Anomalous noise events considerations for the computation of road traffic noise levels: The DYNAMAP's Milan case study. In Proceedings of the 24th International Congress on Sound and Vibration, London, UK, 23–27 July 2017; pp. 23–27.
34. Ausejo, M.; Simón, L.; García, R.; Esteban, M.; Arias, R.; Puga, J.; Rivera, J. Dynamic Noise Map based on permanent monitoring network and street categorization. In Proceedings of the INTERNOISE 2019, Madrid, Spain, 16–19 June 2019.
35. Zambon, G.; Angelini, F.; Salvi, D.; Zanaboni, W.; Smiraglia, M. Traffic noise monitoring in the City of Milan: Construction of a representative statistical collection of acoustic trends. In Proceedings of the 22nd International Congress on Sound and Vibration, ICSV 2015, Florence, Italy; 2015.
36. R Core Team. *R: A Language and Environment for Statistical Computing*; R Foundation for Statistical Computing: Vienna, Austria, 2015; Available online: https://www.r-project.org/ (accessed on 7 January 2020).
37. Brock, G.; Pihur, V.; Datta, S. clValid: An R Package for Cluster Validation. *J. Stat. Softw.* **2008**, *25*, 1–22. [CrossRef]
38. Package 'clValid' version 0.6-4. 2013. Available online: https://cran.r-project.org/web/packages/clValid/clValid.pdf (accessed on 7 January 2020).
39. Ward, J.H. Hierarchical Grouping to Optimize an Objective Function. *J. Am. Stat. Assoc.* **1963**, *58*, 236–244. [CrossRef]
40. Zambon, G.; Benocci, R.; Brambilla, G. Statistical Road Classification Applied to Stratified Spatial Sampling of Road Traffic Noise in Urban Areas. *Int. J. Environ. Res.* **2016**, *10*, 411–420.
41. Smiraglia, M.; Benocci, R.; Zambon, G.; Roman, H.E. Predicting hourly traffic noise from traffic ow rate model: Underlying concepts for the DYNAMAP project. *Noise Mapp.* **2016**, *3*, 130–139.
42. Barrigón Morillas, J.M.; Gomez, V.; Mendez, J.; Vilchez, R.; Trujillo, J. A categorization method applied to the study of urban road traffic noise. *J. Acoust. Soc. Am.* **2005**, *117*, 2844–2852. [CrossRef] [PubMed]
43. Zambon, G.; Roman, H.E.; Smiraglia, M.; Benocci, R. Monitoring and Prediction of Traffic Noise in Large Urban Areas. *Appl. Sci.* **2018**, *8*, 251. [CrossRef]
44. Zambon, G.; Benocci, R.; Bisceglie, A.; Roman, H.E. Milan dynamic noise mapping from few monitoring stations: Statistical analysis on road network. In Proceedings of the INTERNOISE 2016, Hamburg, Germany, 21–24 August 2016.
45. Benocci, R.; Molteni, A.; Cambiaghi, M.; Angelini, F.; Roman, H.E.; Zambon, G. Reliability of DYNAMAP traffic noise prediction. *Appl. Acoust.* **2019**, *156*, 142–150. [CrossRef]
46. AMAT. Available online: https://www.amat-mi.it/it/ (accessed on 13 December 2019).
47. Good Practice Guide for Strategic Noise Mapping and the Production of Associated Data on Noise Exposure. 13 January 2006. Available online: http://sicaweb.cedex.es/docs/documentacion/Good-Practice-Guidefor-Strategic-Noise-Mapping.pdf (accessed on 13 December 2019).

Article

Application of the Intermittency Ratio Metric for the Classification of Urban Sites Based on Road Traffic Noise Events

Giovanni Brambilla [1], Chiara Confalonieri [2] and Roberto Benocci [2,*]

[1] CNR-INM Dept. Acoustics and Sensors "O.M. Corbino", via del Fosso del Cavaliere 100, 00133 Rome, Italy; giovanni.brambilla@artov.inm.cnr.it

[2] Department of Earth and Environmental Sciences (DISAT), University of Milano-Bicocca, Piazza della Scienza 1, 20126 Milano, Italy; c.confalonieri12@campus.unimib.it

* Correspondence: roberto.benocci@unimib.it

Received: 21 October 2019; Accepted: 21 November 2019; Published: 23 November 2019

Abstract: Human hearing adapts to steady signals, but remains very sensitive to fluctuations as well as to prominent, salient noise events. The higher these fluctuations are, the more annoying a sound is possibly perceived. To quantify these fluctuations, descriptors have been proposed in the literature and, among these, the intermittency ratio (*IR*) has been formulated to quantify the eventfulness of an exposure from transportation noise. This paper deals with the application of *IR* to urban road traffic noise data, collected in terms of 1 s A-weighted sound pressure level (SPL), without being attended, monitored continuously for 24 h in 90 sites in the city of Milan. *IR* was computed on each hourly data of the 251 time series available (lasting 24 h each), including different types of roads, from motorways to local roads with low traffic flow. The obtained hourly *IR* values have been processed by clustering methods to extract the most significant temporal pattern features of *IR* in order to figure out a criterion to classify the urban sites taking into account road traffic noise events, which potentially increase annoyance. Two clusters have been obtained and a "non-acoustic" parameter x, determined by combination of the traffic flow rate in three hourly intervals, has allowed to associate each site with the cluster membership. The described methodology could be fruitfully applied on road traffic noise data in other cities. Moreover, to have a more detailed characterization of noise exposure, *IR*, describing SPL short-term temporal variations, has proved to be a useful supplementary metric accompanying L_{Aeq}, which is limited to measure the energy content of the noise exposure.

Keywords: road traffic noise; noise events; intermittency ratio; urban sites classification

1. Introduction

Noise pollution has been estimated as the second major environmental health risk after air pollution in Europe [1]. The noise health effects may emerge directly via autonomous stress reactions to the physical exposure or indirectly via negative affective states, for example the evoked annoyance. Noise annoyance may interfere with daily activities, rest or sleep, and can be accompanied by negative emotional and behavioral responses such as anger, displeasure, exhaustion and by stress-related symptoms [2–4].

There is clear evidence in the literature that annoyance and sleep effects depend not only on sound energy, described by metrics like L_{Aeq}, but also by the characteristics of noise events, which can be quantified by different metrics proposed in the literature, as those reviewed in [5]. It is well known that human hearing is able to adapt to steady noise easier than to the sound pressure level (SPL) fluctuations, as well as to prominent, salient noise events [6,7]. The higher these fluctuations are, the more annoying a sound is possibly perceived. Road traffic noise is typically characterized by the noise events due to the single vehicle pass-by, where the temporal structure of SPL varies between local one-lane city roads, showing highly intermittent noise, up to wide multi-lane motorways, producing a nearly continuous noise with very limited SPL fluctuations. To quantify these SPL fluctuations, common approaches either apply thresholds to detect events exceeding such thresholds and count number and duration of these events, or use SPL statistics, like percentile levels L_{A1}, L_{A5} and L_{A10}, namely the A-weighted SPL exceeded for 1%, 5% and 10% of the measurement time, respectively.

Recently, a new descriptor has been proposed [8], describing the eventfulness (or intermittency) of transportation noise exposure, taking into account both number and magnitude of noise events during a certain time period. The metric, named intermittency ratio (*IR*) and introduced within the framework of the SiRENE project, can be derived either directly from acoustic measurements or calculated from traffic and geometric data for any transportation noise source and any time period. A recent survey, performed on a stratified random sample of 5592 residents exposed to transportation noise all over Switzerland, has shown that for road traffic noise *IR* has an additional effect on the percentage of highly annoyed people and can explain shifts of the exposure-response curve of up to about 6 dB between low *IR* and high *IR* exposure situations, possibly due to the effect of different durations of noise-free intervals between events [9]. Moreover, a parameter study, based on calculations, has showed the dependency of *IR* on source–receiver distance, traffic volume, the percentage of heavy vehicles and travelling speed [10].

The metric *IR* has been determined on the 1 s A-weighted SPL from road traffic, without being attended, monitored continuously for 24 h in 90 sites in the city of Milan. It was computed on each hourly data of the 251 time series available (lasting 24 h each), including different types of roads, from motorways to local roads with low traffic flow. The obtained hourly *IR* values have been processed by clustering methods to extract the most significant temporal pattern features of *IR*, in order to figure out a criterion to classify the urban sites considering road traffic noise events, which potentially increase annoyance. Two clusters have been determined and a "non-acoustic" parameter x, calculated by combination of the traffic flow rate in three hourly intervals, has allowed us to associate each site with the cluster membership. Furthermore, binomial logistic regression has been applied to develop a model to predict the cluster membership on the basis of the *IR* time patterns. The performance of the model, determined comparing the predicted classification of the test data subset with that obtained by the cluster analysis, was satisfactory.

2. Materials and Methods

2.1. Acoustic Data Set

In the framework of the LIFE DYNAMAP project, a large road traffic noise monitoring survey was carried out in the entire area of Milan to collect a database containing noise data related to the city road network [11]. From this database a set of 90 sites have been considered to represent the different types of roads according to the Italian functional road classification, that is motorway (class "A"), thoroughfare roads (class "D"), urban district roads (class "E") and urban local roads ("F"). The distribution of the sites among these classes is given in Table 1, together with the number of 24-hour time series of 1 s A-weighted sound pressure level (SPL) from road traffic monitored continuously by a class 1 sound level meter with the microphone placed at 4 m above the road. In some sites the unattended monitoring has been performed on more consecutive days. The monitoring has been performed on weekdays only (Monday to Friday), without rain and with wind speed less than 5 m/s. Noise events not associated with road traffic have been visually detected and manually masked before further data processing. The microphone was placed close to the road to reduce the influence of noise from other sources. Thus, the noise data are not representative of the real exposure of residents, living at greater distances from the road, especially where the road has a multi-lane geometrical configuration and even further away if the lanes are separated by a median strip area. In particular, the inhabitants' exposure is most likely overestimated because the values of *IR* and L_{Aeq} are greater than those at the road facing building façades (i.e., for *IR* dependency on source–receiver distance see Figure 4 in [10]). Details on the numbers of lanes for each direction and the average distance of the microphone from the roadside are given in Table 1.

Table 1. Distribution of the 90 sites included in the road traffic noise monitoring.

Road Type	N. of Sites	N. of 24-h Time Series	N. of Lanes for Each Direction	Average Distance Microphone-Roadside [m]
A	2	15	3 (50.0%) 4 (50.0%)	18.5 12.5
D	7	18	1 (28.6%) 2 (14.3%) 3 (42.8%) 4 (14.3%)	5.5 7.0 7.0 4.5
E	29	83	1 (32.1%) 2 (35.7%) 3 (28.6%) 4 (3.6%)	6.8 5.8 4.4 8.5
F	52	135	1 one way (30.0%) 1 (56.0%) 2 one way (2.0%) 2 (12.0%)	9.0 10.5 6.5 13.0
Total	90	251		

2.2. Intermittency Ratio IR Formulation

A noise event can be characterized by its maximum level, its sound exposure level (SEL), its "emergence" from background noise, its duration, or by the slope of rise of the level. For the characterization of the "eventfulness" of a noise exposure, the event continuous equivalent level $L_{eq,T,Events}$ is introduced in the *IR* formulation, which accounts for all sound energy contributions that exceed a given threshold, that is clearly stand out from background noise. This parameter is referred to the overall continuous equivalent level $L_{eq,T,tot}$ for the measurement time T to give the following formulation of *IR* [8]:

$$IR = \frac{10^{0.1L_{eq,T,Events}}}{10^{0.1L_{eq,T,tot}}} \cdot 100\ [\%]. \tag{1}$$

A single pass-by only contributes to $L_{eq,T,Events}$ if its SPL exceeds a given threshold K determined by:

$$K = L_{eq,T,tot} + C\ [\text{dB}], \tag{2}$$

where C might be between 0 and 10 dB. For low values of C, almost any situation produces a high *IR*, whereas high values of C almost always produce low *IR*. The balance between these extreme cases was investigated by numerical simulations of various traffic situations and resulted in C = 3 dB [8]. This value has not been set based on any verified psychoacoustic principle, but was derived empirically. As pointed out in [8], "The question of how much an event really has to stand out from background noise in order to be termed "event" by normal listeners depends on various other parameters", like the attentional, cognitive and emotional situation of the listener [6]. By definition, *IR* only takes values between 0% and 100%. An *IR* > 50% means that more than half of the sound exposure is caused by "distinct" pass-by events. In situations with only events that clearly emerge from background noise (e.g., a receiver very close by a road), *IR* yields values close to 100%. For example, Figure 1 shows the A-weighted SPL time history (measurement time T = 1 h) for an urban local road together with the corresponding $L_{Aeq,T,tot}$ and the threshold K used to detect the events (all the SPLs above K), which determine $L_{Aeq,T,Events}$.

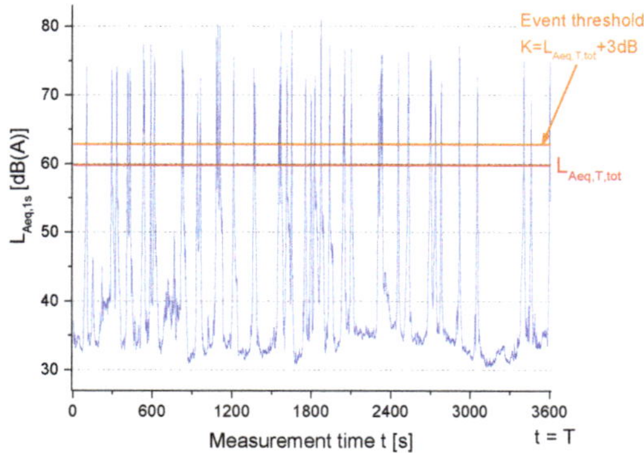

Figure 1. A-weighted sound pressure level versus time t (T = 1 h) for an urban local road. The sound pressure levels (SPLs) above the threshold K are events contributing to determine $L_{eq,T,Events}$. For the plotted hourly SPL time history intermittency ratio (IR) = 93.3%, $L_{Aeq,T,tot}$ = 59.8 dB(A), number of events above K threshold = 45 and $L_{Aeq,T,Events}$ = 59.5 dB(A).

2.3. Data Processing and Analysis

A script running in the "R" environment, version 3.5.1 [12], has been written to import each of the 24-h time series as input in terms of text file (four columns with date, time, SPL in dB(A) at 1 s intervals and a code to indicate the corresponding source, which is road traffic noise or something else). The reference measurement time T was chosen equal to 1 h, as this time frame is established by the Italian legislation for road traffic noise measurement. Besides this requirement, the chosen measurement time T of 1 h was considered a reasonable compromise between longer time (i.e., 24 h, day and night periods, etc.) and shorter ones (i.e., 30 min or even shorter). For each T of 1 h, the output data were exported to an Excel file, including:

- The overall L_{Aeq} hourly value (dB(A));
- The hourly value of intermittency ratio IR (%) and the corresponding number of events;
- For each detected noise event the corresponding start time, the duration (s) and the sound exposure level (SEL; dB(A)).

In addition for each site the hourly traffic flow was provided by the Municipal Agency of Mobility, Environment and Land of Milan (AMAT). The data were calculated by a model of traffic applied to the city road network.

The statistical analysis of the collected data was carried out by the software "R" [12]. For the sites where the noise monitoring lasted more days the median value of IR for each hour was determined, as this parameter is less influenced by the presence of outliers. Thus a matrix of 90 (sites) × 24 (hours) = 2160 values of hourly IR was used as input of the subsequent cluster analysis performed to find out the similarities in the IR time patterns.

To fulfill such an objective, hierarchical clustering, an unsupervised machine learning method for data classification, was applied. This method does not require to pre-specify the number of clusters to be generated and the output is a tree-based representation of the observations (dendrogram) showing the sequence of cluster formation and the distance at which each fusion takes place. Previously, for each hour the IR values have been scaled (mean = 0 and standard deviation = 1). The Euclidean distance has been considered to represent the similarity between pairs of observations. Complete-linkage clustering was considered: at the beginning of the process, each element is in a cluster of its own and, afterwards, the clusters are sequentially combined into larger clusters until all elements end up being in the same cluster. Different clustering methods available in the "clValid" R package, version 0.6-6 [13], were applied. In particular, six methods were considered, that is hierarchical, partitioning around medoids

(PAM), k-means, divisive analysis clustering (DIANA), model-base clustering and self-organizing tree algorithm (SOTA). For the sake of simplicity, minimal discrimination was considered, that is two clusters for both the sites and the hourly time intervals. The clustering performance of the methods was ranked according to seven parameters, namely connectivity, silhouette width and Dunn index (combining measures of compactness and separation of the clusters), the average proportion of non-overlap (APN), the average distance (AD), the average distance between means (ADM) and the figure of merit (FOM). The method selected as "optimal" on the basis of the above parameters was applied to obtain two clusters of *IR* patterns for both the sites and the hourly time intervals.

Afterwards, a model was developed to predict the cluster membership on the basis of the *IR* time patterns. For this purpose the "caret" R package, acronym for "Classification And REgression Training" [14], was used. The dataset needed to be randomly divided into two subsets, one for training the model and the other to test it and evaluate its classification performance. The binomial logistic regression was applied to develop the model because the dependent variable (cluster membership) was categorical with two categories. The classification performance of the model was determined comparing the predicted classification of the test data subset with that obtained by the cluster analysis.

3. Results

Figure 2 shows an example of the obtained 24-h pattern of hourly values of L_{Aeq} and corresponding *IR* for two different types of roads, namely a motorway (class "A") and a local street (class "F"). The plot reports the median of the hourly values ± the median absolute deviation (MAD) because the monitoring included more than one day, namely 12 days for road "A" and 9 days for road "F". It can be seen that road "A" was always much noisier than road "F" (hourly L_{Aeq} average differences across the hours of about 6 dB) and shows always lower *IR* values than road "F" (hourly *IR* average difference across the hours of about −50%, and less pronounced (−30%) during the night). The lower *IR* values observed for road "A" were due to the high traffic flow rate and speed on the motorway, resulting in a high background SPL above which the noise events did not stand out too much. This feature was clearly present in the day period from 6 to 18 h, whereas for the period from 2 to 4 h the highest values of *IR* were observed, when the reduced traffic flow allowed the increase of speed and more prominent noise events occurred above the lower background level. Road "F" shows the same behavior in the night, whereas the lowest *IR* values occurred at the traffic peak hours (8 and 18 h), when the traffic flow was highest and the increased background SPL reduced the prominence of noise events. Thus, given the very different temporal patterns of urban road traffic noise, from relative continuity to high intermittency, it would be worth to consider the *IR* metric as a supplementary quantity to L_{Aeq}.

Regarding clustering, the DIANA method was selected as the "optimal" clustering algorithm to divide the data set into two clusters of *IR* patterns for both the sites and the hourly time intervals. The dendrogram of the scaled *IR* hourly values obtained for the sites (matrix with rows = sites and columns = hours) is given in Figure 3. Table 2 reports the distribution of the sites across the road type and clusters. Cluster 2, on the right hand side in Figure 3, includes the majority of all the road types, whereas Cluster 1, on the left hand side in Figure 3, includes the remaining roads and all those in class "A".

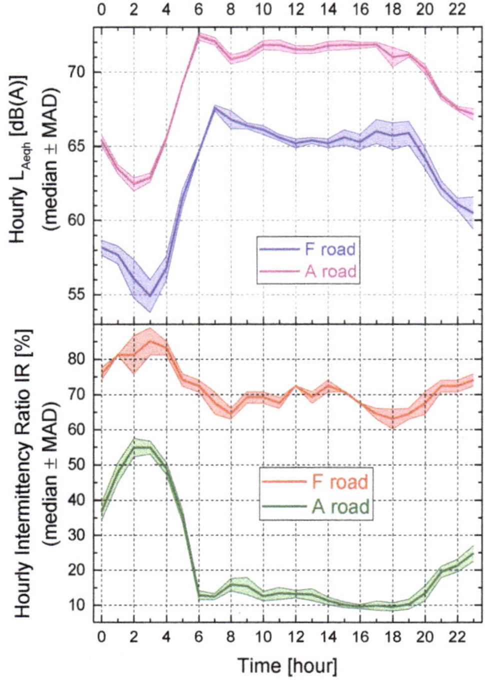

Figure 2. Example of the obtained 24-h pattern of hourly values of L_{Aeq} and corresponding *IR* for two different types of roads, namely a motorway (class "A") and a local street (class "F").

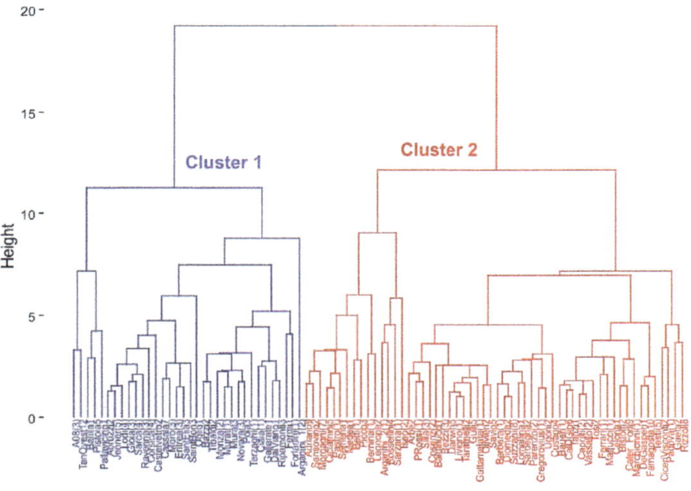

Figure 3. Dendrogram of the scaled *IR* hourly values obtained for the 24-h road traffic noise data monitored in the 90 sites in Milan.

Table 2. Distribution of the 90 sites across the two clusters and type of road for the classification based on *IR* hourly time patterns.

Cluster		1	2
N. of sites (%)		34 (37.8)	56 (62.2)
Road Type	A	2 (100)	0 (0)
	D	4 (57.1)	3 (42.9)
	E	16 (55.2)	13 (44.8)
	F	12 (23.1)	40 (76.9)

The multidimensional scaling (MDS) applied to the data provided the bi-dimensional plot given in Figure 4, where the two clusters appeared satisfactorily separated and the variance explained by the two dimensions was 88.3%.

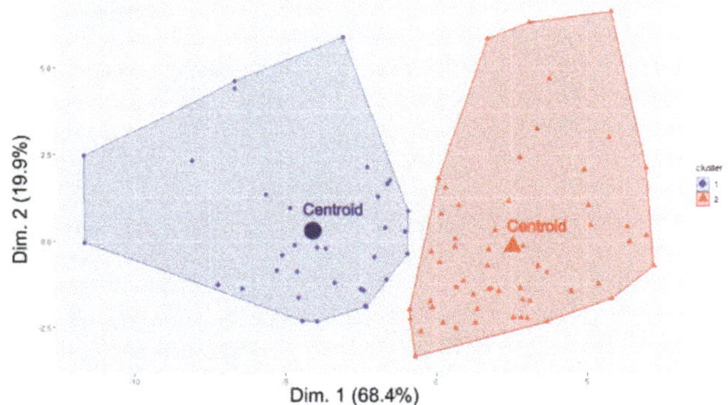

Figure 4. Bi-dimensional plot of the two clusters obtained by multidimensional scaling (MDS). Dimension 1 and 2 explain 68.4% and 19.9% of the variance, respectively.

The dendrogram in Figure 5 shows the clustering in terms of hourly intervals, obtained after the transposition of the matrix containing the 2160 values of hourly *IR* (rows = hours and columns = sites). The night period (from 22 to 7 h) was clearly separated from the day-time. Regarding the *IR* time pattern for each cluster, Figure 6 reports the hourly median *IR* values ± the median absolute deviation (MAD) and the three hourly intervals showing the biggest differences between the two clusters (green rectangles). In the night period the *IR* values were the highest for both clusters because of the presence of noise events clearly emerging above the background noise. In this time period there was an overlapping between *IR* values corresponding to the two clusters. Similar median *IR* time patterns were also observed from 7 to 24 h, with Cluster 1 having lower *IR* values. As expected the night period was the most critical due to prominent noise events, which could produce an increasing of annoyance, considering also the affected activities (mainly sleep).

The above results of clustering were also plotted in terms of a heatmap, reported in Figure 7, a rectangular tiling of the data matrix with cluster trees appended to its margins, where the rows and columns of the matrix are ordered to highlight patterns [15]. The color key legend on the top left in the figure shows also the distribution of the 2160 hourly *IR* values.

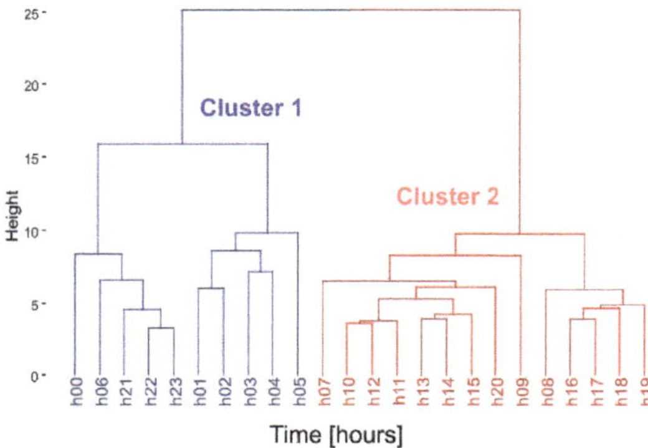

Figure 5. Dendrogram of the scaled *IR* hourly values as a function of the hours.

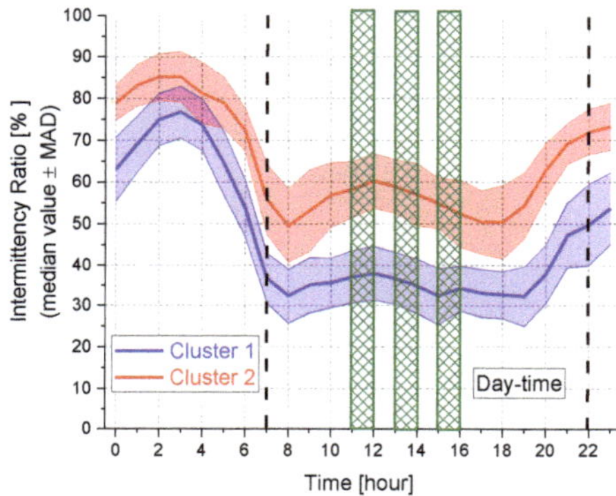

Figure 6. *IR* time pattern for each cluster. Green rectangles correspond to the hourly intervals showing the biggest differences between the two *IR* time patterns.

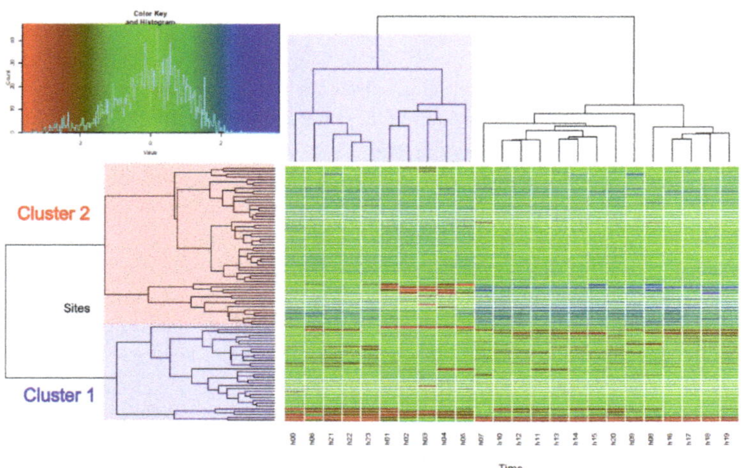

Figure 7. Cluster heatmap of the scaled hourly values of *IR*. On the *y* axis the sites divided into Cluster 1 (blue rectangle) and 2 (red rectangle). On the *x* axis the clustering across the hourly intervals with the night period, from 22 to 7 h, in the blue rectangle.

The obtained *IR* time pattern for each cluster cannot be applied in a straightforward way without any linking to a specific feature of either the road or the corresponding traffic flow. As shown in Table 2, the road type was useless because each cluster included different road types. Thus, to find a "non-acoustic" parameter suitable to predict the cluster membership, the Mann-Whitney U test was performed on the hourly *IR* values to detect the hourly intervals where their differences between the two clusters were biggest. The rank descending order of these differences showed that they corresponded to the hourly intervals 15–16 h, 13–14 h and 11–12 h (see Figure 7). Thus, the traffic flows F in these three hours were combined according to the following relationship, similar to that previously proposed in [16]:

$$x = \sqrt{[lg(F_{15-16})]^2 + [lg(F_{13-14})]^2 + [lg(F_{11-12})]^2}. \quad (3)$$

Having a separation of the sites into two clusters, binomial logistic regression was applied to develop a model to predict this classification. This is a statistical model that in its basic form uses a logistic function (known as "S" shape or sigmoid curve) to model a binary dependent variable, having only two possible values. In such a model, the cluster membership was considered as a dependent variable, in particular Cluster 1 was labeled "0" and Cluster 2 was labeled "1", and the "non-acoustic" parameter x was taken as an independent variable (predictor). The split ratio = 0.7 was used for randomly sub-setting the data set for training the classification model (63 sites) and, afterwards, to test it (27 sites). At the end of the training process, the model equations in terms of probability P of an observation to belong to Cluster 2 (Y = 1) was obtained as follows:

$$P(Y=1) = \frac{1}{1+e^{(-6.84+1.26x)}} \quad (4)$$

The classification model was applied to the test dataset in order to evaluate its classification performance and the obtained confusion matrix, a table counting how often each combination of known categories (the clusters) occurred in combination with each prediction type, is reported in Figure 8. The results were satisfactory, being the model accuracy (fraction of correct predictions) equaled to 0.83, the precision (the ratio of true positives to predicted positives) and recall (the ratio of true positives over all positives) equaled to 0.88 and the Cohen's kappa |ê = 0.60 (moderate agreement). Table 3 reports additional performance parameters. Figure 9 shows the comparison between the cluster membership (blue dots = Cluster 1 and red dots = Cluster 2) obtained by the DIANA clustering and the probabilities predicted by the logistic regression (blue curve obtained by Equation (4)). The proportion of correctly classified observations by the model was equal to 0.74.

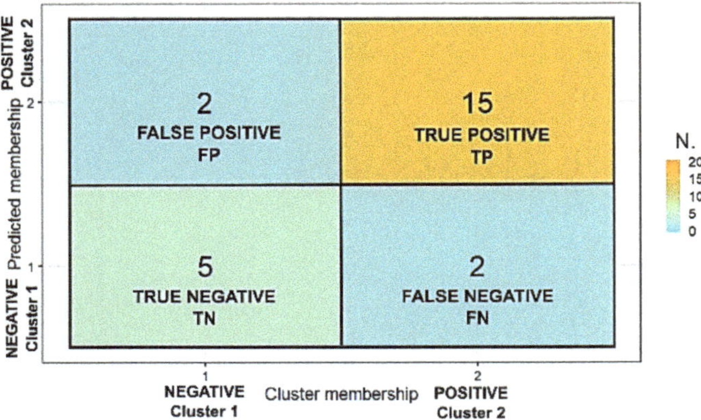

Figure 8. Confusion matrix of the classification model applied to the test dataset.

Table 3. Classification performance of the logistic model.

Parameter	Value
Sensitivity	0.71
Specificity	0.88
Detection rate	0.21
Balanced accuracy	0.80

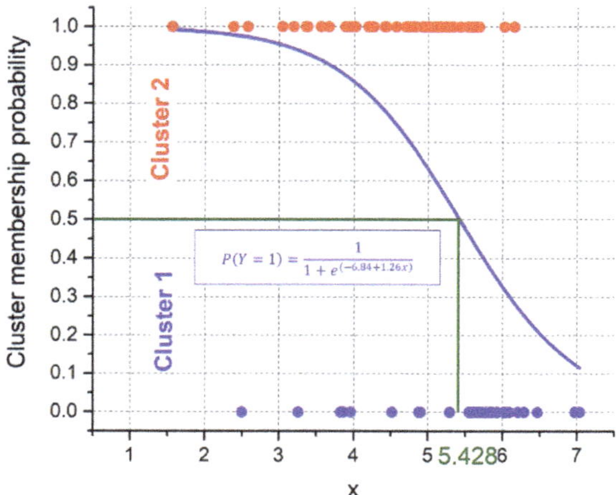

Figure 9. Cluster membership (blue dots = Cluster 1 and red dots = Cluster 2) obtained by the divisive analysis (DIANA) clustering compared with the probabilities predicted by the logistic regression (blue curve obtained by Equation (4)). Probabilities $p \leq 0.5$ and $p > 0.5$ correspond to Cluster 1 and 2, respectively. The threshold for the "non-acoustic" parameter x to discriminate between the cluster membership is reported in green.

Regarding the effective application of the above two clusters, it is essential to determine a threshold for the "non-acoustic" parameter x able to discriminate between the cluster membership. Such a threshold (x = 5.24) was empirically determined as shown in the box plot of the x values reported according to the cluster membership of sites (Figure 10). This value was comparable with that obtained from the intersection of the logistic model curve with the cluster membership probability value of 0.5, shown in Figure 9 (x = 5.428).

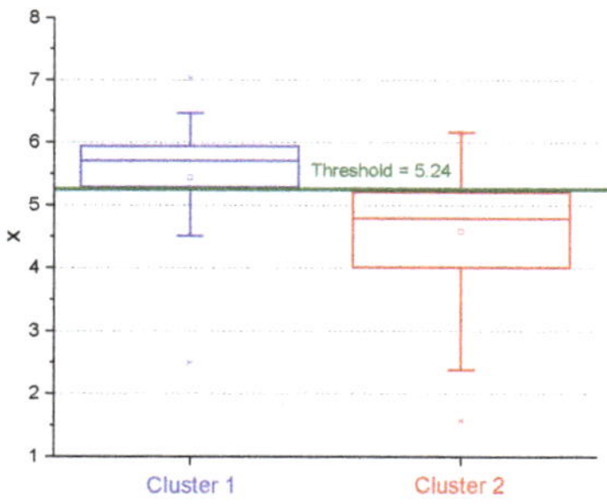

Figure 10. Empirical threshold value of the "non-acoustic" parameter x obtained for the discrimination between the two clusters (x = 5.24).

4. Discussion

It has to be pointed out that the *IR* values calculated from the noise data provided by the noise monitoring network in Milan have some drawbacks due to some factors, like the different distance microphone-longitudinal axis of the road, the microphone proximity to the road and not where the residents live and so forth. In addition, the results of the clustering and classification model were strongly dependent on the local situation and could not be generalized to other contexts. Besides these limitations, the methodology applied could be fruitful applied in other cities and some general considerations could be drawn. For instance, the hourly *IR* and L_{Aeq} time patterns, shown by the example in Figure 2, highlight the complementarity of these two metrics, the former describing SPL short-term temporal variation, the latter measuring the energy content of the noise exposure. In particular, for the available experimental dataset, Figure 11 reports the logistic fitting of the hourly values of these two descriptors for the centroids of cluster 1 and 2.

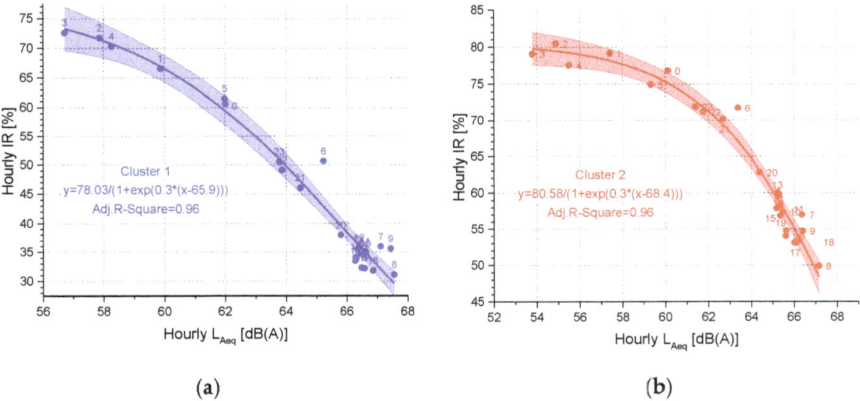

Figure 11. Logistic fitting of the hourly values of *IR* and L_{Aeq} for the centroids of cluster 1 (**a**) and 2 (**b**). The area around the regression line represents the confidence bands at a 95% confidence level. The symbol labels represent the hourly intervals.

Due to its definition, the *IR* value ranges between the following two opposite sonic environments:

1. Sound events with low energy, not so much "emerging" from high background SPL, corresponding to a low value of *IR*;
2. Sound events with high energy, clearly "predominating" above low background SPL, corresponding to a high value of *IR*.

The sonic environment (1) occurs usually at roads with high traffic road rate, such as motorways and thoroughfare roads (road classes "A" and "D") especially during the day-time, whereas the sonic environment (2) is usually observed at roads either with low traffic road rate, such as local roads (road class "F") during the day-time or during the night for all the roads with the exception of motorways.

However, there might be particular cases, indeed very frequent in the urban context, where the local road is very close to a busy street whose noise is clearly influencing the sonic environment in the local road itself. In these circumstances, the low energy noise events, produced by small number of vehicle pass-by at low speed, do not emerge so much above the high background SPL produced by the nearby busy road. In the data set herewith considered there were a few sites with this feature, like the two ones shown in Figure 12. The *IR* time pattern in these sites is similar to those observed for thoroughfare roads. This is, most likely, the reason why a marginal percentage (23.1%) of local roads (class "F") have not been grouped in the cluster containing busy roads. Thus, in the selection of sites to be monitored it is important to avoid, as much as possible, this situation, which, nevertheless, is often present in urban road network.

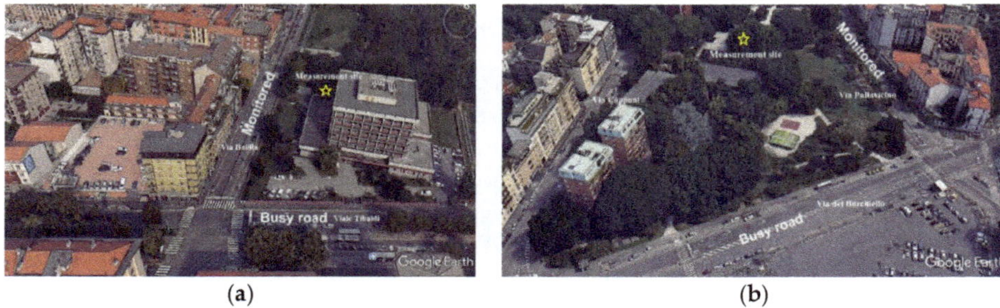

(a) (b)

Figure 12. Examples of two local roads (**a** and **b**) monitored nearby a busy street (adapted from Google Earth images).

The above remarks should not be considered a weakness of the *IR* metric, but rather a reliable representation of the time pattern of the sonic environment and of the potential annoyance it might evoke. In addition, a comparison has been performed between the classification based on *IR* hourly time patterns and that provided by hourly L_{Aeq} time patterns, the latter obtained according to the procedure detailed in [17,18]. The two classifications, as shown in Figure 13, are somewhat different, as they overlap for 64% only.

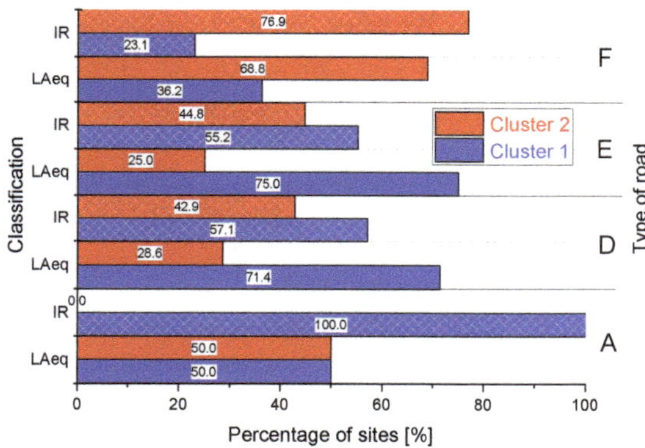

Figure 13. Comparison between the classifications based on *IR* and L_{Aeq} hourly time patterns for the type of roads.

Despite the observed mismatch between the above two classifications, the difference between the hourly L_{Aeq} patterns corresponding to the clusters obtained by the two classifications was not statistically significant at 95% confidence level for any hourly interval, even in the night period, as shown in Figure 14 where the hourly L_{Aeq} median values ± the median absolute deviation (MAD) are reported. However, it has to be pointed out that the two classifications have different aims: the one based on L_{Aeq} pattern is mainly focused on noise mapping, according to the standards issued by the European Directive 2002/49/EC [19], whereas that based on *IR* pattern could be aimed at discriminating the sites according to the potential annoyance their sonic environment might evoke. Thus, these two approaches are not alternative with one another but shall be considered complementary. Furthermore, both the classifications are rather different from the categorization based on the type of road, as established by the Italian legislation, which defines the noise limits as a function of the road category. Thus, this approach did not seem appropriate for an effective protection against road traffic noise pollution.

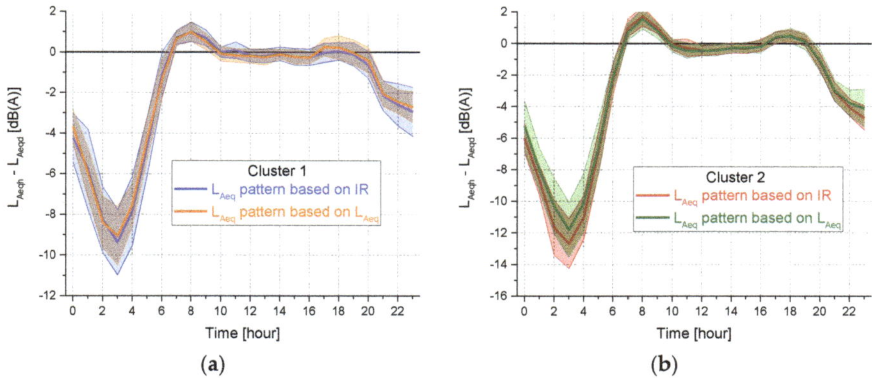

Figure 14. Comparison of the hourly L_{Aeq} patterns corresponding to the two clusters (**a**, cluster 1 and **b**, cluster 2) obtained by the two classifications based on *IR* and L_{Aeq}, respectively.

5. Conclusions

The intermittency ratio *IR* metric was applied to a database of road traffic noise, without being attended, monitored for 24 h in 90 sites in the city of Milan. The reference measurement time T was set at 1 h and the obtained *IR* values were processed by clustering methods. Two clusters were determined, providing hourly *IR* temporal patterns enabling us to classify the urban sites on the basis of the observed noise events, which, potentially, increase the annoyance. A "non-acoustic" parameter *x*, determined by combination of the traffic flow rate in three hourly intervals, was allowed to associate each site with the cluster membership. Furthermore, binomial logistic regression was applied to develop a model to predict the cluster membership on the basis of the *IR* time patterns. The performance of the model, determined comparing the predicted classification of the test data subset with that obtained by the cluster analysis, was satisfactory.

However, the *IR* values calculated from the noise data provided by the road traffic noise monitoring network in Milan, mainly used for a noise mapping update, had some drawbacks due to some factors, like different distances microphone-longitudinal axis of the road and microphone position close to the road and not where the residents live. The reference measurement time *T* chosen, equal to 1 h, had also affected the *IR* values. In addition, the results of clustering and classification model were strongly dependent on the local situation and could not be generalized to other contexts. However, the study showed that data collected for noise monitoring and mapping purposes could be processed to evaluate the occurrences of noise events produced by a vehicle pass-by. Besides the above limitations, the described methodology could be fruitfully applied on road traffic noise data in other cities and some general considerations could be drawn. In particular, *IR* could be a supplementary metric accompanying L_{Aeq}, as the former describes SPL short-term temporal variation and the latter measures the energy content of the noise exposure. Indeed, *IR* could explain deviations of highly annoyed people percentage from that estimated by the classical exposure–response curves that only rely on L_{Aeq} [4], like those in [20].

Furthermore, the two classifications based on *IR* and L_{Aeq} hourly time patterns are rather different from that based on the type of road, as established by the Italian legislation, which defines the noise limits as function of the road category. Thus, this approach does not seem appropriate for an effective protection against road traffic noise pollution.

Further steps of this research are already planned and they include the statistics of errors in the estimate of *IR* values derived by the application of the above time patterns, as well as the potential of *IR* to detect correctly the noise events produced by road traffic, identified by an automatic recognition algorithm already developed within the DYNAMAP project [21,22].

Author Contributions: Conceptualization, G.B.; methodology, G.B. and R.B.; software, G.B.; validation, C.C.; investigation, G.B. and R.B.; data curation, C.C.; writing—original draft preparation, G.B.; writing—review and editing, G.B., R.B. and C.C.

Funding: This research received no external funding.

Acknowledgments: The authors thank Giovanni Zambon to grant the use of the noise monitoring data.

Conflicts of Interest: The authors declare no conflict of interest.

References

1. Hanninen, O.; Knol, A.B.; Jantunen, M.; Lim, T.A.; Conrad, A.; Rappolder, M.; Carrer, P.; Fanetti, A.C.; Kim, R.; Buekers, J.; et al. Environmental burden of disease in Europe: Assessing nine risk factors in six countries. *Environ. Health Perspect.* **2014**, *122*, 439–446. [CrossRef] [PubMed]
2. European Environment Agency. *Good Practice Guide on Noise Exposure and Potential Health Effects*; EEA Technical Report: Copenhagen, Denmark, November 2010. [CrossRef]
3. Stansfeld, S.A.; Matheson, M.P. Noise pollution: Non-auditory effects on health. *Br. Med. Bull.* **2003**, *68*, 243–257. [CrossRef] [PubMed]
4. Fritschi, L.; Brown, A.L.; Kim, R.; Schwela, D.H.; Kephalopoulos, S. (Eds.) *Burden of Disease from Environmental Noise*; World Health Organization: Bonn, Germany, 2011.
5. International Institute of Noise Control Engineering. *I-INCE Supplemental Metrics for Day/Night Average Sound Level and Day/Evening/Night Average Sound Level*; Final Report of I-INCE Technical Study Group 9; 2015.
6. De Coensel, B.; Botteldooren, D.; De Muer, T.; Berglund, B.; Nilsson, M.E.; Lercher, P. A model for the perception of environmental sound based on notice-events. *J. Acoust. Soc. Am.* **2009**, *126*, 656–665. [CrossRef] [PubMed]
7. Bockstael, A.; De Coensel, B.; Lercher, P.; Botteldooren, D. Influence of temporal structure of the sonic environment on annoyance. In Proceedings of the 10th International Congress on Noise as a Public Health Problem (ICBEN), London, UK, 24–28 July 2011; pp. 945–952.
8. Wunderli, J.M.; Pieren, R.; Habermacher, M.; Vienneau, D.; Cajochen, C.; Probst-Hensch, N.; Röösli, M.; Brink, M. Intermittency ratio: A metric reflecting short-term temporal variations of transportation noise exposure. *J. Expo. Sci. Environ. Epidemiol.* **2015**, 1–11. [CrossRef] [PubMed]
9. Brink, M.; Schäffer, B.; Vienneau, D.; Forasterc, M.; Pieren, R.; Eze, I.C.; Cajochen, C.; Probst-Hensch, N.; Röösli, M.; Wunderli, J.-M. A survey on exposure-response relationships for road, rail, and aircraft noise annoyance: Differences between continuous and intermittent noise. *Environ. Int.* **2019**, *125*, 277–290. [CrossRef] [PubMed]
10. Wunderli, J.M.; Pieren, R.; Vienneau, D.; Cajochen, C.; Probst-Hensch, N.; Röösli, M.; Brink, M. Parameter study on IR, a metric reflecting short-term temporal variations of transportation noise exposure. In Proceedings of the 45th INTERNOISE, Hamburg, Germany, 21–24 August 2016; pp. 5710–5719.
11. Zambon, G.; Benocci, R.; Bisceglie, A.; Roman, H.E.; Bellucci, P. The life dynamap project: Towards a procedure for dynamic noise mapping in urban areas. *Appl. Acoust.* **2017**, *124*, 52–60. [CrossRef]
12. R Core Team. *R: A Language and Environment for Statistical Computing*; R Foundation for Statistical Computing: Vienna, Austria, 2018; Available online: https://www.R-project.org/ (accessed on 22 November 2019).
13. Brock, G.; Pihur, V.; Datta, S.; Datta, S. clValid: An R Package for Cluster Validation. *J. Stat. Softw.* **2008**, *25*, 1–22. [CrossRef]
14. Kuhn, M. Building predictive models in R using the caret package. *J. Stat. Softw.* **2008**, *28*, 1–26. [CrossRef]
15. Wilkinson, L.; Friendly, M. The History of the Cluster Heat Map. *Am. Stat.* **2009**, *63*, 179–184. [CrossRef]
16. Zambon, G.; Benocci, R.; Bisceglie, A.; Roman, H.E. Milan dynamic noise mapping from few monitoring stations: Statistical analysis on road network. In Proceedings of the 45th INTERNOISE, Hamburg, Germany, 21–24 August 2016; pp. 6350–6361.

17. Smiraglia, M.; Benocci, R.; Zambon, G.; Roman, H.E. Predicting hourly traffic noise from traffic flow rate model: Underlying concepts for the DYNAMAP project. *Noise Mapp.* **2016**, *3*, 130–139. [CrossRef]
18. Zambon, G.; Roman, H.E.; Smiraglia, M.; Benocci, R. Monitoring and Prediction of Traffic Noise in Large Urban Areas. *Appl. Sci.* **2018**, *8*, 251. [CrossRef]
19. Directive, E.U. Directive 2002/49/EC of the European Parliament and of the Council of 25 June 2002 relating to the assessment and management of environmental noise. *Off. J. Eur. Communities L* **2002**, *189*, 12–25.
20. European Commission. *Position Paper on Dose Response Relationships Between Transportation Noise and Annoyance*; Office for Official Publications of the European Communities: Luxembourg, 2002.
21. Alsina-Pagès, R.M.; Alías, F.; Socoró, J.C.; Orga, F.; Benocci, R.; Zambon, G. Anomalous events removal for automated traffic noise maps generation. *Appl. Acoust.* **2019**, *151*, 183–192. [CrossRef]
22. Orga, F.; Socoró, J.C.; Alías, F.; Alsina-Pagès, R.M.; Zambon, G.; Benocci, R.; Bisceglie, A. Anomalous noise events considerations for the computation of road traffic noise levels: The dynamap's Milan case study. In Proceedings of the 24th International Congress on Sound and Vibration, ICSV 2017, London, UK, 23–27 July 2017.

Article

EAgLE: Equivalent Acoustic Level Estimator Proposal

Claudio Guarnaccia

Department of Civil Engineering, University of Salerno, I-84084 Fisciano, Italy; cguarnaccia@unisa.it

Received: 9 December 2019; Accepted: 23 January 2020; Published: 27 January 2020

Abstract: Road infrastructures represent a key point in the development of smart cities. In any case, the environmental impact of road traffic should be carefully assessed. Acoustic noise is one of the most important issues to be monitored by means of sound level measurements. When a large measurement campaign is not possible, road traffic noise predictive models (RTNMs) can be used. Standard RTNMs present in literature usually require in input several information about the traffic, such as flows of vehicles, percentage of heavy vehicles, average speed, etc. Many times, the lack of information about this large set of inputs is a limitation to the application of predictive models on a large scale. In this paper, a new methodology, easy to be implemented in a sensor concept, based on video processing and object detection tools, is proposed: the Equivalent Acoustic Level Estimator (EAgLE). The input parameters of EAgLE are detected analyzing video images of the area under study. Once the number of vehicles, the typology (light or heavy vehicle), and the speeds are recorded, the sound power level of each vehicle is computed, according to the EU recommended standard model (CNOSSOS-EU), and the Sound Exposure Level (SEL) of each transit is estimated at the receiver. Finally, summing up the contributions of all the vehicles, the continuous equivalent level, L_{eq}, on a given time range can be assessed. A preliminary test of the EAgLE technique is proposed in this paper on two sample measurements performed in proximity of an Italian highway. The results will show excellent performances in terms of agreement with the measured L_{eq} and comparing with other RTNMs. These satisfying results, once confirmed by a larger validation test, will open the way to the development of a dedicated sensor, embedding the EAgLE model, with possible interesting applications in smart cities and road infrastructures monitoring. These sites, in fact, are often equipped (or can be equipped) with a network of monitoring video cameras for safety purposes or for fining/tolling, that, once the model is properly calibrated and validated, can be turned in a large scale network of noise estimators.

Keywords: noise control; sensor concept; road traffic noise model; dynamic model

---

1. Introduction

The problem of road traffic noise in urban and non-urban areas is becoming more and more important nowadays. The effect of noise on human health is well established [1]. The recent publication of the European Environment Agency (EEA) about "The European environment—state and outlook 2020. Knowledge for transition to a sustainable Europe" [2] lists environmental noise among the most dangerous phenomena, dedicating a full chapter to this issue. In this document, the delay in implementing the actions suggested by the Environmental Noise Directive (END) [3] are claimed, underlining how at least 20% of the EU's population is still exposed to noise levels unsafe for health. Due to society and human habits, such as to existing infrastructures, road traffic noise is the most important source of noise in the EU, with more than 100 million of people affected by long-term daily average noise levels greater than 55 dBA and with about 80 million of people exposed to night-time levels above 50 dBA [3].

In order to cope with this issue, many municipalities introduced fixed or temporary monitoring stations and implemented mitigation actions based on the results of the measurements. Expensive and not always accepted acoustic barriers are the most widespread solution to mitigate the noise produced by the main sources [4]. Pavement plays a key-role in noise emitted by road, as recently studied by many authors aiming to integrate noise reduction with green economy by recycling rubber from old tires into asphalts (rubber asphalts) [5–7]. Preventing is also mitigating, thus, innovative solutions, like real time monitoring, are actually studied using a wireless sensor network [8,9].

On the other hand, road infrastructures companies are obliged to perform environmental impact analysis, including noise monitoring and estimation. In addition, when critical situations are highlighted, action plans must be performed, according to the END [2] and to the national regulations of each country.

In any case, measurements are expensive and cannot be performed all over large areas, thus, road traffic noise predictive models (RTNMs) can be adopted to assess noise produced by vehicles. Extensive reviews of the standard statistical RTNMs can be found for instance in [10,11], also in comparison with field measurements [12]. In [13], a brief review of advanced techniques for road traffic noise assessment is reported, including cellular automata [14], Time Series Analysis [15], Poisson models [16], etc. Can et al. in [17] reported a review of the models to estimate the source power level of the single vehicle.

The usage of advanced computing techniques is somehow growing in literature, even though it must still be demonstrated that the adoption of computationally demanding procedures introduces a widespread benefit in the predictions. In "non-standard" conditions (such as traffic jams or congestions or intersections) usually the common RTNMs fail. Therefore, in large areas case studies, such as big municipalities and big road infrastructures, and for long term average (such as L_{den} evaluation), the need of a fast and effective model is more important than having an extreme precision (for instance lower than 1 dBA).

Neglecting the predictive models based on data analysis, such as Time Series Analysis models and Poisson models, it can be affirmed that in order to implement a RTNM it is compulsory to know at least the number of vehicles that pass in a certain time range (flow) and the classification of each vehicle (at least light and heavy categories), together with the geometrical detail of the source and receiver positions. Many other parameters can be included to take into account second order corrections, such as road pavement typology, gradient of the road, temperature, humidity, etc. All these data, compulsory and additional ones, are not always available, and thus the possibility to detect the inputs of any RTNM automatically is an important challenge. In this paper, the author presents the design of a new methodology, the Equivalent Acoustic Level Estimator (EAgLE), based on vehicles detection, counting and tracking, by means of a video processing tool, with a single vehicle noise emission and propagation model.

There are several studies in literature about image processing, video analysis, object detection and tracking. In [18], a detailed study on vehicle recognition based on deep neural network is presented. Huang in [19] presented traffic speed estimation from surveillance video recordings, highlighting the difficulties related to crowded lanes and perspective corrections. Similar research has been presented by Hua et al. in [20], focusing on the tracking and speed estimation from traffic videos. Biswas et al. in [21] presented a speed estimation performed on video recordings taken by unmanned aerial vehicles. Several other studies are reported in literature, focusing mainly on counting, detection and category assignment, tracking, and speed estimation of vehicles from video recordings. Basically, the real time traffic monitoring systems have been deeply innovated in the last years, leading to the development, and sometimes the installation, of very intelligent sensors on road infrastructures and urban areas. In any case, these sensors are somehow limited, since usually they just detect, count, and track vehicles, in order to help in traffic management, for instance in signalized intersections. Sometimes the video cameras are used to control the restricted areas and, in some cases, also for environmental issues (see, for instance, the Ultra Low Emission Zones (ULEZ) in London), but they usually do not produce an assessment of any environmental parameter, such as air and/or noise pollution.

The research presented in this paper aims to partially fill this gap, proposing a methodology to embed these sensor networks with a noise level estimator. The proposed approach starts from the recognition of a moving object on the road. Once the object is tracked, it can be counted and categorized according to its dimension. Its speed can be estimated as well, allowing to assess the sound power level. This assessment can be performed in many ways, according to several noise emission models (as presented by Can et al. in [17]). In this paper, the proposed approach is to adopt the CNOSSOS-EU emission model [22]. Once the noise emission, i.e., the source sound power level, of each vehicle is estimated, the overall continuous equivalent level over a given time range can be calculated, summing up all the contributions coming from the vehicles flowing in that time range. This technique, named after "EAgLE: Equivalent Acoustic Level Estimator", in honor of one of the animals that have the best visual capacities, can be implemented in a new sensor to be developed for road traffic noise assessment purposes.

The above brief description of the EAgLE methodology is detailed in Section 2, while in Section 3 a preliminary application is presented, showing the results obtained in a preliminary comparison with sound levels recorded on an Italian highway. It will be highlighted that this methodology can be implemented in an existing or under development sensor network. In fact, many road and railway infrastructures, such as many municipalities, have already implemented a video recording network, mainly for safety reasons, that can be easily integrated to become an environmental monitoring network. The integration between existing video recordings and the proposed methodology is the starting point for transforming standard video cameras in smart sensors. In fact, the new proposed sensor, based on video recording, is able to give a quantitative estimation of the noise levels in many points, without the adoption of sound level meters and extensive (and expensive) measurement campaigns. Of course, this methodology being a concept, with only a small dataset for validation, it has several shortcomings at the moment that are reported in the discussion section. The EAgLE efficacy must be tested on a large dataset. This is the reason why a long term validation should be run, for instance, using it in parallel with existing monitoring stations. At any rate, EAgLe seems to be very promising, giving the chance to produce large noise maps, with the only aid of existing, or to be installed, video cameras.

2. Materials and Methods

The EAgLE technique adopted in this paper is based on the recognition of any moving object by means of background subtraction, defining a moving "blob" (Figure 1). The blob is bounded in a box (yellow box) and a centroid is applied in the center of the box (red dot). This centroid is tracked and when it passes a given line (green horizontal line in Figure 1), the vehicle is counted and assigned to light or heavy vehicle category according to the box's diagonal length (see counter on top right of Figure 1). The time each centroid takes for going from the green line to the white line (or vice versa) is used to estimate the speed of the vehicle, after a conversion from frame per second to meter per second.

Figure 1. Image analysis on the video frame. (**a**) Moving object are bounded in a yellow box and a red dot (centroid) is applied. Green and white line are used for counting and speed estimation; (**b**) blob detection after background subtraction.

The algorithm has been developed in the "Microsoft Visual Studio 2015" framework, with the aid of the "Open Source Computer Vision Library" (OpenCV). The code is written in Python. The source code of the recognition part has been created starting from codes shared in the "GitHub" platform [23].

The principal functions of the code are

- *Main* function: it performs calculations needed to count the vehicles and separate the categories. It is also used to estimate the speed of each vehicle;
- *Blob* auxiliary function: it initializes the parameters of the blob, determining the bounding box and diagonal dimensions. In addition, it includes the "*predictNextPosition*" function described below;
- *Header* function: the blob class is defined here, applying the parameters defined in the "blob" function to each continuous mass detected in the frame.

The input file is a MPEG-4 file. At this stage, the algorithm works only in offline mode, analyzing the single frames after having processed the input video. The detection is performed by subtracting the background into two following frames. The tracking of the blob centroid is performed with an improved algorithm, proposed in [23]. The classic approach suggests minimizing the distance between all the centroid positions and the referenced one in two following frames. In this algorithm, a prediction of the position in the next frame is performed for each centroid, on the basis of the trajectory that followed in the previous close frames. Then, a weighted mean between previous positions, with weights varying according to the time distance, is performed and this position is proposed for the following frame. This calculation is done on 4 previous positions, as a compromise between tracking efficiency and computing time. Then, the distance between the predicted position and the real one is minimized, assigning the position to each blob, in all the frames of the video. This is useful to avoid multiple recognition due to several vehicles moving close each other.

When a centroid crosses a chosen line (in our case, the green line in Figure 1), that can be horizontal or vertical, the counting is increased by one. The category is assigned according to the length of the diagonal of the box. A short video sample of this procedure (Video S1) is proposed in the supplementary material of the paper. A time stamp (frame number) of the crossing is recorded and used for evaluation of the speed, by combining this time with the time of crossing the white line.

Once the vehicle has been detected and classified, and its speed has been assigned to the velocity vector, the sound power level can be estimated. In this preliminary stage, the following procedure has been implemented in Matlab©, but it can be implemented in the same framework of the video processing.

As mentioned in the introduction, among the several emission models that are presented in the literature (see Can et al. [17]), EAgLE implements the CNOSSOS-EU emission model that suggests calculating the sound power level as follows:

$$L_{W,i,m}(v_m) = 10log\left(10^{\frac{L_{WR,i,m}(v_m)}{10}} + 10^{\frac{L_{WP,i,m}(v_m)}{10}}\right) \quad (1)$$

where, i is the index related to the frequency band of octave, m is the index related to the type of vehicle, v_m is the average speed of the flow of the m-th category of vehicles, $L_{w,R,i,m}$ is the rolling noise, and $L_{w,P,i,m}$ is the propulsion noise, given by:

$$L_{WR,i,m}(v_m) = A_{R,i,m} + B_{R,i,m}\, log\left(\frac{v_m}{v_{ref}}\right) + \Delta L_{WR,i,m}(v_m) \quad (2)$$

$$L_{WP,i,m} = A_{P,i,m} + B_{P,i,m}\left(\frac{v_m - v_{ref}}{v_{ref}}\right) + \Delta L_{WP,i,m}(v_m) \quad (3)$$

with v_{ref} being the reference speed (70 km/h), A and B table coefficients, and ΔL_w the correction terms. Of course, other emission models can be easily implemented, according to the needs and the country of application of the EAgLE system.

Once the L_w is obtained for each vehicle, the instantaneous sound pressure level at the receiver $L_p(t)$ can be estimated using the pointlike source propagation formula, and the single event Sound Exposure Level (SEL) of each pass-by, i.e., the amount of acoustic energy of each transit "compressed" in 1 s, at the fixed receiver, is calculated:

$$SEL = 10 \log \frac{1}{t_0} \int_{t1}^{t2} 10^{\frac{L_p(t)}{10}} \, dt \tag{4}$$

where t_0 = 1 s, t_1, and t_2, respectively, are the beginning and the end of the transit. This step is fundamental in order to make all the transits comparable, since they have strong differences in terms of duration, according to the speed of the vehicles [24]. This procedure is done for each vehicle and for each category, in particular for light and heavy duty vehicles. Then, the overall SEL is calculated with a log sum for light and heavy vehicles. The continuous equivalent level L_{eq} evaluated in the time range Δt is finally obtained with the following formula:

$$L_{eq}^{(\Delta t)} = 10 \log \frac{1}{\Delta t} + 10 \log \left(\sum_{i=1}^{N_L} 10^{0.1 SEL_i^{light}} + \sum_{i=1}^{N_H} 10^{0.1 SEL_i^{heavy}} \right) \tag{5}$$

A résumé of the main steps of the EAgLE methodology is

1. To acquire the video from cameras;
2. To run the counting and recognition algorithm (in real time or in post processing analysis);
3. To remove fake counts and adjust category recognition (only in offline analysis);
4. To feed the noise level estimator with input data;
5. To calculate noise emission levels (according to CNOSSOS-EU);
6. To calculate the SEL of each vehicle;
7. To calculate the overall SEL for light and heavy duty vehicles' categories;
8. To estimate the L_{eq} on the required time basis (it should coincide with the video duration).

Of course, once the EAgLE methodology is embedded in existing sensors for video recording and validated with on-site measurements and calibration, the choice of time basis and time range to calculate the L_{eq} can be tuned according to the needs of the case study. For instance, for urban planning purposes, in urban areas with specific limits, the L_{den} (i.e., equivalent level evaluated on the day, evening, and night periods, with penalties for evening and night) can be calculated by running the algorithm on the video recordings of one year. Several other applications are possible, changing and tuning the parameters of the EAgLE methodology, depending on the aim of the investigation and on the case study.

Preliminary Application on a Case Study on an Italian Highway: Case Study Description

A preliminary application of the EAgLE methodology has been performed on a site located along the Italian highway A2 "Autostrada del Mediterraneo". This highway is managed by ANAS S.p.a. and goes from the crossing between A30 and RA2, in Fisciano, to Reggio Calabria. The video recording and the measurements have been performed in the city of Baronissi (Figure 2a), in the segment between Fisciano and Salerno, from the sidewalk of a bridge (Figure 2b,c), in safety conditions (Figure 2d). In this segment, the highway is made of two lanes per direction, with an entering lane coming from a gas station, in the south-north direction. Anyway, the entering flow recorded during the measurements was negligible. Furthermore, the traffic on the bridge was negligible. No unusual events have been recorded, such as noisy motorcycles, airplanes passing by, honking, etc., meaning that the conditions of test are quite ideal for the application of the methodology.

Figure 2. Measurement location: (**a**) Position of Baronissi (red mark), in the Campania region (courtesy of Google Earth©); (**b**) 3D aerial view of the bridge from Google Earth©; (**c**) lateral view of the bridge from Google Street View©; (**d**) picture of the instruments during the measurement collection.

The instruments used for the measurements are a class 1 sound level meter Fusion by 01 dB and a video camera embedded in a mobile phone. Two measurements of 15 minutes have been collected around lunch time on Friday, 17 November, 2017. All the acoustic parameters, in particular $L_{pA,F}$, $L_{eq,A}$, percentile levels, acoustic spectrum in third of octaves, etc., and the video of the vehicles passing-by have been recorded in parallel. Temperature was approximately in the range 11 °C–14 °C and wind speed was below 5 m/s on average. Furthermore, to protect the sound level meter from sudden wind peaks, the wind cover was used (see Figure 2d). The flow was running almost freely, with little variations of speed. The average number of vehicles flowing in 15 minutes is 1091 vehicles, with a percentage of heavy vehicles of about 15% in both the measurements. Details about the manual counts performed on the videos are reported in Table 1.

Table 1. Details of the manual counts results.

Measurement ID	Starting Time [hh:mm:ss]	Light Vehicles Flow [veh/15 min]	Heavy Vehicles Flow [veh/15 min]	Percentage of Heavy Vehicles [%]
1	12:52:37	930	168	15.3
2	13:16:14	917	167	15.4

The detection algorithm is obviously strongly influenced by the stability of the image that is affected by vibrations of the bridge and wind. Since in this sample application a simple camera with a tripod has been used, the overall recognition efficiency is affected by the vibration of the image. Without any post processing and offline analysis, the detection error is greater than 200%. For this reason, in order to check the complete EAgLE technique, a sampling of the two videos was tested, choosing the time ranges in which the camera was more stable, in order to find subsections of the videos less affected by image movements. Two video subsections, each of them made by 5 cuts collected at the beginning, at the end, and in the middle of the video, were extracted, one per each measurement. The overall duration of each subsection is around 300 seconds. Moreover, an offline analysis was run, removing the counts due to the moving of the frames. The periods chosen for the videos' cuts are summarized in Table 2.

Table 2. Starting and ending time of the 5 cuts sampled in Video 1 and Video 2.

	Period				
	Period 1 [mm:ss] From–to	**Period 2 [mm:ss] From–to**	**Period 3 [mm:ss] From–to**	**Period 4 [mm:ss] From–to**	**Period 5 [mm:ss] From–to**
Video 1 cut	00:00–01:02	03:40–04:41	06:03–07:03	10:30–11:30	13:55–15:00
Video 2 cut	00:00–01:18	02:38–03:34	06:29–07:40	10:48–11:51	14:08–15:02

3. Preliminary Results

The results of vehicle counting and detection is reported in Table 3, for the two video cuts, approximately five minutes long each, after post processing of the videos and moving frames counts removal.

Table 3. Results of the manual and Equivalent Acoustic Level Estimator (EAgLE) counting and recognition, after the post processing of the video, in the two video cuts.

	Manual counts		**EAgLE Counts**		**Error Percentage**	
	Light Vehicles [counts]	**Heavy Vehicles [counts]**	**Light Vehicles [counts]**	**Heavy Vehicles [counts]**	**Light Vehicles [%]**	**Heavy Vehicles [%]**
Video 1 cut (308 s)	334	52	342	55	+2%	+6%
Video 2 cut (322 s)	349	64	355	67	+2%	+5%

The efficiency achieved after the removal of moving frames counts is good. Moreover, the recognition is performed with satisfying results. The mistakes in the category are usually overestimated due to the fact that some slightly moving frames could not be removed. That led to the creation of fake moving blobs due to the difference in the background between two following frames. When these fake blobs appeared close to the counting line, they were counted (usually as light vehicles because of the little variation between the two images in the following frames). In addition, it occurred that in some cases two light vehicles moving very closely to each other were recognized as a single heavy vehicle, leading to a small overestimation in this category. The author believes that these problems can be solved by means of a more stable video camera, an optimized angle of view, and a more advanced recognition tool.

The distributions of the speed estimated with the EAgLE algorithm are reported in Figure 3. It can be noticed that the distribution of light vehicles' speeds is very close to a normal distribution, as suggested in literature for free flows. For the heavy vehicles, the different shape of the distribution is probably influenced by the mixing of medium and heavy vehicles, which in principle have different average speeds. The EAgLE algorithm run in this preliminary application, in fact, did not distinguish between vans (medium vehicles) and buses or trucks (heavy vehicles). The mean values of the two distributions are of course different, due to the different speed limits and run conditions.

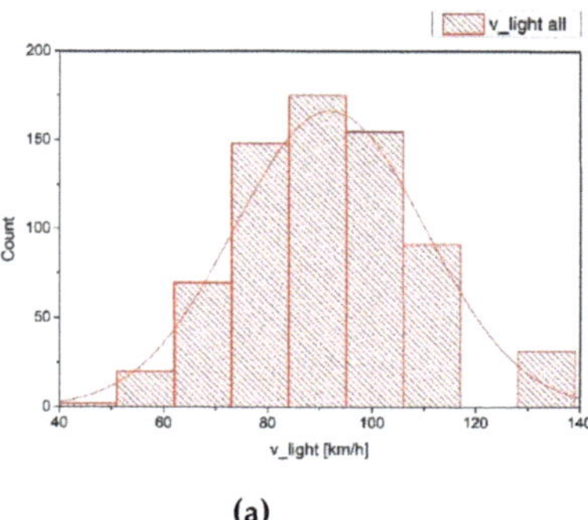

(a)

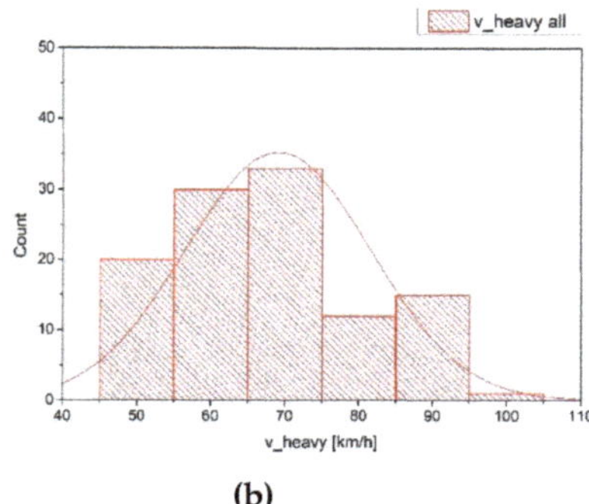

(b)

Figure 3. Speeds distributions for light (**a**) and heavy vehicles (**b**) summing the speeds estimated in both the video cuts.

The missing bins in light vehicles' speeds distribution figures are due to the discretization in detection of the speed. The frame rate of the camera (30 fps), in fact, influences the speed estimation, that is performed converting the number of frames per second needed to go from the trigger line to the "arrival" line. In particular, the discretization due to the frame rate introduces a discretization in the speed estimated as well. The resulting "delta" is a function of the speed itself (it grows according to the growth of the speed), of the frame rate and of the position of the lines. This position is the result of a compromise between a distance large enough to estimate the speed in a sufficiently large range, and the best location for vehicles pass-by detection. The delta ranges from about 4 km/h in the low speeds part of the distribution to about 14 km/h in the high speeds zone. It is expected that a more advanced camera, with a higher frame rate, will lead to a more precise estimation of the speed, with a consequently better distribution plot. Additional error sources can be the uncertainty on the centroid position, for instance, due to the shadow effect and the resolution of the image, since it influences the bounding box shape.

Due to the results obtained in the first phase with the detection algorithm, basically, once the identification and the speeds vectors for light and heavy vehicles have been detected, the noise levels estimation has been performed in Matlab framework. As already described in Section 2, the sound power level of the sources has been estimated with the CNOSSOS-EU approach, and the propagation to the receiver has been done with the standard pointlike source propagation formula. The measured continuous equivalent level L_{eq} on the 15 minutes time range, the levels predicted with some predictive statistical models, the levels predicted with CNOSSOS-EU model, and the L_{eq} simulated with the EAgLE technique, are resumed in Table 4. The predictive models selected for the comparison are a fully statistical and simple model, i.e., the Burgess model [25], that includes just the traffic flow, the percentage of heavy vehicles, and the distance between source and receiver, and a "semi-dynamical" model, i.e., CNOSSOS-EU, that, in addition to the previous inputs, includes the mean speed of the flow and some correction factors, such as road gradient, temperature, etc.

Table 4. Summary of measured L_{eq} over 15 minutes compared with predictive model results and with the L_{eq} simulated with the EAgLE methodology.

Measurement ID	Measured L_{eq} [dBA]	L_{eq} Burgess [dBA]	L_{eq} Cnossos [dBA]	L_{eq} EAgLE [dBA]
1	75.9	77.9	76.2	76.0
2	75.9	77.9	76.1	75.9

It can be immediately noticed that the statistical models overestimate the measured L_{eq}, while the models that consider the speed of the flow (as a mean value, such as CNOSSOS, or for the single vehicle, such as EAgLE) give a much better estimation of the noise levels.

4. Discussion

The preliminary results reported in Section 3 are very encouraging and the comparison performed on the two test videos present a very good agreement between EAgLE simulated levels and the measured L_{eq}. Furthermore, it should be underlined that at the moment the methodology presents some limitations and shortcomings.

First of all, the EAgLE technique is strongly affected by the video recording. In particular, the critical points seem to be the angle of recording, which affects the parallax and the conversion between frames and real world distances, the resolution of the camera, which influences the speed estimation, the light conditions, and the shadow effects. The former two points can be quite easily solved with a calibration of the system and with the adoption of high resolution cameras. In regards to the latter two points, of course, a dark image is not feasible for EAgLE at the moment. Problems can occur during the first and last hours of the day, when the sun is barely perpendicular to the road and shadows can modify the size of the bounding boxes, leading to a misclassification of the vehicle. This means that the proposed methodology can be used continuously, during day and night, only in places with artificial lights, but by calibrating the angle of view and the sensibility of the bounding box can include effects of the shadows. It should be also underlined that the actual video recording sensors are always placed on illuminated sites, since it makes no sense to place a video camera on dark sites. For this reason, the EAgLE methodology is still interesting to be embedded in existing sensors, and, for new installation, should be designed in proper locations to avoid the night (or little light) issues and the shadow effects. Moreover, tests with the light projectors of the cars should be performed to see if the recognition efficiency can be kept using the moving lights. Furthermore, tests at different hours of the day have to be performed, to assess the effect of the sunrays inclination on the recognition performance.

Another important issue concerning the video recording is the detection ability in crowded and congested roads. While the exclusion of other "non-noisy" moving objects (pedestrians, animals, etc.) can be performed with a proper placement of the video camera, the possibility of giving bad results in congested situations is a critical point, especially in urban areas. In highways, in fact, congestions are quite rare, especially out of the rush hours. In the author's opinion, this is a problem that can be solved by improving the detection and classification code of EAgLE. As mentioned in the introduction, several techniques have been developed for this purpose, much more advanced that the one implemented in this preliminary application, based on machine learning, deep learning, neural network, etc. (see for instance [19]). For this reason, the author is confident that great improvements can be done on this issue, by tuning the detection algorithm on the case study under investigation.

Another limitation of the EAgLE methodology lies in the estimation of non-standard events, such as honking, sirens, extremely noisy vehicles, external sources, etc. It must be underlined that none of the predictive models present in literature nowadays can predict such events, thus, from this point of view, EAgLE, at the moment, is somehow aligned with the other models. In any case, trigger events could be implemented in the recognition code, for instance, using the lights of the ambulances or of the police cars, to tag these events as non-standard and treat them in a proper way.

The preliminary application presented in previous sections is limited due to the small number of measurements and to the free flow condition. More on this part must be done in future researches in order to validate the technique on a larger sample of measurements, with different traffic conditions and geometric features of the sites.

Looking at the comparison in Table 4, it could be argued that such a strong computation effort is not needed, since the CNOSSOS model gives very similar results. The key point is that the CNOSSOS model needs several inputs to run, while the EAgLE methodology produced excellent results just using the video recordings. Moreover, EAgLE includes a fully dynamic model, since it considers the speed and the kinematics of the single vehicle.

Even with the above mentioned limitations of this study, the EAgLE methodology is really promising, because of its easiness in application in any place controlled by video camera recordings. The actual algorithm is quite easy and can be implemented in real time monitoring, to produce raw estimations of noise levels. Of course, for a more reliable estimation, an offline analysis is mandatory in order to clean the raw data from mistakes in counting or the classification of vehicles, such as in estimating the speed. The integration of this system in a complex sensor, including video recording, online analysis, data transmission, and offline processing, is encouraged by the preliminary results obtained in the case study application. The author believes that with a more powerful video camera network and an improved data processing system, this methodology can be extremely useful in qualitative noise monitoring systems, especially in urban areas and big infrastructures, where usually video recording is already present for safety reasons or for tolling/fining systems.

Future studies should include the production of a test sensor that embeds the EAgLE methodology, with a video camera, a sound level meter, and a processor able to run the algorithm at a local site. In this way, the sensor can be tested on a large scale validation, with a continuous recording of pressure levels and video images in order to test the online performances and the criticisms. A sensibility analysis of the sensor can be performed, testing the variations according to the detection and propagation critical elements (such as angle of view, distance, geometry of the site, etc.) and to the source parameters (such as flow volume, typology and dynamics, pulsing conditions, and/or congestions). Moreover, the non-standard events, such as honking, ambulances, police sirens, etc., should be investigated, since the noise produced is due to both the vehicle and to external loudspeakers.

Once the EAgLE methodology will be validated, a large spatial scale can be tested, with the aim to produce a noise map of a city or of a transportation infrastructure, taking advantage of the existing video camera networks. When long term recordings are available, for instance, in more than a year, the L_{den} estimation can be performed, using real traffic data, instead of simulating ideal conditions in noise predictive software. This could help local policy makers and infrastructure managers in finding the critical points of their networks and, if needed, in committing to implement further investigations, based on standard noise level measurements or other tools.

5. Conclusions

In this paper, the EAgLE (Equivalent Acoustic Level Estimator) technique has been presented. This technique, based on image analysis, vehicle tracking, and dynamic noise modeling, aims at producing a robust estimation of the continuous equivalent noise level on given time ranges, by using just a video camera recording.

A preliminary application of the technique, in a short time range (630 s) related to a case study along a highway in South Italy, has been presented, showing how, with a good recognition efficiency, the noise levels estimated with EAgLE are extremely close to the measured levels in this reduced sample of measurements performed in free flow and standard conditions.

More tests are needed to validate the EAgLE procedure. Moreover, beside the shortcomings discussed in the previous sections, several strength points arise from the first tests. In particular, the possibility to provide reliable qualitative estimations of the noise level in any place embedded with a video camera, in cities, or along transportation infrastructures, is definitively the key point of the proposed sensor. These estimations can be used on one side to cope with the need of a large spatial monitoring, and on the other side to provide first level alarms of exceeding limit thresholds, to be checked with follow-up interventions at specific sites.

Supplementary Materials: The following are available online at http://www.mdpi.com/1424-8220/20/3/701/s1, Video S1: 1-minute video of the EAgLE counting algorithm running.

Funding: This research received no external funding.

Acknowledgments: The author is grateful to Joseph Quartieri for providing support for this research. This research would not be possible without the efforts of Antonio Marino who developed the code during his undergraduate thesis period and who helped in the analyses. The author thanks Valentina Salzano and Angela Raimondo for participating in the field measurement campaign, in the framework of their undergraduate thesis. The author expresses gratitude to the editors and to the reviewers for the valuable suggestions and comments.

Conflicts of Interest: The authors declare no conflict of interest.

References

1. Bodin, T.; Albin, M.; Ardö, J.; Stroh, E.; Östergren, P.O.; Björk, J. Road traffic noise and hypertension: Results from a cross-sectional public health survey in southern Sweden. *Env. Heal.* **2009**, *8*. [CrossRef] [PubMed]
2. European Environment Agency. *The European Environment—State and Outlook 2020. Knowledge for Transition to a Sustainable Europe*; Publications Office of the European Union: Luxembourg, 2019. [CrossRef]
3. Directive 2002/49/EC of the European Parliament and of the Council of relating to the assessment and management of environmental noise. *Offi. J. Eur. Commun.* **2002**, *189*, 18.07.
4. Fredianelli, L.; Del Pizzo, A.; Licitra, G. Recent Developments in Sonic Crystals as Barriers for Road Traffic Noise Mitigation. *Environments* **2019**, *6*, 14. [CrossRef]
5. Praticò, F.G. On the dependence of acoustic performance on pavement characteristics. *Transp. Res. Part Transp. Environ.* **2014**, *29*, 79–87.
6. Praticò, F.G.; Anfosso-Lédée, F. Trends and issues in mitigating traffic noise through quiet pavements. *Procedia Soc. Behav. Sci.* **2012**, *53*, 203–212.
7. Licitra, G.; Cerchiai, M.; Teti, L.; Ascari, E.; Bianco, F.; Chetoni, M. Performance assessment of low-noise road surfaces in the Leopoldo project: comparison and validation of different measurement methods. *Coatings* **2015**, *51*, 3–25. [CrossRef]
8. Zambon, G.; Roman, H.E.; Smiraglia, M.; Benocci, R. Monitoring and prediction of traffic noise in large urban areas. *Appl. Sci.* **2018**, *8*, 251. [CrossRef]
9. Zambon, G.; Benocci, R.; Bisceglie, A.; Roman, H.E.; Bellucci, P. The LIFE DYNAMAP project: Towards a procedure for dynamic noise mapping in urban areas. *Appl. Acoust.* **2017**, *124*, 52–60. [CrossRef]
10. Steele, C. A critical review of some traffic noise prediction models. *Appl. Acoust.* **2001**, *62*, 271–287. [CrossRef]
11. Quartieri, J.; Iannone, G.; Guarnaccia, C.; D'Ambrosio, S.; Troisi, A.; Lenza, T.L.L. A Review of Traffic Noise Predictive Models, in Recent Advances in Applied and Theoretical Mechanics. In Proceedings of the 5th WSEAS International Conference on Applied and Theoretical Mechanics (MECHANICS'09), Puerto de la Cruz, Tenerife, Spain, 14–16 December 2009; pp. 72–80.
12. Guarnaccia, C.; Lenza, T.L.L.; Mastorakis, N.E.; Quartieri, J. A Comparison between Traffic Noise Experimental Data and Predictive Models Results. *Int. J. Mech.* **2011**, *5*, 379–386.
13. Guarnaccia, C. Advanced Tools for Traffic Noise Modelling and Prediction. *WSEAS Trans. Syst.* **2013**, *12*, 121–130.
14. Quartieri, J.; Mastorakis, N.E.; Guarnaccia, C.; Iannone, G. Cellular Automata Application to Traffic Noise Control. In Proceedings of the 12th International Conference on "Automatic Control, Modelling & Simulation" (ACMOS '10), Catania, Italy, 29–31 May 2010; pp. 299–304.
15. Guarnaccia, C.; Quartieri, J.; Mastorakis, N.E.; Tepedino, C. Development and Application of a Time Series Predictive Model to Acoustical Noise Levels. *WSEAS Trans. Syst.* **2014**, *13*, 745–756.
16. Guarnaccia, C.; Quartieri, J.; Tepedino, C.; Rodrigues, E.R. An analysis of airport noise data using a non-homogeneous Poisson model with a change-point. *Appl. Acou.* **2015**, *91*, 33–39.
17. Can, A.; Aumond, P. Estimation of road traffic noise emissions: The influence of speed and acceleration. *Transp. Res. D* **2018**, *58*, 155–171. [CrossRef]
18. Huttunen, H.; Yancheshmeh, F.S.; Chen, K. Car Type Recognition with Deep Neural Networks. In Proceedings of the IEEE Intelligent Vehicles Symposium (IV), Gothenburg, Sweden, 19–22 June 2016; pp. 1115–1120.
19. Huang, T. Traffic Speed Estimation from Surveillance Video Data. Proceedings of IEEE/CVF Conference on Computer Vision and Pattern Recognition Workshops (CVPRW); Salt Lake City, UT, USA: 18–22 June 2018; pp. 161–165.
20. Hua, S.; Kapoor, M.; Anastasiu, D. Vehicle Tracking and Speed Estimation from Traffic Videos. Proceedings of IEEE/CVF Conference on Computer Vision and Pattern Recognition Workshops (CVPRW), Salt Lake City, UT, USA, 18–22 June 2018; pp. 153–160.

21. Biswas, D.; Hongbo, S.; Wang, C.; Stevanovic, A. Speed Estimation of Multiple Moving Objects from a Moving UAV Platform. *Int. J. Geo-Inf.* **2019**, *8*, 259. [CrossRef]
22. Kephalopoulos, S.; Paviotti, M.; Anfosso-Lédée, F. Common Noise Assessment Methods in Europe (CNOSSOS-EU). Available online: https://ec.europa.eu/jrc/en/publication/reference-reports/common-noise-assessment-methods-europe-cnossos-eu (accessed on 24 January 2020).
23. GitHub, Microcontrollers and More; OpenCV_3_Car_Counting_Cpp. Available online: https://github.com/MicrocontrollersAndMore/OpenCV_3_Car_Counting_Cpp (accessed on 5 December 2019).
24. Iannone, G. Improvements in the acoustical modeling of traffic noise prediction: theoretical and experimental results. Ph.D Thesis, University of Salerno, Fisciano, Italy, April 2011.
25. Burgess, M.A. Noise prediction for Urban Traffic Conditions. Related to Measurement in Sydney Metropolitan Area. *Appl. Acou.* **1977**, *10*, 1–7. [CrossRef]

Article

Sound Levels Forecasting in an Acoustic Sensor Network Using a Deep Neural Network

Juan M. Navarro [1,*], Raquel Martínez-España [1], Andrés Bueno-Crespo [1], Ramón Martínez [1] and José M. Cecilia [2]

[1] Escuela Politécnica. Universidad Católica de Murcia (UCAM), Campus de los Jeronimos, 30107 Guadalupe, Spain

[2] Computer and Systems Department (DISCA). Universitat Politècnica de València (UPV), Camino de Vera, s/n, 46022 Valencia, Spain

* Correspondence: jmnavarro@ucam.edu

Received: 13 December 2019; Accepted: 4 February 2020; Published: 7 February 2020

Abstract: Wireless acoustic sensor networks are nowadays an essential tool for noise pollution monitoring and managing in cities. The increased computing capacity of the nodes that create the network is allowing the addition of processing algorithms and artificial intelligence that provide more information about the sound sources and environment, e.g., detect sound events or calculate loudness. Several models to predict sound pressure levels in cities are available, mainly road, railway and aerial traffic noise. However, these models are mostly based in auxiliary data, e.g., vehicles flow or street geometry, and predict equivalent levels for a temporal long-term. Therefore, forecasting of temporal short-term sound levels could be a helpful tool for urban planners and managers. In this work, a Long Short-Term Memory (LSTM) deep neural network technique is proposed to model temporal behavior of sound levels at a certain location, both sound pressure level and loudness level, in order to predict near-time future values. The proposed technique can be trained for and integrated in every node of a sensor network to provide novel functionalities, e.g., a method of early warning against noise pollution and of backup in case of node or network malfunction. To validate this approach, one-minute period equivalent sound levels, captured in a two-month measurement campaign by a node of a deployed network of acoustic sensors, have been used to train it and to obtain different forecasting models. Assessments of the developed LSTM models and Auto regressive integrated moving average models were performed to predict sound levels for several time periods, from 1 to 60 min. Comparison of the results show that the LSTM models outperform the statistics-based models. In general, the LSTM models achieve a prediction of values with a mean square error less than 4.3 dB for sound pressure level and less than 2 phons for loudness. Moreover, the goodness of fit of the LSTM models and the behavior pattern of the data in terms of prediction of sound levels are satisfactory.

Keywords: acoustics; wireless sensor networks; smart cities; deep learning; long short-term memory; temporal forecast

1. Introduction

Noise pollution is one of the main environmental concerns of modern cities because of its effects on the quality of life, health and livability of cities. The European Commission adopted the European Noise Directive (END) [1], which focuses on the monitoring of environmental noise by generating noise maps of the main population centers and elaborating action plans [2,3]. Noise measurements in urban areas are typically carried out by designated officers that collect data at a few accessible spots, where sound level meters are installed during short time intervals. Collected noise data is often input into a model that attempts to predict noise levels for a temporal long-term throughout the landscape to be evaluated. As a result, noise maps are generated using sound sources and propagation models leveraging geographic information systems to improve the accuracy and quality of the results [4,5]. Specifically, road [6–10], railway [11–13] and aerial [14,15] traffic models are used, among others. However, according to Maisonneuve [16], this approach presents several limitations since noise maps are actually generated from synthetic data. Even though these models allow to gain a first insight into the noise pollution problem, they are mainly focused on long-term acoustic parameters prediction and require auxiliary data such as source definition, traffic flow, street geometry, day period, urban topology, etc.

Wireless Acoustic Sensor Networks (WASN) [17,18] are becoming an indispensable tool for monitoring and assessment of short-term noise levels. WASN are a balanced technology regarding the characteristics of cost, scalability, flexibility, reliability and accuracy [19]. Such networks are supported by recent advances in low-power wireless communications technology as well as the integration of several functionalities in electronic devices, including sensing, communication and processing, even allowing the implementation of neural networks in the nodes [20]. They are being extensively used in smart city applications in recent years. This trend has led to intensive deployments in numerous cities such as New York [21], Barcelona [22] or Monza [23]. WASN can be deployed over an area of interest to operate continuously by creating a real-time monitoring system, which collects historical data related to the sound environment over longer periods of time, operating unattended and requiring human intervention only for network installation, maintenance and removal. This data is transmitted to a central sink node, then could be stored and subsequently be used, for instance, to dynamically update noise maps [24]. Indeed, all these information acquired by WASN can be analyzed to obtain useful information for the city [25]. Moreover, it is very interesting and relevant to predict the short-term behavior of the acoustic parameters that evaluate the sound environment. For instance, it allows the ability to detect behavior patterns depending on different times of day and, furthermore, in the event of failure or error in sending information from a sensor, this information can be estimated with precision. In addition, by being able to know these unique level values several days in advance, preventive measures could be taken if necessary to avoid the population from being exposed to risk levels. Therefore, in this work, a novel approach based on deep neural networks is introduced to forecast the near-time short-term sound level values using only historical sound level data from the location of study. In this way, the approach that is presented in this paper can be applied to every node of the sensor network, where the inputs of the model are the past and actual sound level values and the outputs are the future values.

To achieve this objective, in this paper the use of the Long Short-Term Memory (LSTM) deep neural network technique is proposed to model the behavior pattern of the acoustic parameters which has demonstrated very good results in prediction of time series [26,27]. Sound sources, specifically those concerning a sound environment in this work, can be considered as time variant functions, i.e., time series, both the audio signal and the corresponding calculated parameters. Time series data analysis has been actively researched for decades and is considered one of the ten most difficult problems in data mining due to its unique properties. In this work, the capability of LSTM networks to estimate short-term future values of sound levels in a certain location using historical data is explored. In particular, several models are obtained by training the LSTM networks with sound pressure level and loudness level values captured by a node of a WASN. Comparison with ARIMA technique results together with some experiments are presented to evaluate the proposed approach.

The paper is structured as follows. After this introduction, Section 2 presents a review of related work and the difference with the proposed approach. Then, Section 3 describes the deployed sensor network, designed LSTM networks and the collection and pre-processing of the data-set handled to train and evaluate them. In Section 4 different results obtained from the experiments to evaluate the implemented LSTM networks are shown and discussed. Finally, Section 5 presents the general conclusions of this study and proposes future work.

2. Related Work

A significant amount of information generated by sound sources is carried by acoustic signals, and this information can be used to describe and understand human and social activities. Sound signal acquired by acoustic sensors can be processed in two ways: (i) capturing and processing the audio signal (e.g., event detection [28,29], classification of sound sources [30,31], sound source location [32], etc.) and (ii) calculating values of acoustic parameters from the captured audio signal (e.g., sound pressure level [33], loudness [34], etc.) that are the data collected to generate sound maps.

Several works have been developed in applying artificial neural networks to estimate sound source features and/or acoustic parameters values in a certain location for a given period of time, using data obtained through WASN or other information data base. In what follows we introduce differences between the proposed work and these previous works. Regarding audio signal processing, in publications [35,36] a WASN is proposed to monitor and analyze urban noise pollution, deploying a network of sensors to measure sound pressure level and using convolutional neural networks to classify sound sources from captured audio. In other work, Socoró et al. [37] introduced an anomalous noise event detector to remove sound frames unrelated to road traffic sound sources to provide more reliable data captured by a WASN. In [38], a convolutional recurrent neural network in a dilated spiral is used as a classifier fed by the energy recording feature in the mel band for the detection of sound events. Regarding to parameters calculation, some published papers introduce neural networks to estimate advanced acoustic parameters values. Yu and Kang [39] explored the feasibility of using machine learning models to predict the sound landscape quality in urban open spaces by correlating various physical, behavioral, social, demographic and psychological factors. In [40], a convolutional neural network was implemented to estimate the psycho-acoustic annoyance Zwicker's model from an input audio signal. In contrast with these related works, in our research a neural network approach is used to predict future time values of acoustic parameters instead of estimating current time values.

There are some studies that apply neural networks to create a prediction model in order to estimate sound pressure levels emitted by sound sources across a spatial domain but using also geospatial and description information as input parameters. Specifically in [41], a multi-layer perceptron neural network model trained with the Levenberg–Marquardt algorithm was used to predict the equivalent sound level from road traffic noise. In another publication [42], a system proposition is presented that uses an ensemble of machine learning techniques to estimate both environmental sound levels and uncertainty in model predictions by taking geospatial data as input. In addition to making use of auxiliary information, these neural network-based models predict long-term values and do not take into account the temporal composition of the short-term sound environment. An attempt to predict the temporal component of traffic noise levels is presented in [43] through the use of back-propagation neural networks, however it only estimates index values describing temporal variability and impulsiveness in addition to using auxiliary data as input. Although noise sources are mainly non-stationary, statistical techniques such as AutoRegressive Integrated Moving Average (ARIMA) [44] have been also used in the literature to model traffic noise pollution.

Finally, it is worth highlighting that there are several works in the literature that predict other pollution factors through deep neural networks, considering the data of these variables as time series. Specifically, the most common pollution problem studied is air pollution, particulate matter and carbon monoxide concentrations among others [45,46]. However, the use of deep learning models such as LSTM require an optimized configuration and settings for each type of problem, as it is carried out in Section 3.5, considering the inputs and its behavior in time.

3. Materials and Methods

3.1. Wireless Acoustic Sensor Network

In this work, data captured from a node of a deployed WASN was used to train and validate the designed neural network prediction models. This WASN is a scalable and extensible system used to monitor sound levels in a certain environment. This is a static and homogeneous WASN allowing continuous monitoring indoors and outdoors. This network was composed of ten acoustic nodes deployed in the campus of the Catholic University of Murcia. In this WASN, each acoustic node [47] collected and processed the audio signal and after that, it calculated and sent data every minute to the sink node. The low-cost acoustic node design included two main parts: the audio acquisition system and the processing core. The former consisted of an array of the four-microphones of a Sony PlayStation Eye camera. Regarding the processing core, a Raspberry Pi 3 Model B computer [48] was selected for the processing, acquisition and publishing stages. Although a node is able to compute results related to diverse acoustic parameters, see [47] for details, this research is focused on the equivalent sound pressure level (L_p) and loudness level (N) values [49] in a one-minute period. A sink node plays the additional role of transmitting the data to an Internet of Things (IoT) platform to store and to perform analysis of the overall data. The audio signal was not stored nor transmitted from the node to keep public privacy. Concerning the network design, acoustic nodes transmit data via Wi-Fi technology using two communications protocols: TCP for communication between nodes and HTTP for communication between the sink node and the IoT platform. Further in-depth control and maintenance of the deployed nodes was provided via a virtual private network that provides a method for remote Secure SHell (SSH) access to each node. The virtual private network also enhances the wireless transmission security of the sensor as all data and control traffic was routed through this secure network.

Specifically for this research, a data-set with these acoustic parameters, L_p and N, was built, as it is explained in detail in the following section.

3.2. Acoustic Data-Set

In this research, the acoustic data acquired on a continuous basis with a temporal period, i.e., a time step of 1 min by a node of the described WASN in the previous section was used to train a LSTM network. This data-set was collected from the beginning of October to the end of November 2019 and it contains quantitative and temporal data related to two acoustic parameters: the equivalent sound pressure level in decibels (dB) and loudness level in phons in one-minute of integration time. The selected node was located in-door in an open-office room where lecturers and researchers work. Working days are mainly from Monday to Friday but Saturday is also open. This data-set is representative of a random noise, of which the main sound sources are speech and human activities. This long-period study can help to analyze and predict the temporal behavior pattern of this type of soundscape.

From the principal data-set, a total of ten data-sets have been generated, five for each parameter, computing a temporal average of the data for the following periods: 1, 5, 15, 30 and 60 min. The following average has been used for time intervals:

$$X = 10log\left(\frac{1}{n}\sum_{i=1}^{n}10^{\frac{X_i}{10}}\right), \tag{1}$$

where X can be either L_p or N, and X_i corresponds respectively to the equivalent sound pressure level (L_{p_i}) and loudness level (N_i) for each time step i. For example, the data-set denoted as noise15 in Table 1 indicates that the 1-min values have been averaged over 15 min, generating one value for L_p and other for N. A description of the quantity of samples used for each data-set can be seen in Table 1. The number of samples in each data-set corresponds to approximately 50 days.

Table 1. Number of samples per data-set for each of the pressure level and loudness parameters.

Data-Sets	Total Instances L_p	Total Instances N
noise01	72,300	72,300
noise05	14,460	14,460
noise15	4820	4820
noise30	2410	2410
noise60	1205	1205

3.3. Deep Learning: Long Short-Term Memory

A Recurrent Neural Network (RNN) in very powerful for everything that has to do with sequence analysis, such as text, sound or video analysis. The main feature of an RNN is that information can persist by looping into the network diagram, so they can basically "remember" previous states and use this information to decide what will be next. This feature makes them very suitable for managing time series. However, a conventional RNN presents problems in training because retro-propagated gradients tend to grow enormously or fade over time because the gradient depends not only on the present error but also on past errors. The accumulation of errors makes it difficult to memorize long-term dependencies. These problems are solved by the Long Short-Term Memory neural networks (LSTM), for which it incorporates a series of steps to decide which information will be stored and which erased. The LSTM networks are composed of LSTM modules which are a special type of recurrent neural network described in 1997 by Hochreiter and Schmidhuber [50]. The LSTM module contains three internal gates, known as input, forgotten and output (as can be seen in more detail in the Figure 1), consisting basically of a sigmoid layer and a multiplication operation, and in the case of the forgetting door, it also incorporates a hyperbolic tangent layer. These gates allow to remove or add information to the cell state, which is a connection that transfers information from one LSTM module to the next. The input gates controls when new information can enter memory. Forgotten gates control when a piece of information is forgotten, allowing the cell state to discriminate between important and superfluous data, leaving room for new data, for this, a hyperbolic tangent layer is added which is combined with the sigmoid layer. Output gate controls when used in the result of memories stored in the cell state. The cell state has a weighting optimization mechanism based on the resulting network output error, which controls each gate. The output and the cell state value generated by the LSTM module are transferred to the next LSTM module. Figure 1 shows the gates and operations of an LSTM module graphically for L_p (for N it would be the same scheme), and in which it can be observed that the input for a unit, is the output of the previous one. This way, each LSTM module transmits to the next one its prediction that together with the current input of the module, generate the output that is sent as input to the next LSTM module.

The network proposed in this work is univariate, that is, it takes a single input variable and obtains a single output variable, given that the objective of the work is to predict both the L_p sound levels and the loudness N. Thus, for the prediction of each one of these values, a different LSTM model will be made for each data-set.

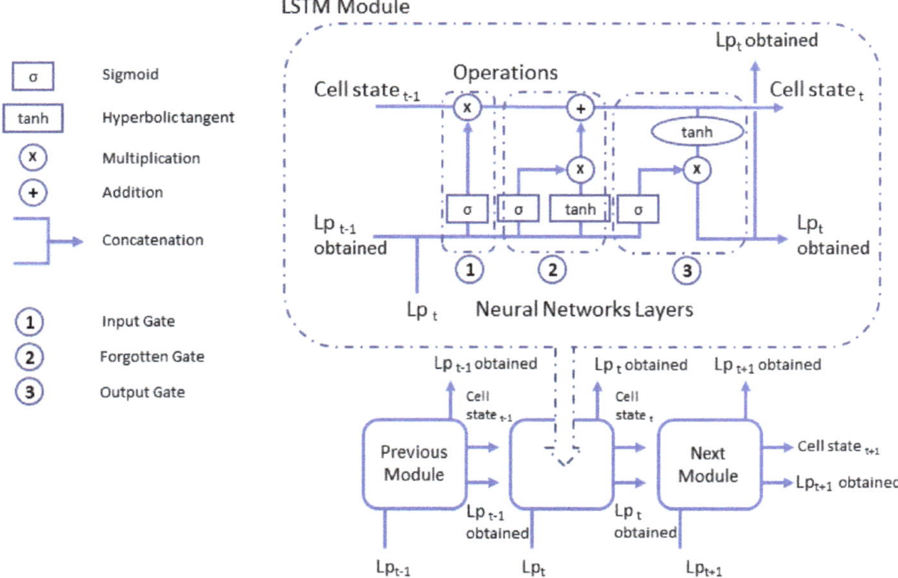

Figure 1. General scheme of an Long Short-Term Memory neural networks (LSTM) for L_p. The interaction between LSTM modules can be observed, as well as the three types of gates that make up an LSTM module.

3.4. Statistical Approach: Auto Regressive Integrated Moving Average

Classical approach to predict time-series is based in statistics. The Auto Regressive Integrated Moving Average technique [51] is a statistical model that uses variations and regressions of statistical data in order to find patterns for a prediction into the future. It has been also applied to sound level parameters prediction [44], as it has been introduced in Section 2. ARIMA is a dynamic time series model, i.e., future estimates are explained by past data rather than independent variables. This model was developed in the late 1960s. Box and Jenkins (1976) systematized it [52]. An ARIMA model is characterized by 3 terms: $(p, d, q,)$ where, p is the order of the Auto Regressive (AR) term, q is the order of the Moving Average (MA) term and d is the number of differences needed to make the time series stationary. In this work, an ARIMA model has been created using the same data-set described in Section 3.2 to compare with quality metrics of the proposed LSTM models.

3.5. Experiment Configuration

The viability and suitability of the proposed LSTM technique is assessed using two types of experiments. On the one hand, an experiment was executed using 80% of the data-set to train the model and 20% to test it. This experiment was applied to the five data-sets (different time intervals) described in Section 3.2, for each acoustic parameter. In addition, to validate the LSTM model, we performed a comparison with the Auto Regressive Integrated Moving Average (ARIMA) technique [51]. On the other hand, to analyze the robustness and adaptability of the proposed LSTM model, we performed several types of validation for the 30 and 60 min data-sets, which are the best results obtained globally. Specifically on the proposed LSTM model; comparisons will be made using the validations of 60%, 70% and 80% to train and 40%, 30% and 20% to test respectively. Thus, depending on the results, the response capacity of the model presented can be analyzed in the absence of training data.

For the ARIMA model, used in the comparison, the parameter (p, d, q) used for the for the estimation of the acoustic parameter L_p were (1,1,14) and for the acoustic parameter N were (1,1,10). In the LSTM model proposed in this paper, the optimal parameters that have been chosen, after a previous adjustment carried out to obtain the optimum parameters, are shown in the Table 2. For the number of neurons, intervals are shown depending on the acoustic parameter.

Table 2. Optimal parameters for LSTM execution experiments.

Parameter	Value
Number of input neurons	L_p[50:100] N[17:70]
Batch size	32
Number of epochs	100
Learning factor	0.001
Optimizer	Adam
Activation function	hyperbolic tangent
Loss Function	quadratic mean error
Delay Sequence	6

The quality evaluation of the model proposed is performed by measuring the goodness of the prediction by the following metrics:

- the Root Mean Square Error (RMSE)
- the Mean Absolute Error (MAE)
- the Pearson Correlation Coefficient (PCC)
- Determination Coefficient (R^2)

Experiments were been carried out in a GPU-based platform. This platform was composed of an Intel(R) Xeon(R) CPU E5-2640 v4 @ 2.40GHz, 128 GB of RAM, 1 TB SSD Hard Disk and a NVIDIA GeForce GTX 780 GPU (Kepler).

4. Results and Discussion

In this section, the behavior of the LSTM model proposed for the prediction of the sound pressure level and loudness values is discussed and analyzed. The evaluation and analysis is detailed in two subsections. First, a comparison with a technique to predict the time series of ARIMA was made by performing an experiment with 80% of the data-set to train and 20% to test. Then, to validate the robustness of the proposed LSTM technique, several validations increasing the test percentage and reducing the train percentage were performed. It should be noted that the predictions were estimated for the values L_p and N, therefore for each of these values a different model was made.

4.1. Comparing the LSTM Model with the ARIMA Model

This section presents the results obtained by the LSTM models for the prediction of the parameters L_p and N for the different data-sets described in Section 1. In addition, LSTM models are compared with the ARIMA technique models for both parameters to validate the results. The validation carried out for both LSTM and ARIMA models was using 80% of the data-sets to train and 20% to test. The number of days is equivalent to about 40 days for training and about 10 consecutive days of prediction for testing.

Table 3 shows the values of RMSE, MAE, PCC and R^2 for each of the data-set of L_p parameter for the LSTM and ARIMA models. For the LSTM models, the calculated metrics are very satisfactory in general, obtaining a RMSE lower than 4.3 dB for L_p in all the data-sets. Regarding to the fit of the model, R^2, the better is this fit the greater the temporal amplitude of the interval is. This may be caused by the smoothing obtained by the averaging of punctual noise peaks. The best fit of the model, 0.75, is obtained for L_p when the prediction period is 60 min. With respect to ARIMA models, the RMSE values increase considerably, which indicates that the ARIMA technique is not adequate for estimating the behavior of the L_p parameter in short-term intervals. For all data-sets the ARIMA model fit is very low and the errors much higher than for the LSTM model. It must be taken into account that ARIMA may need more days of training to be able to reduce the error and improve the fit of the predicted time series. This is one of the advantages of the LSTM technique.

Table 3. Representation of Root Mean Square Error (RMSE), Mean Absolute Error (MAE), Pearson Correlation Coefficient (PCC) and R^2 of the five data-sets of sound pressure level values (L_p) for the LSTM proposed models and the ARIMA models.

	L_p-noise60		L_p-noise30		L_p-noise15		L_p-noise05		L_p-noise01	
	ARIMA	**LSTM**	**ARIMA**	**LSTM**	**ARIMA**	**LSTM**	**ARIMA**	**LSTM**	**ARIMA**	**LSTM**
RMSE	9.3000	3.9400	112.0704	4.2700	9.5734	4.2500	78.2656	3.9500	6.0694	3.5900
MAE	6.6400	2.7500	2.2050	2.8500	7.7755	2.5500	5.8934	2.0700	4.4851	1.7100
PCC	0.1732	0.8600	0.0131	0.8300	0.2798	0.8100	0.0335	0.8100	0.0521	0.8000
R^2	0.0300	0.7500	0.0002	0.6900	0.0783	0.6600	0.0011	0.6400	0.0027	0.5800

Table 4 shows the values of RMSE, MAE, PCC and R^2 for each of the data-set of N parameter for the LSTM and ARIMA models. For the LSTM models, the calculated metrics are very satisfactory in general, obtaining a RMSE lower than 2 phons for N in all the data-sets. Particularly, metrics show that the RMSE of N is similar for all time intervals. In addition, the value of adjustment of the model, R^2, of N is very similar in all the cases, which indicates that it is less affected by the time interval considered to predict sound levels. For ARIMA models, the behavior and results for predicting the N parameter is similar to the L_p parameter. In this case, the error does not increase as significantly as for the L_p parameter. However, the error is always more than double that obtained by the LSTM technique. Moreover, as far as the model's adjustment is concerned, the result is not at all satisfactory. This indicates that the ARIMA models are not able to adapt to the non-stationary behavior of the sound level parameters in short-term intervals.

Table 4. Representation of RMSE, MAE, PCC and R^2 of the five data-sets of loudness values (N) for the LSTM proposed models and the ARIMA models.

	N-noise60		N-noise30		N-noise15		L_p-noise05		N-noise01	
	ARIMA	**LSTM**	**ARIMA**	**LSTM**	**ARIMA**	**LSTM**	**ARIMA**	**LSTM**	**ARIMA**	**LSTM**
RMSE	3.1100	1.9900	14.6412	2.0100	8.4290	1.9600	12.3481	1.8700	3.1400	1.7900
MAE	2.7100	0.9900	2.2050	1.0500	2.0675	1.0000	1.8617	0.8900	0.1563	0.7400
PCC	0.2000	0.7900	0.0198	0.7800	0.3769	0.7800	0.0031	0.7800	0.0011	0.7600
R^2	0.0400	0.6000	0.0004	0.5900	0.1420	0.6000	0.0000	0.6100	0.0000	0.5700

In summary, results show that the LSTM technique outperforms the ARIMA technique for creating temporal short-term models and predicts the behavior of the L_p and N parameters. One aspect to consider about the obtained LSTM models is the difference between the RMSE and MAE values for both N and L_p levels. The MAE value is almost double the RMSE value, indicating that there are outliers in the data [53]. These outliers data are usually reflected by the peaks. In this case, the outliers can be observed in Figures 2 and 3, for both N and L_p levels, in the eventually impulsive sound events that occur throughout the day.

Figures 2 and 3 represent a temporal graph for a ten days interval of the captured data, i.e., real data from the test-subset, along with the estimated data using the obtained LSTM models for both N and L_p. The test-subset begins on Sunday and ends on Tuesday of the following week. Therefore, it can be observed that the minimum noise level on Sunday because the open-office room where the data has been collected is closed. However, the acoustic level increases over the next five working days on the day-period and decreases on the night-period. On Saturday, the activity of people in the office is reduced, thus the noise level is quieter than a regular working day. Then, the time sequence starts again with a Sunday having the lowest noise levels. In general, the model obtained by the LSTM technique, as a pattern of sound level behavior for both L_p and N, adequately follows the trend of sound level. The greater the interval in time averages, the peaks of short event high noises are smoothed, obtaining a better prediction and adjustment of the model comparing with models of shorten intervals.

In order to explore in detail the obtained LSTM models, Figure 4a shows a zoomed view of graph of Figures 2d and 4b shows a zoomed view of graph of Figure 3d for a two days interval with a time average of 30 minutes. It can be observed that the LSTM model has difficulties in precisely estimate short-time events where the sound level increase and decrease drastically, i.e., when sound level suddenly rise or decay. However, the behavior of the LSTM model is much more stable when the peaks are less relevant, e.g., during Saturdays.

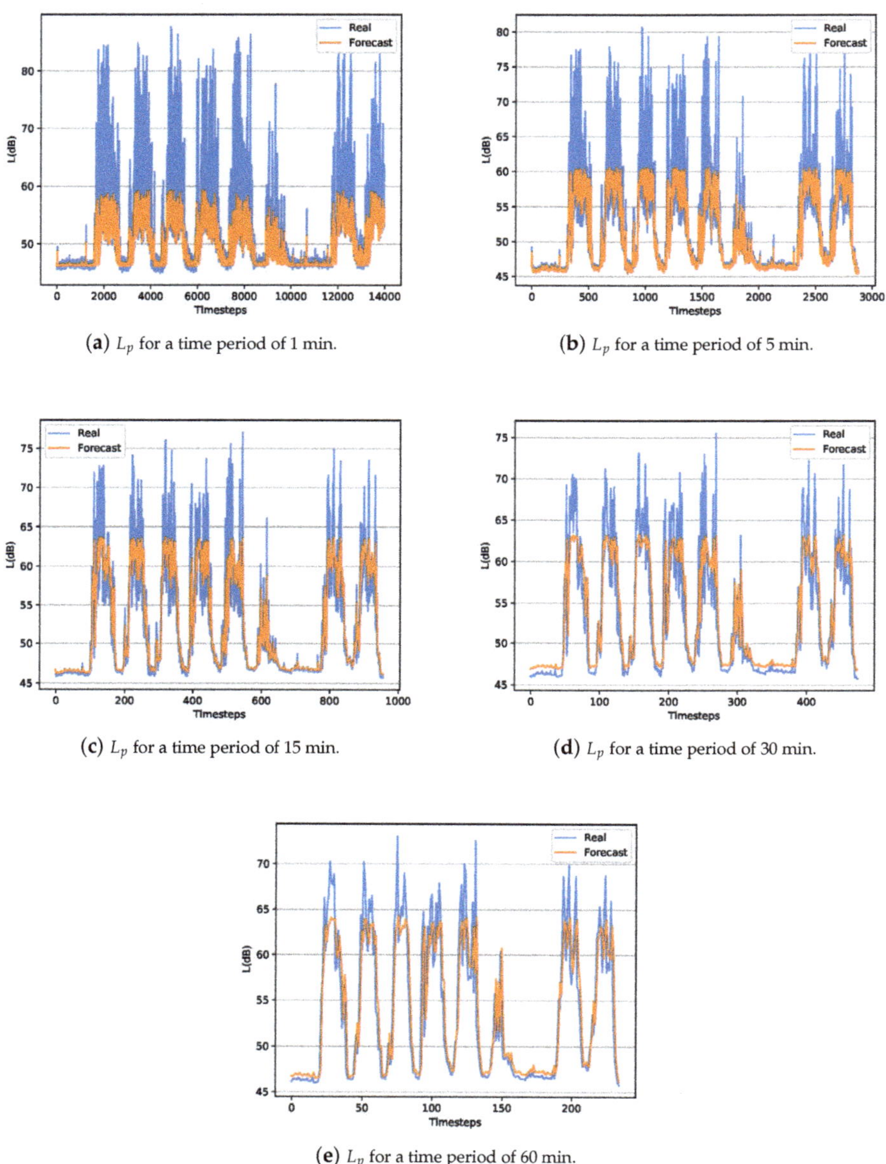

(**a**) L_p for a time period of 1 min.

(**b**) L_p for a time period of 5 min.

(**c**) L_p for a time period of 15 min.

(**d**) L_p for a time period of 30 min.

(**e**) L_p for a time period of 60 min.

Figure 2. Representation of captured and estimated LSTM data during approximately ten days test-interval (20% of the data) of L_p.

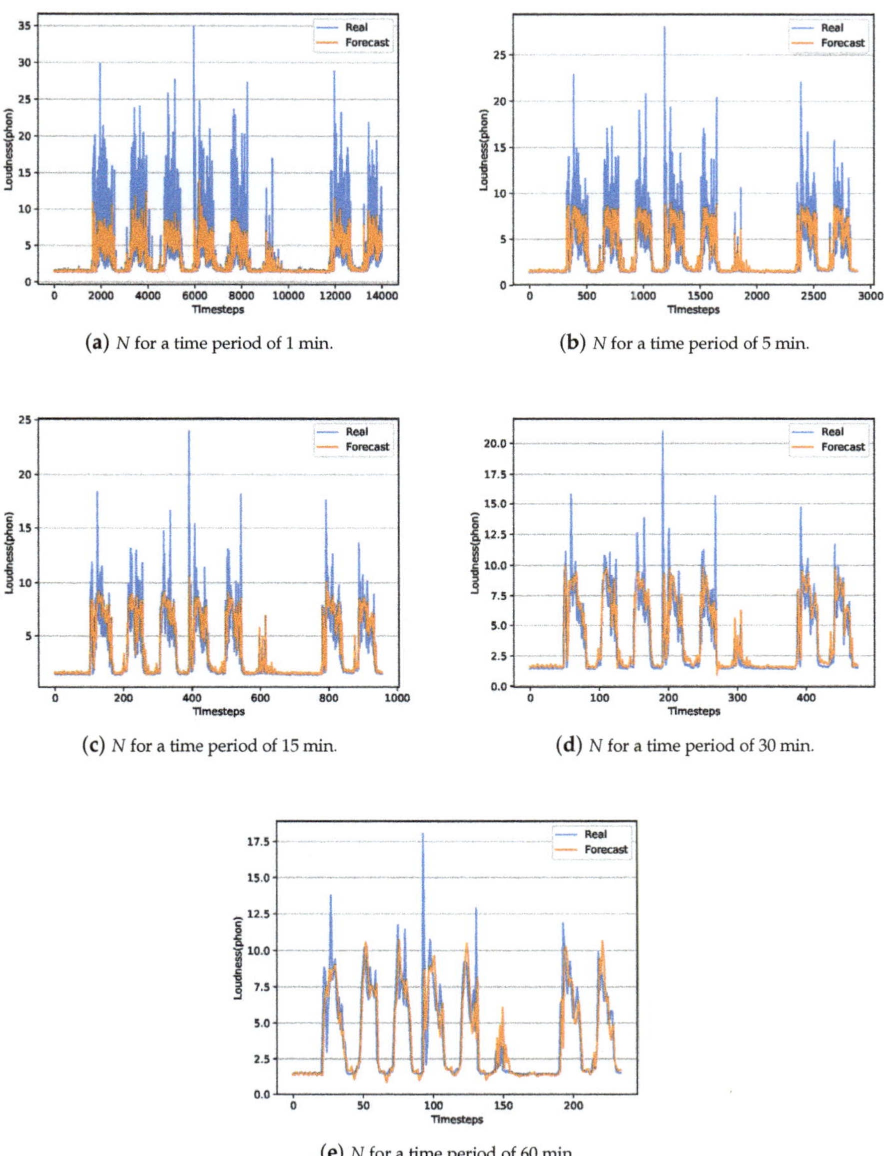

(**a**) N for a time period of 1 min.

(**b**) N for a time period of 5 min.

(**c**) N for a time period of 15 min.

(**d**) N for a time period of 30 min.

(**e**) N for a time period of 60 min.

Figure 3. Representation of captured and estimated LSTM data during approximately ten days test-interval (20% of the data) of N.

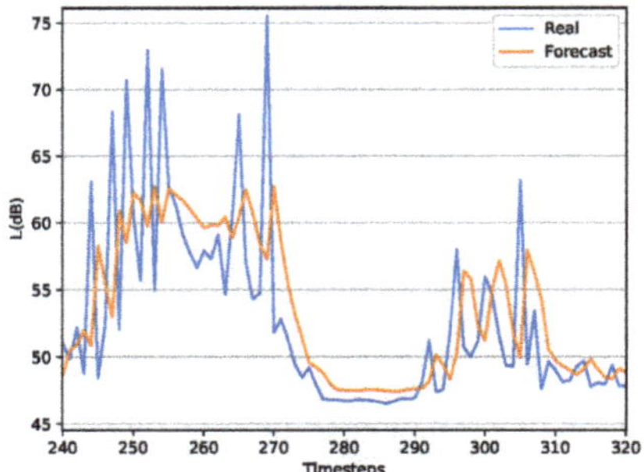

(**a**) Detail for two days of L_p for a time period of 30 min.

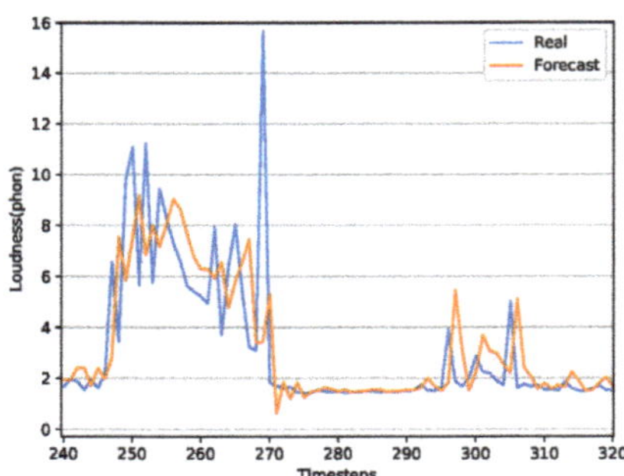

(**b**) Detail for two days of N for a time period of 30 min.

Figure 4. Detailed representation for two days, Friday and Saturday, for a time period of 30 min of captured and estimated LSTM data.

4.2. Assessing the Robustness of the Proposed LSTM Model

In the previous section, it was concluded that the LSTM technique can develop precise models for predicting the sound parameters L_p and N in short-term. In this section, a validation of the behavior, the stability and the robustness of the LSTM technique is carried out throughout different types of tests. The objective is to analyze the variability of the LSTM models when a greater amount of samples are predicted having a smaller amount of training samples. The validations that have been made are as follows:

- 80% train and 20% test (80/20)—approximately 40 days to train and 10 days to test (validation already done in the previous experiment, used to analyze and compare).
- 70% train and 30% test (70/30)—approximately 35 days to train and 15 days to test.
- 60% train and 40% test (60/40)—approximately 30 days to train and 20 days to test.

Table 5 shows the values of RMSE, MAE, PCC and R^2 of the validations indicated for noise60 and noise30 data-sets. Analyzing the results for the parameter L_p, it can be appreciated how independently of the type of validation the RMSE error is, around 4 dB for the noise60 data-set and around 4.3 dB for the noise30 data-set. The variations of the LSTM models for both data-sets are minimal when the type of validation performed is changed. These minimum variations can be seen with the value of R^2 that hardly suffers variations of 0.04 points. Regarding the N parameter, the results are very similar to the L_p parameter in terms of model variability. Analyzing the RMSE value of the N parameter, it is observed that it is around 2 dB for any of the two data-sets and any of the validations. The same happens with the determination coefficient R^2 where the differences between models of different validations and data-sets do not exceed 0.05 points. A remarkable aspect of the N parameter for the 60/40 validation is that it gets the best result than the other validations for both the noise30 and noise60 data-sets. The explanation for this situation can be that by obtaining more test days, these days include more weekends where the noise is more stable and there are fewer punctual peaks, hence the model fit is better.

Table 5. Representation of RMSE, MAE, PCC and R^2 for different training and test percentages of L_p and N values.

		Sound Pressure Level				Loudness			
Data-Set	**Train/Test**	**RMSE**	**MAE**	**PCC**	R^2	**RMSE**	**MAE**	**PCC**	R^2
noise30	**80/20**	4.27	2.85	0.83	0.69	2.01	1.05	0.78	0.59
	70/30	4.32	2.74	0.82	0.68	2.08	1.09	0.77	0.59
	60/40	4.51	3.15	0.84	0.65	2.00	1.05	0.80	0.63
noise60	**80/20**	3.94	2.75	0.86	0.75	1.99	0.99	0.79	0.60
	70/30	4.13	2.92	0.85	0.72	2.03	1.17	0.79	0.61
	60/40	4.05	3.13	0.86	0.74	1.97	1.14	0.82	0.65

After detailing and analyzing the results of the various performed validations together with the comparison with the ARIMA technique in the previous experiment, it can be concluded that the LSTM technique obtains a considerably stable and satisfactory performance for the problem posed. It must be taken into account that the challenges presented by the LSTM technique have allowed us to make reliable models regarding the error and the adjustment of the model using very few training samples and allowing a prediction of 20 consecutive days. Although the LSTM models created follow the trend of sound with a stable behavior, they present limitations in detecting impulsive short events, i.e., high peak noises at certain times.

5. Conclusions and Future Work

Wireless acoustic sensor networks are an important tool for monitoring and managing noise pollution in cities. In addition to economic cost savings as compared to traditional procedure to create a noise map, these networks are helping in the design of new noise maps with extended sound sources information and enabling existing noise maps to be updated dynamically. However, it must be taken into account that sensors within a network can fail or that network signal coverage may drop in certain situations, producing missing values in the IoT platform. Moreover, it would be helpful for local administrations to know in advance the trend in noise levels in cities in the temporal short-term. As a support to address these issues and even to decrease the number of necessary nodes in a network, the techniques of artificial intelligence can help through the execution of its different algorithms.

This paper proposes the use of a deep neural network, specifically a Long Short-Term Memory neural network (LSTM) to forecast future time values creating a model that represents the behavior of an acoustic environment in a certain location, specifically sound pressure level (L_p) and loudness values (N) parameter are contemplated. To create this model, values taken from a node of a deployed acoustic sensor network that collects information every minute have been used. Different models have been designed for L_p and N applying several time periods varied up to 60 min, in order to assess and analyze the behavior of the acoustic environment at different time intervals. To validate the model, it has been compared with the Auto Regressive Integrated Moving Average (ARIMA) time series technique, to evaluate and discuss the benefits and limitations of the proposed LSTM. Besides, to analyze the stability of the LSTM technique, several types of validations have been made. The results indicate that LSTM models obtain a lower prediction error and a better model fit than ARIMA. In general, the results achieved through the application of the LSTM technique are satisfactory since all the created models predict in a correct way the rising and falling trends of the sound levels. Moreover, obtained root mean square error values are lower than 4.3 dB for L_p and lower than 2 phons for N all considered models. Analyzing the parameters separately, using the N level more robust models than L_p are obtained, resulting in smaller error values and no significance differences between considered time periods. Regarding the L_p level models, a more reliable model is achieved when a higher time period is considered. Although L_p is a parameter with higher variance than N, the trend of the behavior pattern estimated by the model is satisfactory in terms of determination coefficient. Regarding the results of the different validations, these indicate that the proposed LSTM technique has little variability and needs little training data to obtain good predictions, therefore, the technique could be applied in any city, without the need to obtain long previous historical data. Regarding the limitations of the proposed LSTM technique, the difficulty of the model to follow the trend of high sound levels of the L_p and N parameters has been observed.

As a future work, an evaluation of the implementation of LSTM models within the nodes of the network of acoustic sensors is proposed. Moreover, a study to determine the influence of other climatic parameters or variables in predicting acoustic pollution through a multivariate neural network is of interest.

Author Contributions: Conceptualization, J.M.N. and R.M.; data curation, J.M.N. and R.M.; investigation, R.M.-E., A.B.-C. and J.M.C.; methodology, R.M.-E. and A.B.-C.; resources, J.M.N. and R.M.; software, R.M.-E. and A.B.-C.; supervision, J.M.N.; validation, J.M.N., R.M.-E., A.B.-C., R.M. and J.M.C.; writing—original draft, J.M.N., R.M.-E., A.B.-C. and R.M.; writing—review and editing, J.M.N., R.M.-E., A.B.-C., R.M. and J.M.C. All authors have read and agreed to the published version of the manuscript.

Funding: This work was partially supported by the Fundación Séneca del Centro de Coordinación de la Investigación de la Región de Murcia under Project 20813/PI/18.

Conflicts of Interest: The authors declare no conflict of interest.

References

1. European Commission. *END, Directive 2002/49/EC of the European Parliament and of the Council of 25 June 2002 relating to the Assessment and Management of Environmental Noise*; European Commission: Brussels, Belgium, July 2002.
2. Hornikx, M. Ten questions concerning computational urban acoustics. *Build. Environ.* **2016**, *106*, 409–421. [CrossRef]
3. Murphy, E.; King, E. A. Strategic environmental noise mapping: Methodological issues concerning the implementation of the EU Environmental Noise Directive and their policy implications. *Environ. Int.* **2010**, *36*, 290–298. [CrossRef] [PubMed]
4. Murphy, E.; Rice, H. J.; Meskell, C. Environmental noise prediction, noise mapping and GIS integration: The case of inner Dublin, Ireland. In Proceeding of the 8th International Transport Noise and Vibration Symposium, St. Petersburg, Russia, 4–6 June 2006.
5. Arana, M.; Martín, R. S.; Martin, M. L. S.; Aramendía, E. Strategic noise map of a major road carried out with two environmental prediction software packages. *Environ. Monit. Assess.* **2010**, *163*, 503–513. [CrossRef] [PubMed]
6. Garg, N; Maji, S. A critical review of principal traffic noise models: Strategies and implications. *Environ. Impact Assess. Rev.* **2014**, *46*, 68–81. [CrossRef]
7. Steele, C. A critical review of some traffic noise prediction models. *Appl. Acoust.* **2001**, *62*, 271–287. [CrossRef]
8. Guarnaccia, C. Advanced tools for traffic noise modelling and prediction. *WSEAS Trans. Syst.* **2013**, *12*, 121–130.
9. Barry, T; Reagan, J.A. *FHWA Highway Traffic Noise Prediction Model*; U.S. Department of Transportation: Washington, DC, USA, 1978.
10. Li, B; Tao, S.; Dawson, R.W.; Cao, J.; Lam, K. A GIS based road traffic noise prediction model. *Appl. Acoust.* **2002**, *63*, 679–691. [CrossRef]
11. van Leeuwen, H.J.A. Railway noise prediction models: A comparison. *J. Sound Vib.* **2000**, *231*, 975–987. [CrossRef]
12. Lui, W.K.; Li, K.M.; Ng, P. L.; Frommer, G. A comparative study of different numerical models for predicting train noise in high-rise cities. *Appl. Acoust.* **2006**, *67*, 432–449. [CrossRef]
13. van Leeuwen, H. J. A. Noise Predictions Models to Determine the Effect of Barriers Placed Alongside Railway Lines. *J. Sound Vib.* **1996**, *193*, 269–276. [CrossRef]
14. Oerlemans, S.; Schepers, J. G. Prediction of wind turbine noise and validation against experiment. *Int. J. Aeroacoust.* **2009**, *8*, 555–584. [CrossRef]
15. Tadamasa, A.; Zangeneh, M. Numerical prediction of wind turbine noise. *Renew. Energy* **2011**, *36*, 1902–1912. [CrossRef]

16. Maisonneuve, M.; Stevens, M.; Ochab, B. Participatory noise pollution monitoring using mobile phones. *Inf. Polity* **2010**, *15*, 51–71. [CrossRef]
17. Akyildiz, I.; Su, W.; Sankarasubramaniam, Y.; Cayirci, E. Wireless sensor networks: A survey. *Comput. Netw.* **2002**, *38*, 393–422. [CrossRef]
18. Peckens, C.; Porter, C.; Rink, T. Wireless sensor networks for long-term monitoring of urban noise. *Sensors* **2018**, *18*, 3161. [CrossRef]
19. Basten, T.; Wessels, P. An overview of sensor networks for environmental noise monitoring. In Proceedings of the 21st International Congress on Sound and Vibration (ICSV21), Beijing, China, 13–17 July 2014.
20. Alías, F.; Alsina-Pagés, R. Review of Wireless Acoustic Sensor Networks for Environmental Noise Monitoring in Smart Cities. *J. Sens.* **2019**. [CrossRef]
21. Mydlarz, C.; Salamon, J.; Bello J.P. The implementation of low-cost urban acoustic monitoring devices. *Appl. Acoust.* **2017**, *117*, 207–218. [CrossRef]
22. Camps-Farrés, J. Barcelona noise monitoring network. In Proceedings of the EuroNoise 2015, Maastricht, The Netherlands, 31 May–3 June 2015; pp. 2315–2320.
23. Bartalucci, C.; Borchi, F.; Carfagni, M.; Furferi, R.; Governi, L. Design of a prototype of a smart noise monitoring system. In Proceedings of the 24th International Congress on Sound and Vibration (ICSV24), London, UK, 23–27 July 2017; pp. 1–8.
24. Mietlicki, C.; Mietlicki, F. Medusa, a new approach for noise management and control in urban environment. In Proceedings of the 11th European Congress and Exposition on Noise Control Engineering (Euronoise2018), Crete, Greece, 27–31 May 2018; pp. 727–730.
25. Navarro, J. M.; Tomas-Gabarron, J. B.; Escolano, J. A big data framework for urban noise analysis and management in smart cities. *Acta Acust. United Acust.* **2017**, *103*, 552–560. [CrossRef]
26. Langkvist, M.; Karlsson, L.; Loutfi, A. A review of unsupervised feature learning and deep learning for time-series modeling. *Pattern Recognit. Lett.* **2014**, *42*, 11–24. [CrossRef]
27. Che, Z.; Purushotham, S.; Cho, K.; Sontag, D.; Liu, Y. Recurrent neural networks for multivariate time series with missing values. *Sci. Rep.* **2018**, *8*, 1–12. [CrossRef]
28. Wittenburg, G.; Dziengel, N.; Wartenburger, C.; Schiller, J. A system for distributed event detection in wireless sensor networks. In Proceedings of the 9th ACM/IEEE International Conference on Information Processing in Sensor Networks (IPSN'10), Stockholm, Sweden, 12–16 April 2010; pp. 94–104.
29. Kim, H.G.; Kim, J.Y. Environmental sound event detection in wireless acoustic sensor networks for home telemonitoring. *China Commun.* **2017**, *14*, 1–10. [CrossRef]
30. Luque, A.; Romero-Lemos, J.; Carrasco, A.; Barbancho, J. Improving Classification Algorithms by Considering Score Series in Wireless Acoustic Sensor Networks. *Sensors* **2018**, *18*, 2465. [CrossRef] [PubMed]
31. Zhang, Y.; Fu, Y.; Wang, R. Collaborative representation based classification for vehicle recognition in acoustic sensor networks. *J. Comput. Methods Sci. Eng.* **2018**, *18*, 349–358. [CrossRef]
32. Cobos, M.; Perez-Solano, J.J.; Felici-Castell, S.; Segura, J.; Navarro, J.M. Cumulative-sum-based localization of sound events in low-cost wireless acoustic sensor networks. *IEEE/ACM Trans. Audio Speech Lang. Process.* **2014**, *22*, 1792–1802. [CrossRef]
33. Sevillano, X.; Socoró, J.C.; Alías, F.; Bellucci, P.; Peruzzi, L.; Radaelli, S.; Benocci, R. DYNAMAP—Development of low cost sensors networks for real time noise mapping. *Noise Mapp.* **2016**, *3*, 172–189. [CrossRef]
34. Segura-Garcia, J.; Navarro-Ruiz, J.; Perez-Solano, J.; Montoya-Belmonte, J.; Felici-Castell, S.; Cobos, M. Torres-Aranda, Spatio-Temporal Analysis of Urban Acoustic Environments with Binaural Psycho-Acoustical Considerations for IoT-Based Applications. *Sensors* **2018**, *18*, 690. [CrossRef]
35. Bello, J. P.; Silva, C.; Nov, O.; Dubois, R.L.; Arora, A.; Salamon, J.; Doraiswamy, H. SONYC: A system for monitoring, analyzing, and mitigating urban noise pollution. *Commun. ACM* **2019**, *62*, 68–77. [CrossRef]
36. Jakob, A.; Marco, G.; Stephanie, K.; Robert, G.; Christian, K.; Tobias, C.; Hanna, L. A Distributed Sensor Network for Monitoring Noise Level and Noise Sources in Urban Environments. In Proceedings of the 2018 IEEE 6th International Conference on Future Internet of Things and Cloud (FiCloud), Barcelona, Spain, 6–8 August 2018; pp. 318–324.
37. Socoró, J.; Alías, F.; Alsina-Pagès, R. An anomalous noise events detector for dynamic road traffic noise mapping in real-life urban and suburban environments. *Sensors* **2017**, *17*, 2323. [CrossRef]

38. Li, Y.; Liu, M.; Drosos, K.; Virtanen, T. Sound event detection via dilated convolutional recurrent neural networks. *arXiv* **2019**, arXiv:1911.10888.
39. Yu, L.; Kang, J. Modeling subjective evaluation of soundscape quality in urban open spaces: An artificial neural network approach. *J. Acoust. Soc. Am.* **2009**, *126*, 1163–1174. [CrossRef]
40. Lopez-Ballester, J.; Pastor-Aparicio, A.; Segura-Garcia, J.; Felici-Castell, S.; Cobos, M. Computation of Psycho-Acoustic Annoyance Using Deep Neural Networks. *Appl. Sci.* **2019**, *9*, 3136. [CrossRef]
41. Mansourkhaki, A.; Berangi, M.; Haghiri, M.; Haghani, M. A neural network noise prediction model for Tehran urban roads. *J. Environ. Eng. Landsc. Manag.* **2018**, *26*, 88–97. [CrossRef]
42. Pedersen, K.; Transtrum, M.K.; Gee, K.L.; Butler, B.A.; James, M.M.; Salton, A.R. Machine learning-based ensemble model predictions of outdoor ambient sound levels. *Proc. Meet. Acoust.* **2018**, *35*, 022002.
43. Torija, A.J.; Ruiz, D.P.; Ramos-Ridao, A.F. Use of back-propagation neural networks to predict both level and temporal-spectral composition of sound pressure in urban sound environments. *Build. Environ.* **2012**, *52*, 45–56. [CrossRef]
44. Garg, N.; Soni, K.; Saxena, T.K.; Maji, S. Applications of Autoregressive integrated moving average (ARIMA) approach in time-series prediction of traffic noise pollution. *Noise Control. Eng. J.* **2015**, *63*, 182–194. [CrossRef]
45. Tong, W.; Li, L.; Zhou, X.; Hamilton A.; Zhang, K. Deep learning PM 2.5 concentrations with bidirectional LSTM RNN. *Air Qual. Atmos. Health* **2019**, *12*, 411–423. [CrossRef]
46. Krishan, M.; Jha, S.; Das, J.; Singh, A.; Goyal, M.K.; Sekar, C. Air quality modelling using long short-term memory (LSTM) over NCT-Delhi, India. *Air Qual. Atmos. Health* **2019**, *12*, 899–908. [CrossRef]
47. Noriega-Linares, J.E.; Rodriguez-Mayol, A.; Cobos, M.; Segura-Garcia, J.; Felici-Castell, S.; Navarro, J.M. A Wireless Acoustic Array System for Binaural Loudness Evaluation in Cities. *IEEE Sens. J.* **2017**, *17*, 7043–7052. [CrossRef]
48. Raspberry PI. Available online: https://www.raspberrypi.org (accessed on 6 February 2020).
49. Zwicker, E.; Fastl, H. *Psychoacoustics: Facts and Models*; Springer: Berlin, Germany, 2013; Volume 22.
50. Hochreiter, S.; Schmidhuber, J. LSTM can solve hard long time lag problems. In Proceedings of the 9th International Conference on Neural Information Processing Systems, Denver, CO, USA, 3–5 December 1996; pp. 473–479.
51. Brockwell, P. J.; Davis, R. A. *Introduction to Time Series and Forecasting*; Springer: Berlin, Germany, 2016.
52. Box, G.E.; Jenkins, G.M. *Time Series Analysis. Forecasting and Control*, Revised ed.; In Holden-Day Series in Time Series Analysis; Holden-Day: San Francisco, CA, USA, 1976.
53. Legates, D.R.; McCabe, G.J., Jr. Evaluating the use of "goodness-of-fit" measures in hydrologic and hydroclimatic model validation. *Water Resourc. Res.* **1999**, *35*, 233–241. [CrossRef]

Article

Fault Detection of Electric Impact Drills and Coffee Grinders Using Acoustic Signals

Adam Glowacz

Department of Automatic Control and Robotics, Faculty of Electrical Engineering, Automatics, Computer Science and Biomedical Engineering, AGH University of Science and Technology, al. A. Mickiewicza 30, 30-059 Kraków, Poland; adglow@agh.edu.pl

Received: 20 December 2018; Accepted: 8 January 2019; Published: 11 January 2019

Abstract: Increasing demand for higher safety of motors can be noticed in recent years. Developing of new fault detection techniques is related with higher safety of motors. This paper presents fault detection technique of an electric impact drill (EID), coffee grinder A (CG-A), and coffee grinder B (CG-B) using acoustic signals. The EID, CG-A, and CG-B use commutator motors. Measurement of acoustic signals of the EID, CG-A, and CG-B was carried out using a microphone. Five signals of the EID are analysed: healthy, with 15 broken rotor blades (faulty fan), with a bent spring, with a shifted brush (motor off), with a rear ball bearing fault. Four signals of the CG-A are analysed: healthy, with a heavily damaged rear sliding bearing, with a damaged shaft and heavily damaged rear sliding bearing, motor off. Three acoustic signals of the CG-B are analysed: healthy, with a light damaged rear sliding bearing, motor off. Methods such as: Root Mean Square (RMS), MSAF-17-MULTIEXPANDED-FILTER-14 are used for feature extraction. The MSAF-17-MULTIEXPANDED-FILTER-14 method is also developed and described in the paper. Classification is carried out using the Nearest Neighbour (NN) classifier. An acoustic based analysis is carried out. The results of the developed method MSAF-17-MULTIEXPANDED-FILTER-14 are very good (total efficiency of recognition of all classes—TE_D = 96%, $TE_{CG\text{-}A}$ = 97%, $TE_{CG\text{-}B}$ = 100%).

Keywords: motor; mechanical fault; detection; RMS; sound; drill; safety; pattern; bearing; fan; shaft

1. Introduction

Today rotating machinery is used for a wide variety of industrial applications such as electrical motors, engines, home appliances and electric power tools. It can also find applications in mining, oil, car, energy, and the steel industry. Cost-effective and non-destructive fault detection is profitable for industry. It can be used for rotating machinery. Reliable operation of rotating machinery is essential for many factories, oil refineries, industrial plants. Gas turbines, motors, pumps, aircraft engines, drive trains can be diagnosed by fault diagnosis techniques. Machines must operate safely without interruptions. If faults occur, the consequences can be catastrophic. Damaged machines generate costs, for example replacement of the machine or stopped production lines in the factory. Thus, the benefits of fault detection are maintenance cost savings.

There are lots of studies in the literature related to fault diagnosis and fault detection of rotating machinery. Analysis of electric currents is developed in the articles [1–5]. The results of current recognition are very good. However it can only be used for limited number of electrical faults such as broken bars, shorted rotors, stator coils. Electric current-based methods are usually useless for many mechanical faults such as damaged teeth on sprockets, faulty gears, faulty fans, etc. The next methods developed in the literature are based on vibration analysis [6–13] and acoustic analysis [14–22]. They are very effective. There is no need to connect a measuring sensor with the machine for acoustic-based measurements. Vibration-based measurements require a connection between the sensor and the machine. Vibration signals are less noisy than acoustic signals. Both of them can measure signals immediately. Vibration and acoustic analysis can also detect mechanical and electrical faults of rotating machinery.

The next method of fault detection is thermal analysis. Thermal analysis methods are described in [23–25]. Temperature detection can be performed using thermal imaging cameras, infrared thermometers and portable laser thermometers. If we use a thermal imaging camera or portable laser thermometer, then we can measure from a distance. The next method of fault detection of rotating machinery is oil analysis. It can provide diagnostic information about the condition of rotating machinery. In [26,27] some methods are mentioned: rotating disc electrode spectroscopy, inductively coupled plasma spectroscopy, FPQ-XRF, acid digestion, light blocking, light scattering, laser imaging, laser imaging, ferrography, light blocking, light scattering, laser imaging, fuel sniffer, gas chromatography, gravimetric, Karl Fischer titration, viscosity, etc. Multidimensional prognostics for rotating machinery was also presented [28].

This article describes the application of the acoustic-based approach to an electric impact drill (EID)—Verto 50G515, made in China, and two coffee grinders designated as coffee grinder A (CG-A)—Metrox ME-1497, made in China, and coffee grinder B (CG-B)—Sencor SCG 1050WH, made in China. The EID, CG-A, and CG-B use commutator motors. The commutator motor is a type of electrical motor used for power tools and home appliances such as blenders, coffee grinders and hair driers. The author analysed five electric impact drills (one healthy and four faulty). Each of them generates acoustic signals. Five signals are analysed: healthy (Figures 1 and 2), with 15 broken rotor blades (faulty fan) (Figure 3), with a bent spring (Figure 4), with a shifted brush (Figure 5), with a rear ball bearing fault (Figure 6).

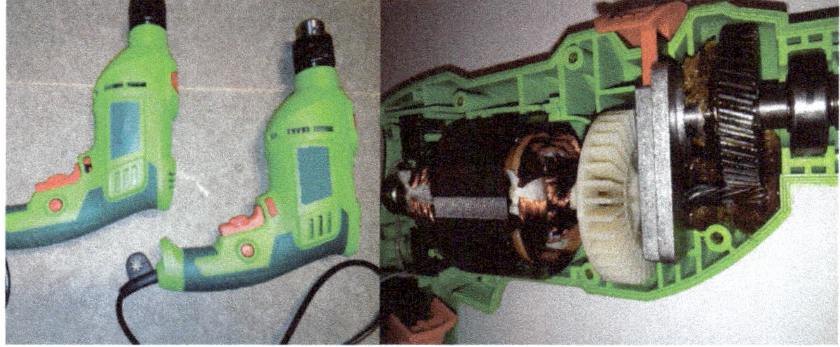

Figure 1. Healthy EID.

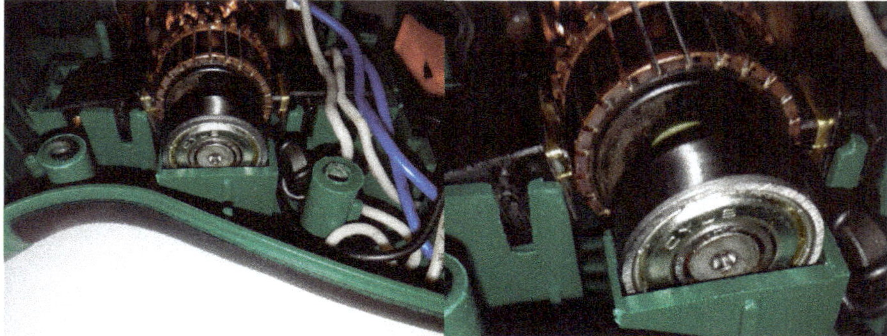

Figure 2. Healthy EID (EID with a healthy rear bearing).

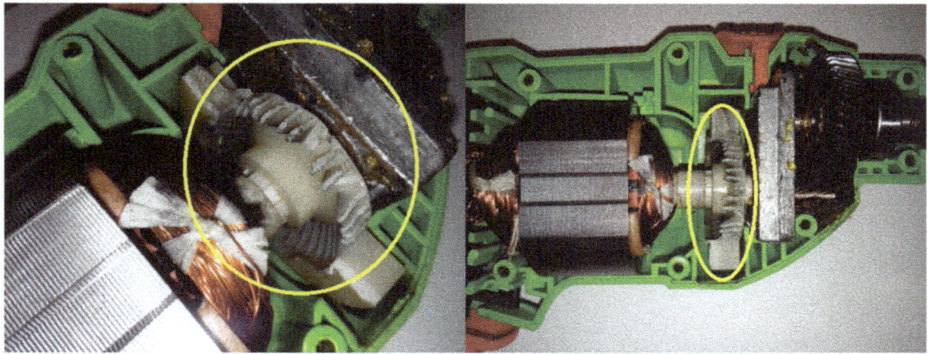

Figure 3. EID with 15 broken rotor blades (indicated by yellow circle).

Figure 4. EID with a bent spring (indicated by yellow circle).

Figure 5. EID with a shifted brush (indicated by yellow circle).

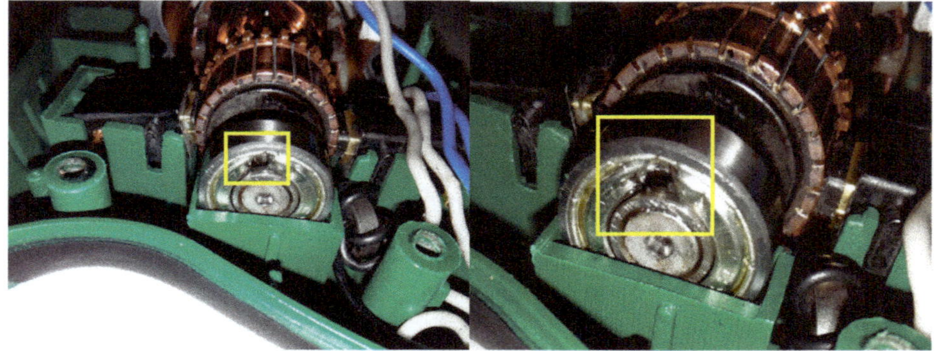

Figure 6. EID with a rear ball bearing fault (indicated by yellow square).

Four signals of the CG-A were analysed: healthy CG-A (Figure 7), CG-A with a heavily damaged rear sliding bearing (Figure 8), CG-A with a damaged shaft and heavily damaged rear sliding bearing (Figure 9), motor off (Figure 10).

Figure 7. Healthy CG-A.

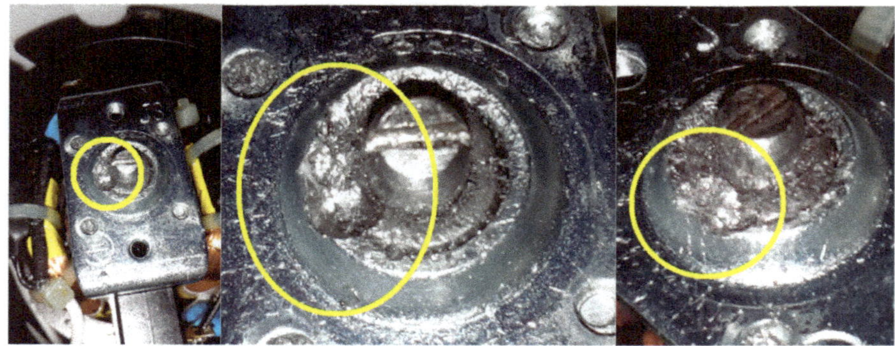

Figure 8. CG-A with a heavily damaged rear sliding bearing (indicated by yellow circle).

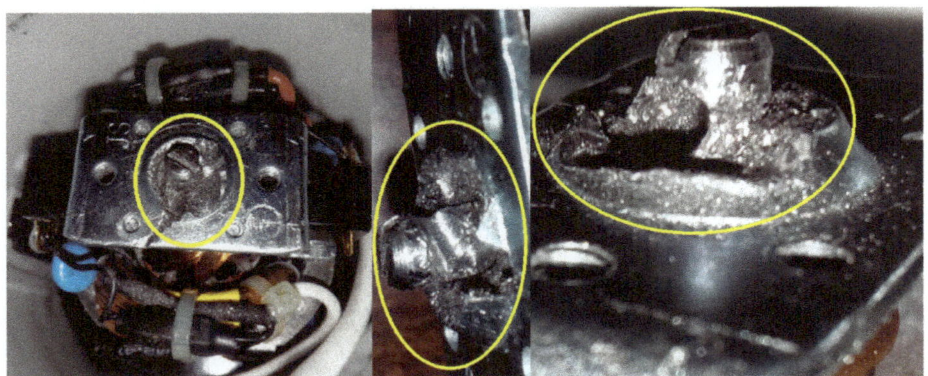

Figure 9. CG-A with a damaged shaft and heavily damaged rear sliding bearing (indicated by yellow circle).

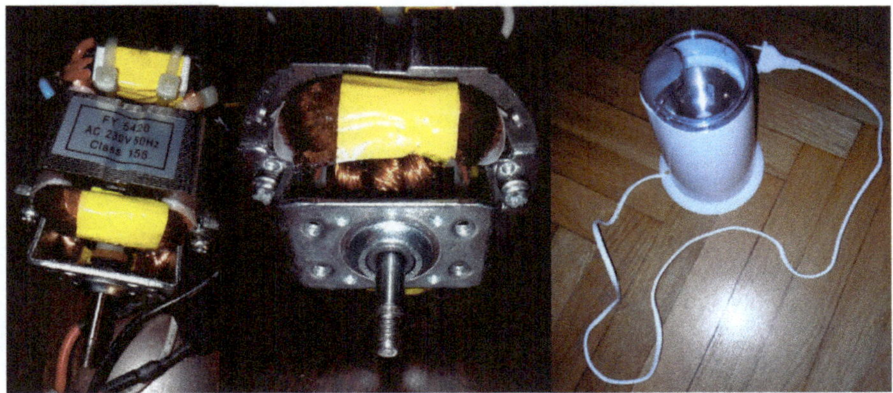

Figure 10. Motor off (CG-A off).

Three signals of the CG-B were analysed: healthy CG-B (Figure 11), CG-B with a light damaged rear sliding bearing (Figure 12), motor off (Figure 13).

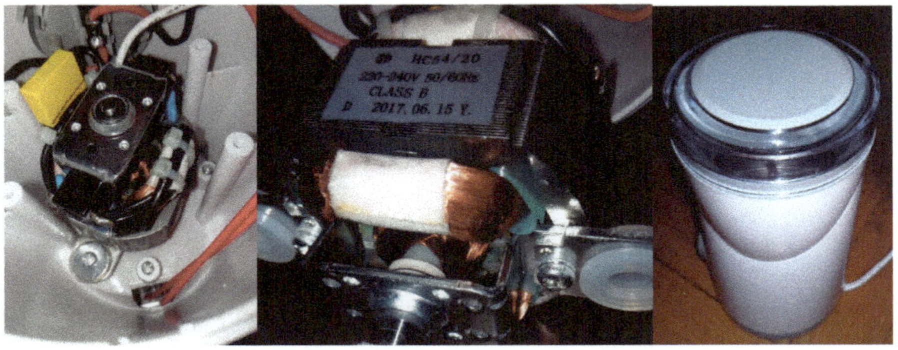

Figure 11. Healthy CG-B.

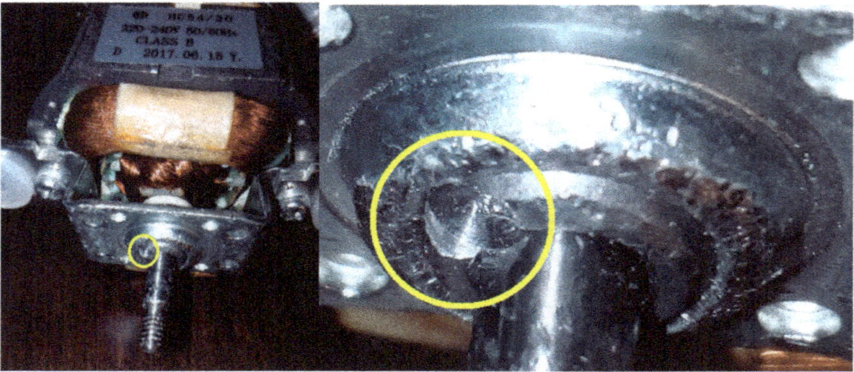

Figure 12. CG-B with a light damaged rear sliding bearing (indicated by yellow circle).

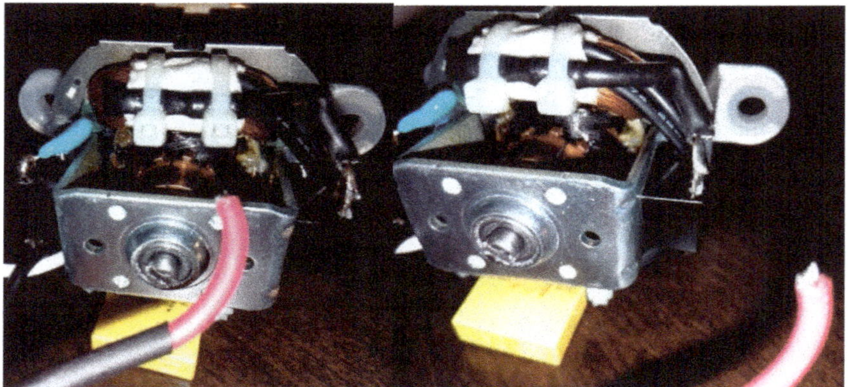

Figure 13. Motor off (CG-B off).

In Section 1, the author presents a review of the fault detection methods. In Section 2, the author describes the acoustic based approach and proposed methods of signal processing. In Section 3, the recognition results of the EID, CG-A, and CG-B are presented. A discussion is presented in Section 4. In Section 5, summary and conclusions are described.

2. Developed Acoustic Based Approach

The developed acoustic-based approach used signal processing methods and the acoustic data of the EID, CG-A and CG-B. Acoustic data were obtained using a HAMA 00057152 microphone. The parameters of the microphone are: frequency response 30–16,000 Hz, rated impedance 1400 Ω, sensitivity −62 dB. The microphone was placed 0.2–0.3 m away from the EID, CG-A and CG-B. Other types of microphones could be also used. Acoustic data were split (using "MPlayer library—The Movie Player"—wav file parameters sampling frequency 44,100 Hz, single channel, 16 bits resolution, stationary signal) and normalized. Normalization of amplitude divided each sample (in the time domain) by the maximum value of the signal (in time domain). After that feature vectors were formed using the RMS or MSAF-17-MULTIEXPANDED-FILTER-14 (the methodology is presented in Section 2.1). Next the Nearest Neighbour (NN) classifier compared feature vectors in the classification step. The developed acoustic based approach is shown in Figure 14.

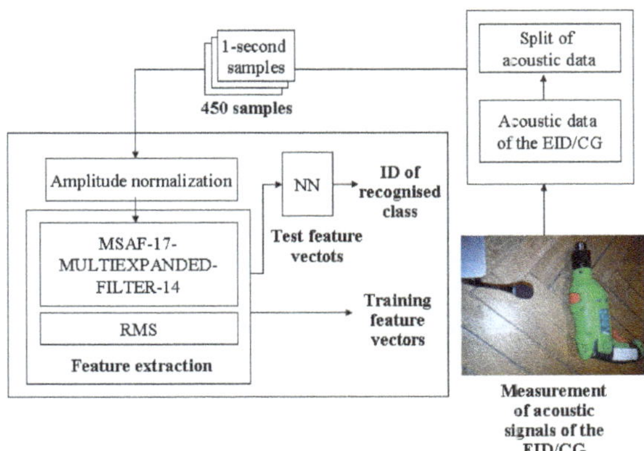

Figure 14. Developed acoustic based approach.

An experimental setup consisted of the microphone and a computer. It was used to analyse the electric impact drill/coffee grinder (Figure 15a). Measurement of acoustic signals is depicted in Figure 15b.

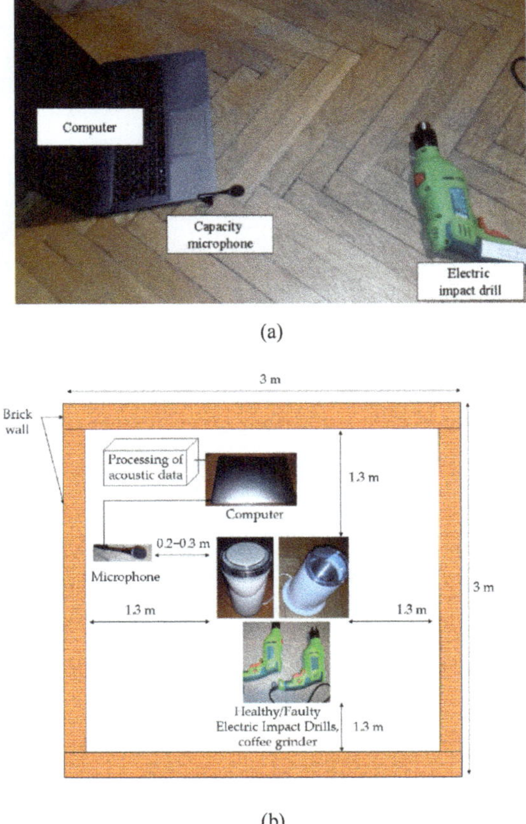

Figure 15. (**a**) Capacity microphone, computer and electric impact drill. (**b**) Measurement of acoustic signals.

2.1. MSAF-17-MULTIEXPANDED-FILTER-14

The Method of Selection of Amplitudes of Frequency Multiexpanded Filter (MSAF-17-MULTIEXPANDED-FILTER-14) was developed and implemented. This feature extraction method used differences between FFT spectra. It consists of seven signal processing steps:

(1) Compute Fast Fourier Transform (FFT) spectra for all states of the EID (for all training vectors). In the presented acoustic based approach the FFT provided a vector of 16384-elements. For 16,384 frequency components, the frequency spectrum is 22,050 Hz. Therefore, each frequency component is every 1.345 Hz. The computed vectors were defined as follows: healthy EID—$\mathbf{h} = [h_1, h_2, ..., h_{16,384}]$, EID with 15 broken rotor blades (faulty fan)—$\mathbf{f} = [f_1, f_2, ..., f_{16,384}]$, EID with a bent spring—$\mathbf{s} = [s_1, s_2, ..., s_{16,384}]$, EID with a rear ball bearing fault—$\mathbf{b} = [b_1, b_2, ..., b_{16,384}]$.

(2) For each training vector compute: $\mathbf{h} - \mathbf{f}$, $\mathbf{h} - \mathbf{s}$, $\mathbf{f} - \mathbf{s}$, $\mathbf{b} - \mathbf{h}$, $\mathbf{b} - \mathbf{f}$, $\mathbf{b} - \mathbf{s}$.

(3) Compute: $|\mathbf{h} - \mathbf{f}|$, $|\mathbf{h} - \mathbf{s}|$, $|\mathbf{f} - \mathbf{s}|$, $|\mathbf{b} - \mathbf{h}|$, $|\mathbf{b} - \mathbf{f}|$, $|\mathbf{b} - \mathbf{s}|$.

(4) Find 1–17 *Common Frequency Components* (*CFCs*) or set a parameter *Threshold of CFCs (ToCFCs)*. If there are no *CFCs*, then set a parameter *ToCFCs*. The parameter is defined as Equation (1):

$$ToCFCs = \frac{Number \quad of \quad required \quad CFCs}{Number \quad of \quad all \quad differences} \tag{1}$$

Let's analyse the following example: three training sets are given. Each of them has four training samples. Eighteen differences are computed (six for the first training set, six for the second training set, six for the third training set). Let's suppose that frequency component 130 Hz is found three times for $|\mathbf{h} - \mathbf{f}|$. Let's suppose that frequency components 110, 160 Hz are found two times for $|\mathbf{h} - \mathbf{s}|$. Let's suppose that frequency components 110, 140 Hz are found two times for $|\mathbf{f} - \mathbf{s}|$. Let's suppose that frequency component 500 Hz is found three times for $|\mathbf{b} - \mathbf{h}|$. Let's suppose that frequency components 600, 610 Hz are found two times for $|\mathbf{b} - \mathbf{f}|$. Let's suppose that frequency components 600, 710 Hz are found two times for $|\mathbf{b} - \mathbf{s}|$. There are no *CFCs*. Only frequency components 110 Hz and 600 Hz are found four times. The MSAF-17-MULTIEXPANDED finds frequency components 110, 130, 140, 160, 500, 600, 610, 710 Hz, if *ToCFCs* is equal to 0.1111 (2/18). The MSAF-17-MULTIEXPANDED method finds 0 frequency components, if *ToCFCs* is equal to 0.2777 (5/18).

(5) Form groups of frequency components for a proper recognition. Considering the presented example, it can be noticed that the frequency component 110 Hz is good for $|\mathbf{h} - \mathbf{s}|$ and $|\mathbf{f} - \mathbf{s}|$. The frequency component 130 Hz is good for $|\mathbf{h} - \mathbf{f}|$. The frequency component 500 Hz is good for $|\mathbf{b} - \mathbf{h}|$. The frequency component 600 Hz is good for $|\mathbf{b} - \mathbf{f}|$ and $|\mathbf{b} - \mathbf{s}|$. The MSAF-17-MULTIEXPANDED-FILTER-14 finds 1 group consisted of 110, 130, 500, 600 Hz.

(6) Form bandwidths of frequency. Considering the presented example, 14 Hz bandwidths are selected. The MSAF-17-MULTIEXPANDED-FILTER-14 uses a value of 14 Hz. The value of 14 Hz is set experimentally. The middle of the first bandwidth is located at 110 Hz. The middle of the second bandwidth is located at 130 Hz. The middle of the third bandwidth is located at 500 Hz. The middle of the fourth bandwidth is located at 600 Hz. Following bandwidths are selected <103–117 Hz>, <123–137 Hz >, <493–507 Hz>, <593–607 Hz>.

(7) Using computed bandwidths, form a feature vector.

In other words, we can say that: 17—means that, we analyse 17 (local) maximum values of analysed difference between FFT spectra of acoustic signals, for example $|\mathbf{h} - \mathbf{f}|$, 14—means that, we set 14 Hz frequency bandwidth, for example for frequency 50 Hz it will be <50 − 7 Hz, 50 + 7Hz>.

A block diagram of the developed method MSAF-17-MULTIEXPANDED-FILTER-14 is presented in Figure 16.

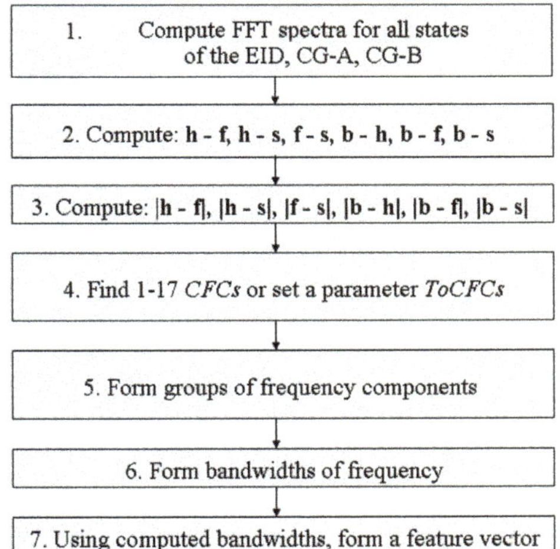

Figure 16. Block diagram of the developed method MSAF-17-MULTIEXPANDED-FILTER-14.

Differences between FFT spectra |**h** − **f**|, |**h** − **s**|, |**f** − **s**|, |**b** − **h**|, |**b** − **f**|, |**b** − **s**| were computed and are presented in Figures 17–22.

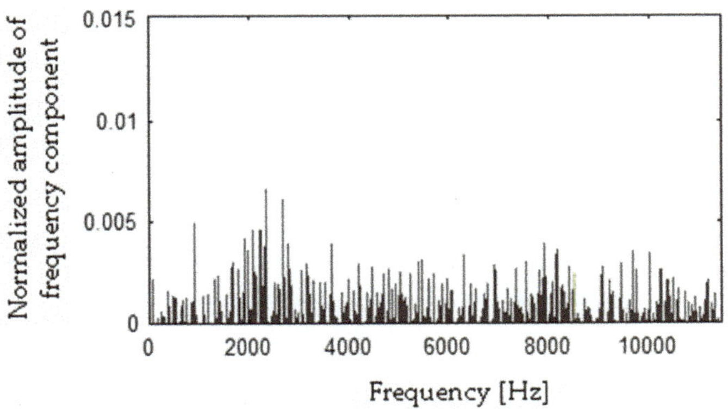

Figure 17. Difference (|**h** − **f**|) using the MSAF-17-MULTIEXPANDED-FILTER-14 method.

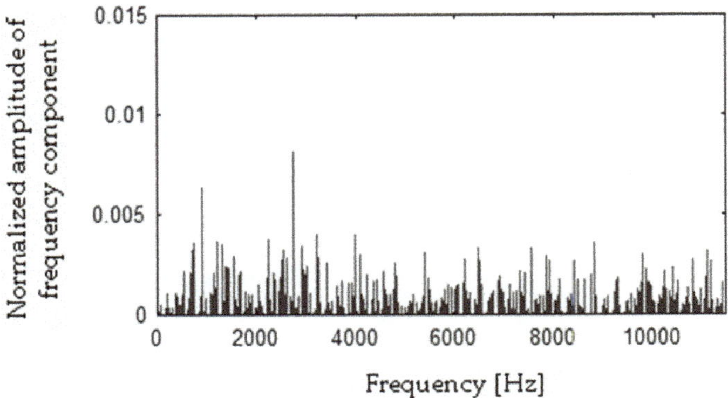

Figure 18. Difference (|**h** − **s**|) using the MSAF-17-MULTIEXPANDED-FILTER-14 method.

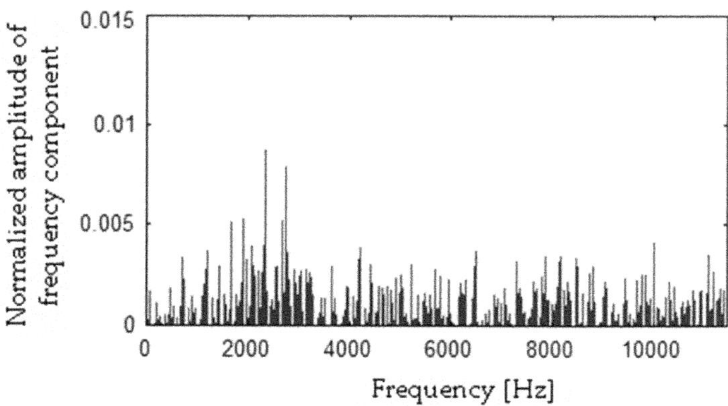

Figure 19. Difference (| **f** − **s** |) using the MSAF-17-MULTIEXPANDED-FILTER-14 method.

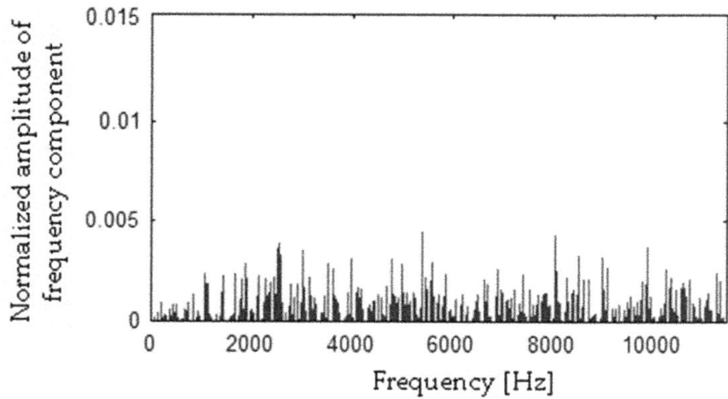

Figure 20. Difference (| **b** − **h** |) using the MSAF-17-MULTIEXPANDED-FILTER-14 method.

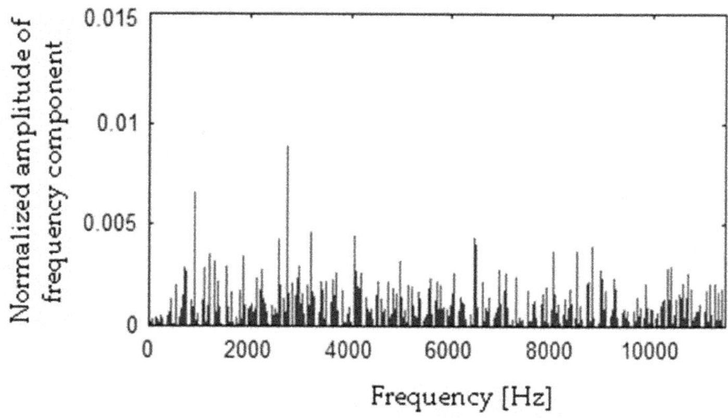

Figure 21. Difference (| **b** − **s** |) using the MSAF-17-MULTIEXPANDED-FILTER-14 method.

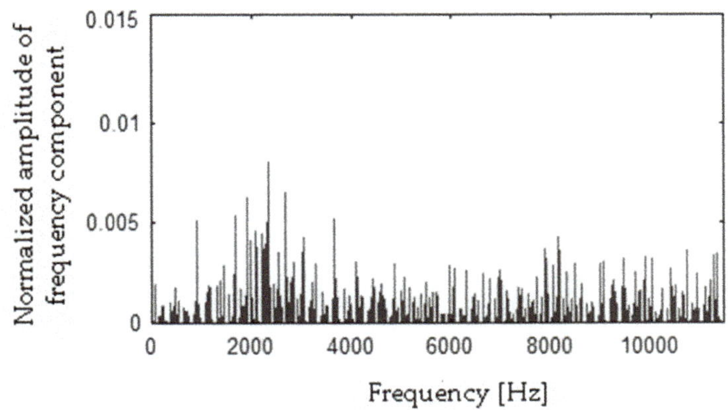

Figure 22. Difference (|**b** − **f**|) using the MSAF-17-MULTIEXPANDED-FILTER-14 method.

The developed method MSAF-17-MULTIEXPANDED-FILTER-14 found the following frequency components: 278, 280, 457, 464, 468, 477, 479, 480, 481, 483, 557, 558, 2297, 2313, 2316, 2317, 11098, 11099, 11103, 11106, 11110, 11111, 11190, 11192, 11193, 11197, 11198, 11205, 11207, 11208, 11209, 11213, 11239, 11240, 11242, 11244, 11246 Hz.

Next the MSAF-17-MULTIEXPANDED-FILTER-14 selected seven frequency bandwidths of the EID: <271–287 Hz>, <450–490 Hz>, <550–565 Hz>, <2290–2324 Hz>, <11091–11118 Hz>, <11183–11220 Hz>, <11232–11253 Hz>.

The frequency component 278 Hz was found, so the first frequency bandwidth is 271–285 Hz (278 − 7 Hz, 278 + 7 Hz). The MSAF-17-MULTIEXPANDED-FILTER-14 method computed frequency bandwidth 14 Hz. It can be noticed that frequency component 280 Hz is within the frequency bandwidth. Thus, the frequency bandwidth is 271–287 Hz etc. The selected frequency bandwidths/features of the EID were depicted in Figures 23–26. The value of the parameter *ToCFCs* was equal to 0.25 for the EID.

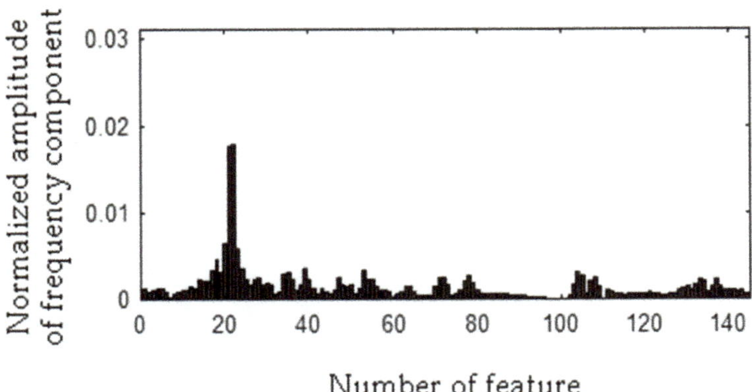

Figure 23. Values of features of healthy EID (145 features, seven frequency bandwidths, <271–287 Hz>, <450–490 Hz>, <550–565 Hz>, <2290–2324 Hz>, <11091–11118 Hz>, <11183–11220 Hz>, <11232–11253 Hz>).

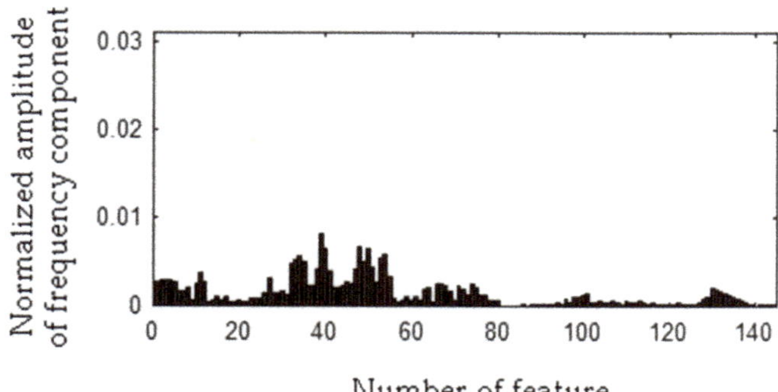

Figure 24. Values of features of the EID with 15 broken rotor blades (faulty fan) (145 features, seven frequency bandwidths, <271–287 Hz>, <450–490 Hz>, <550–565 Hz>, <2290–2324 Hz>, <11091–11118 Hz>, <11183–11220 Hz>, <11232–11253 Hz>).

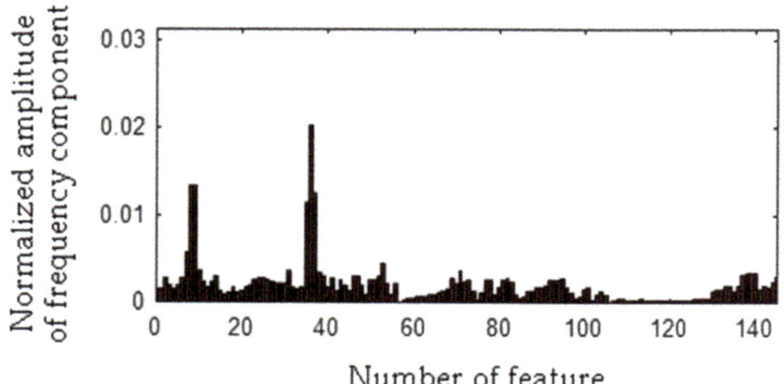

Figure 25. Values of features of the EID with a bent spring (145 features, seven frequency bandwidths, <271–287 Hz>, <450–490 Hz>, <550–565 Hz>, <2290–2324 Hz>, <11091–11118 Hz>, <11183–11220 Hz>, <11232–11253 Hz>).

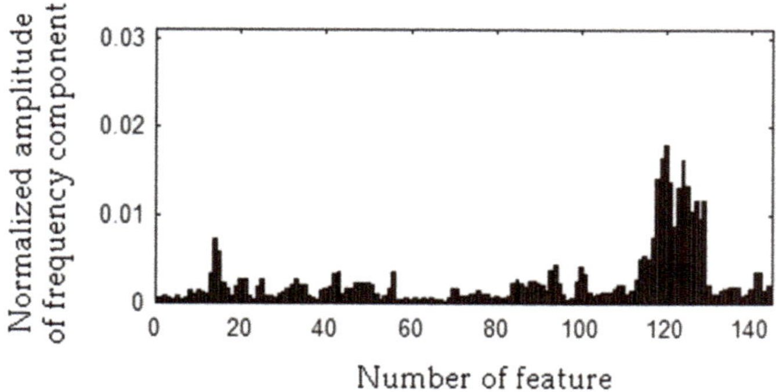

Figure 26. Values of features of the EID with a rear ball bearing fault (145 features, seven frequency bandwidths, <271–287 Hz>, <450–490 Hz>, <550–565 Hz>, <2290–2324 Hz>, <11091–11118 Hz>, <11183–11220 Hz>, <11232–11253 Hz>).

The MSAF-17-MULTIEXPANDED-FILTER-14 selected two frequency bandwidths of the CG-A: <515–537 Hz>, <1560–1575 Hz>. The selected frequency bandwidths/features of the CG-A are depicted in Figures 27–29. The value of the parameter *ToCFCs* was equal to 0.5 for the CG-A.

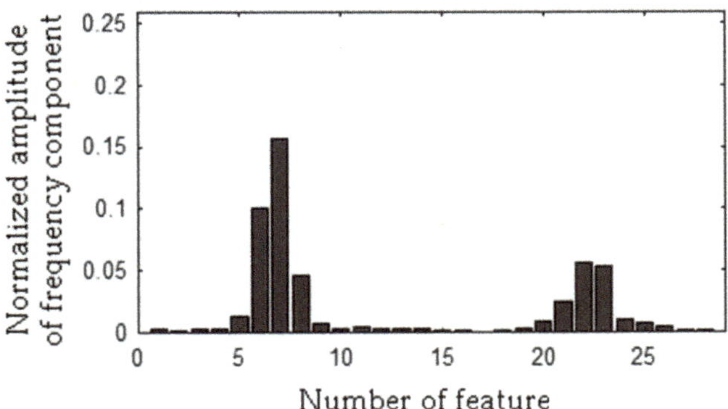

Figure 27. Values of features of the healthy CG-A (29 features, two frequency bandwidths, <515–537 Hz>, <1560–1575 Hz>).

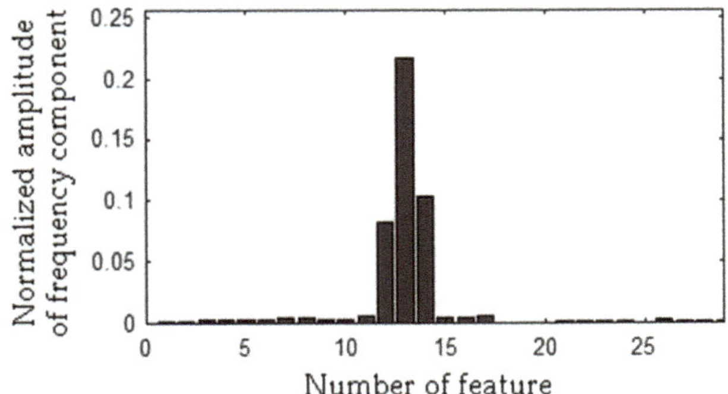

Figure 28. Values of features of the CG-A with a heavily damaged rear sliding bearing (29 features, two frequency bandwidths, <515–537 Hz>, <1560–1575 Hz>).

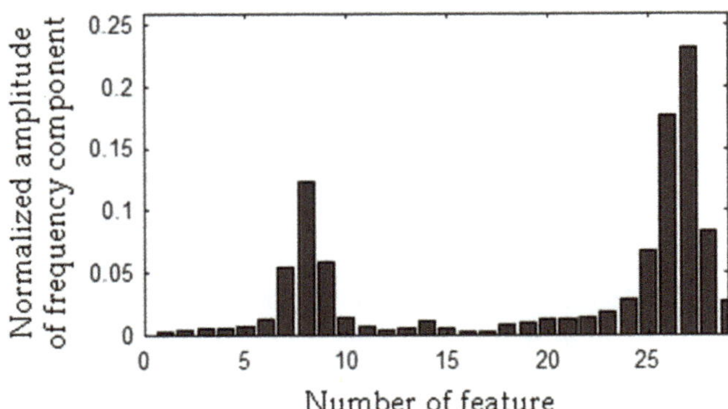

Figure 29. Values of features of the CG-A with a damaged shaft and heavily damaged rear sliding bearing (29 features, two frequency bandwidths, <515–537 Hz>, <1560–1575 Hz>).

The MSAF-17-MULTIEXPANDED-FILTER-14 selected three frequency bandwidths of the CG-B: <94–109 Hz>, <194–207 Hz>, <463–488 Hz>. The selected frequency bandwidths/features of the CG-B are depicted in Figures 30 and 31. The value of the parameter *ToCFCs* was equal to 0.5 for the CG-B.

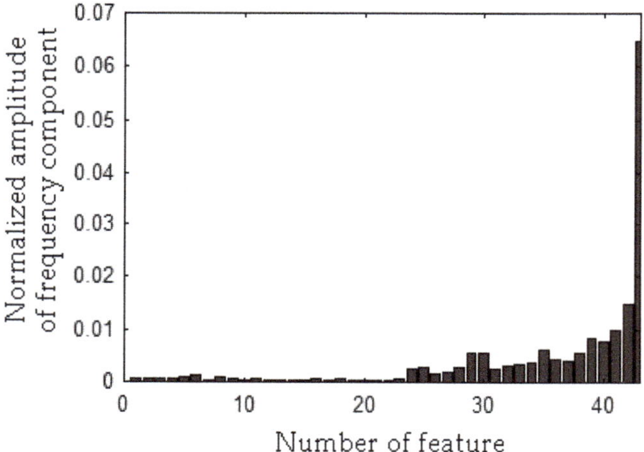

Figure 30. Values of features of the healthy CG-B (43 features, three frequency bandwidths, <94–109 Hz>, <194–207 Hz>, <463–488 Hz>).

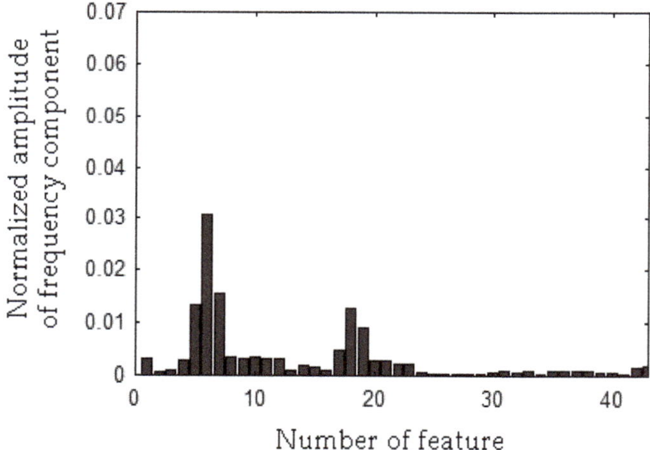

Figure 31. Values of features of the CG-B with a light damaged rear sliding bearing (43 features, three frequency bandwidths, <94–109 Hz>, <194–207 Hz>, <463–488 Hz>).

Next computed features were classified. To classify features the NN classifier [29–31] was used (please see Section 2.3). There are 145 features in the feature vector. It can be noticed that distance classifiers (for example: k-means, Nearest Mean) should have also good results. Fuzzy classifiers [32] and neural network [33–35] can be also suitable for the acoustic-based approach. The NN classifier was selected because of its good recognition efficiency for multi-dimensional vectors.

2.2. RMS

The second method of feature extraction used for the proposed acoustic based approach is the Root Mean Square (RMS). The RMS is a well-known method for feature extraction. It is defined as Equation (2):

$$x_{RMS} = \sqrt{\frac{1}{n}(x_1^2 + x_2^2 + \ldots + x_n^2)} \tag{2}$$

where x_{RMS}—RMS for 1-s sample (44,100 values), n—number of all samples, n = 44,100, x_1, ..., x_n—values of samples 1, ..., n (sampling rate 44,100 Hz).

In the presented analysis (please see Section 3) the author used 50 1-s samples for each class of the EID. Two hundred and fifty 1-s samples were used for five classes (of the EID). There were x_{RMS1}, ..., x_{RMS50}—RMS values of the healthy EID, x_{RMS51}, ..., x_{RMS100}—RMS values of the EID with 15 broken rotor blades (faulty fan), x_{RMS101}, ..., x_{RMS150}—RMS values of the EID with a bent spring, x_{RMS151}, ..., x_{RMS200} – RMS values of the EID with a shifted brush (motor off), x_{RMS201}, ..., x_{RMS250}—RMS values of the EID with a rear ball bearing fault. The computed RMS values of the EID are presented in Tables 1–5.

Table 1. RMS values of the healthy EID.

Number of Samples	RMS Value	Number of Samples	RMS Value
x_{RMS1}	0.237122	x_{RMS5}	0.240819
x_{RMS2}	0.231192	x_{RMS6}	0.236356
x_{RMS3}	0.234878	x_{RMS7}	0.239650
x_{RMS4}	0.238282	x_{RMS8}	0.238406

Table 2. RMS values of the EID with 15 broken rotor blades (faulty fan).

Number of Samples	RMS Value	Number of Samples	RMS Value
x_{RMS51}	0.322252	x_{RMS55}	0.312347
x_{RMS52}	0.316197	x_{RMS56}	0.318529
x_{RMS53}	0.317383	x_{RMS57}	0.310883
x_{RMS54}	0.305535	x_{RMS58}	0.302719

Table 3. RMS values of the EID with a bent spring.

Number of Samples	RMS Value	Number of Samples	RMS Value
x_{RMS101}	0.250579	x_{RMS105}	0.245578
x_{RMS102}	0.244888	x_{RMS106}	0.243813
x_{RMS103}	0.244461	x_{RMS107}	0.246395
x_{RMS104}	0.249611	x_{RMS108}	0.246297

Table 4. RMS values of the EID with a shifted brush.

Number of Samples	RMS Value	Number of Samples	RMS Value
x_{RMS151}	0.006427	x_{RMS155}	0.006478
x_{RMS152}	0.006338	x_{RMS156}	0.007226
x_{RMS153}	0.008981	x_{RMS157}	0.007020
x_{RMS154}	0.009021	x_{RMS158}	0.006644

Table 5. RMS values of the EID with a rear ball bearing fault.

Number of Samples	RMS Value	Number of Samples	RMS Value
x_{RMS201}	0.235278	x_{RMS205}	0.234696
x_{RMS202}	0.236730	x_{RMS206}	0.236078
x_{RMS203}	0.233518	x_{RMS207}	0.237600
x_{RMS204}	0.234478	x_{RMS208}	0.237778

The values of the RMS of acoustic signals "Healthy EID" and "EID with a rear ball bearing fault" were similar. It will be difficult to recognise these two classes. In the presented analysis (please see Section 3) the author used 50 1-s samples for each class of the CG-A. Two hundred 1-s samples were used for four classes (of the CG-A). There were x_{RMS251}, ..., x_{RMS300}—RMS values of the healthy CG-A, x_{RMS301}, ..., x_{RMS350}—RMS values of the CG-A with a heavily damaged rear sliding bearing, x_{RMS351},

..., x_{RMS400} – RMS values of the CG-A with a damaged shaft and heavily damaged rear sliding bearing, x_{RMS401}, ..., x_{RMS450}—RMS values of the motor off (CG-A off). The values x_{RMS401}, ..., x_{RMS450} were the same as RMS values of the EID with a shifted brush (EID off). The computed RMS values of the CG-A are presented in Tables 6–8.

Table 6. RMS values of the healthy CG-A.

Number of Samples	RMS Value	Number of Samples	RMS Value
x_{RMS251}	0.203343	x_{RMS255}	0.209252
x_{RMS252}	0.203521	x_{RMS256}	0.215012
x_{RMS253}	0.201109	x_{RMS257}	0.209241
x_{RMS254}	0.205511	x_{RMS258}	0.205984

Table 7. RMS values of the CG-A with a heavily damaged rear sliding bearing.

Number of Samples	RMS Value	Number of Samples	RMS Value
x_{RMS301}	0.234359	x_{RMS305}	0.234927
x_{RMS302}	0.234860	x_{RMS306}	0.233882
x_{RMS303}	0.231783	x_{RMS307}	0.235229
x_{RMS304}	0.237120	x_{RMS308}	0.229835

Table 8. RMS values of the CG-A with a damaged shaft and heavily damaged rear sliding bearing.

Number of Samples	RMS Value	Number of Samples	RMS Value
x_{RMS351}	0.239449	x_{RMS355}	0.248779
x_{RMS352}	0.246317	x_{RMS356}	0.250027
x_{RMS353}	0.246894	x_{RMS357}	0.250791
x_{RMS354}	0.247325	x_{RMS358}	0.250203

The values of the RMS of acoustic signals "CG-A with a heavily damaged rear sliding bearing" and "CG-A with a damaged shaft and heavily damaged rear sliding bearing" were similar. It will be difficult to recognise these two classes.

In the presented analysis (please see Section 3) the author used 50 1-s samples for each class of the CG-B. One hundred and fifty 1-s samples were used for three classes (of the CG-B). There were x_{RMS451}, ..., x_{RMS500}—RMS values of the healthy CG-B, x_{RMS501}, ..., x_{RMS550}—RMS values of the CG-B with a light damaged rear sliding bearing, x_{RMS551}, ..., x_{RMS600}—RMS values of the motor off (CG-B off). The values x_{RMS551}, ..., x_{RMS600} were the same as RMS values of the EID with a shifted brush (EID off). The computed RMS values of the CG-B are presented in Tables 9 and 10.

Table 9. RMS values of the healthy CG-B.

Number of Samples	RMS Value	Number of Samples	RMS Value
x_{RMS451}	0.248146	x_{RMS455}	0.248331
x_{RMS452}	0.254812	x_{RMS456}	0.259062
x_{RMS453}	0.248951	x_{RMS457}	0.263240
x_{RMS454}	0.240446	x_{RMS458}	0.264600

Table 10. RMS values of the CG-B with a lightly damaged rear sliding bearing.

Number of Samples	RMS Value	Number of Samples	RMS Value
x_{RMS501}	0.131587	x_{RMS505}	0.103367
x_{RMS502}	0.121155	x_{RMS506}	0.095910
x_{RMS503}	0.103567	x_{RMS507}	0.108105
x_{RMS504}	0.094650	x_{RMS508}	0.105756

2.3. NN Classifier

The NN classifier is very known in the literature [29–31]. This type of a classifier is based on lazy learning. It does not generalize the training data. Each training feature vector has a label with a class (ID of the class). The label (ID of the class) is given to the feature vector in the training phase.

An unlabeled test feature vector is used in the classification (testing) phase. The NN classifier assigns the label, which is the closest to the training data. For this reason, distance metric is used. The author used Euclidean distance, although other distance functions could be used. Similar results were obtained using other distance functions (Manhattan distance and Minkowski distance). Euclidean distance was defined as Equation (3):

$$ED(\mathbf{x}-\mathbf{y}) = \sqrt{\sum_{i=1}^{n} |(x_i - y_i)^2|} \quad (3)$$

where **x**—test feature vector, **y**—training feature vector, $ED(\mathbf{x}-\mathbf{y})$—Euclidean distance, n—number of features (it is 1 feature for the RMS).

The NN classifier is useful for classification of feature vectors. It was found application in pattern recognition, speaker recognition, image recognition, text recognition, face recognition etc. The NN classifier is described in detail in [29–31].

3. Recognition Results of the EID, CG-A, CG-B

The analysed EID was powered from the 230 V/50 Hz mains. The author used 50G515 electric impact drills. Other devices could be used. It generated five acoustic signals denoted as: healthy EID, EID with 15 broken rotor blades (faulty fan), EID with a bent spring, EID with a shifted brush (motor off), EID with a rear ball bearing fault. Measurements were carried out in the room 3 m × 3 m. The analysed EID had rated power P_D = 500 W, rotation speed R_D = 3000 rpm and weight M_D = 1.84 kg.

The analysed CG-A was also powered from the 230 V/50 Hz mains. The author used a ME-1498 coffee grinder. Other devices could be used. The analysed CG-A consisted of a FY5420 motor (rated power 140 W). It had rotor speed of 28,000–30,000 rpm. It generated four acoustic signals denoted as: healthy, with a slightly damaged rear sliding bearing, with a moderately damaged rear sliding bearing, motor off.

The analysed CG-B was also powered from the 230 V/50 Hz mains. The author used a SCG 1050WH coffee grinder. The analysed CG-B consisted of a HC5420 motor (rated power 150 W). It had a rotor speed of 11,300 rpm. It generated three acoustic signals denoted as: healthy, with a light damaged rear sliding bearing, motor off.

Patterns were computed using 32 training samples of the EID, 24 training samples of the CG-A, and 24 training samples of the CG-B. Each training sample had 44,100 values. The results of recognition were computed using 250 test samples of the EID, 200 test samples of the CG-A and 150 test samples of the CG-B. Test samples had the same audio parameters (sampling rate 44,100 Hz, single channel) as training samples.

The efficiency of the proposed approach was evaluated using Equation (4). This Equation (4) defined the efficiency of recognition of the EID (E_D):

$$E_{D1} = (N_{D1})/(N_{ALL-D1}) \cdot 100\% \quad (4)$$

where: E_{D1}—the efficiency of recognition for D1 class (in the analysis it is one of five classes, for example healthy EID), N_{D1}—the number of test samples classified as D1 class, $N_{ALL\text{-}D1}$—the number of all test samples in D1 class. The values of $E_{CG\text{-}A}$ and $E_{CG\text{-}B}$ were computed similarly to E_{D1}.

The total efficiency of recognition of all classes (TE_D) was also introduced. It was defined as follows Equation (5):

$$TE_D = (E_{D1} + E_{D2} + E_{D3} + E_{D4} + E_{D5})/5 \quad (5)$$

where TE_D—the total efficiency of recognition of all classes (five states of the EID), E_{D1}—the efficiency of recognition for D1 class (in the presented analysis D1 class—healthy EID), E_{D2}—the efficiency of recognition for D2 class (in the presented analysis D2 class—EID with a bent spring), E_{D3}—the efficiency of recognition for D3 class (in the presented analysis D3 class—EID with 15 broken rotor blades), E_{D4}—the efficiency of recognition for D4 class (in the presented analysis D4 class—EID with a shifted brush), E_{D5}—the efficiency of recognition for D5 class (in the presented analysis D5 class—EID with a rear ball bearing fault). The values of $TE_{CG\text{-}A}$ and $TE_{CG\text{-}B}$ were computed similarly to TE_D. Four acoustic signals were used for $TE_{CG\text{-}A}$. Three acoustic signals were used for $TE_{CG\text{-}B}$. The computed values of E_D and TE_D were presented in Tables 11 and 12. Acoustic signals of the EID were processed by the MSAF-17-MULTIEXPANDED-FILTER-14 method and the NN classifier (Table 11).

Table 11. Computed values of E_D and TE_D of the EID using the MSAF-17-MULTIEXPANDED-FILTER-14 method and the NN classifier.

Type of Acoustic Signal	E_D (%)
Healthy EID	100
EID with a bent spring	92
EID with (15 broken rotor blades) faulty fan	100
EID with shifted brush (motor off)	100
EID with rear ball bearing fault	88
	TE_D (%)
Total efficiency of recognition of the EID	96

Table 12. Computed values of E_D and TE_D of the EID using the RMS and the NN classifier.

Type of Acoustic Signal	E_D (%)
Healthy EID	56
EID with a bent spring	100
EID with (15 broken rotor blades) faulty fan	100
EID with shifted brush (motor off)	100
EID with rear ball bearing fault	60
	TE_D (%)
Total efficiency of recognition of the EID	83.2

Acoustic signals of the EID were processed by the RMS and NN classifier (Table 12).

The computed values of E_D and TE_D of the proposed approach were following: E_D = 88–100%, TE_D = 96% for the MSAF-17-MULTIEXPANDED-FILTER-14 method and E_D = 56–100%, TE_D = 83.2% for the RMS. The computed values of $E_{CG\text{-}A}$ and $TE_{CG\text{-}A}$ were presented in Tables 13 and 14. Acoustic signals of the CG-A were processed by the MSAF-17-MULTIEXPANDED-FILTER-14 method and the NN classifier (Table 13).

Table 13. Computed values of $E_{CG\text{-}A}$ and $TE_{CG\text{-}A}$ of the CG-A using the MSAF-17-MULTIEXPANDED-FILTER-14 method and the NN classifier.

Type of Acoustic Signal	$E_{CG\text{-}A}$ (%)
Healthy CG-A	100
CG-A with a heavily damaged rear sliding bearing	100
CG-A with a damaged shaft and heavily damaged rear sliding bearing	88
Motor off	100
	$TE_{CG\text{-}A}$ (%)
Total efficiency of recognition of the CG-A	97

Table 14. Computed values of $E_{CG\text{-}A}$ and $TE_{CG\text{-}A}$ of the CG-A using the RMS and the NN classifier.

Type of Acoustic Signal	$E_{CG\text{-}A}$ (%)
Healthy CG-A	100
CG-A with a heavily damaged rear sliding bearing	92
CG-A with a damaged shaft and heavily damaged rear sliding bearing	92
Motor off	100
	$TE_{CG\text{-}A}$ (%)
Total efficiency of recognition of the CG-A	96

Acoustic signals of the CG-A were processed by the RMS and NN classifier (Table 14).

The computed values of $E_{CG\text{-}A}$ and $TE_{CG\text{-}A}$ of the proposed approach were following: $E_{CG\text{-}A}$ = 88–100%, $TE_{CG\text{-}A}$ = 97% for the MSAF-17-MULTIEXPANDED-FILTER-14 method and $E_{CG\text{-}A}$ = 92–100%, $TE_{CG\text{-}A}$ = 96% for the RMS. The computed values of $E_{CG\text{-}B}$ and $TE_{CG\text{-}B}$ were presented in Tables 15 and 16. Acoustic signals of the CG-B were processed by the MSAF-17-MULTIEXPANDED-FILTER-14 method and the NN classifier (Table 15).

Table 15. Computed values of $E_{CG\text{-}B}$ and $TE_{CG\text{-}B}$ of the CG-B using the MSAF-17-MULTIEXPANDED-FILTER-14 method and the NN classifier.

Type of Acoustic Signal	$E_{CG\text{-}B}$ (%)
Healthy CG-B	100
CG-B with a light damaged rear sliding bearing	100
Motor off	100
	$TE_{CG\text{-}B}$ (%)
Total efficiency of recognition of the CG-B	100

Table 16. Computed values of $E_{CG\text{-}B}$ and $TE_{CG\text{-}B}$ of the CG-B using the RMS and the NN classifier.

Type of Acoustic Signal	$E_{CG\text{-}B}$ (%)
Healthy CG-B	100
CG-B with a light damaged rear sliding bearing	100
Motor off	100
	$TE_{CG\text{-}B}$ (%)
Total efficiency of recognition of the CG-B	100

Acoustic signals of the CG-B were processed by the RMS and NN classifier (Table 16).

The computed values of $E_{CG\text{-}B}$ and $TE_{CG\text{-}B}$ of the proposed approach were following: $E_{CG\text{-}B}$= 100%, $TE_{CG\text{-}B}$ = 100% for the MSAF-17-MULTIEXPANDED-FILTER-14 method and RMS.

4. Discussion

The acoustic-based fault-detection technique is significant for the recent research area of electrical motors. This approach is useful for inspection of motor condition. It can analyse acoustic signals in places with limited or no access. The novelty of the proposed work was to detect faults of an EID and two coffee grinders. The author focused on feature extraction of five acoustic signals of the EID, four acoustic signals of the CG-A and three acoustic signals of the CG-B. The method MSAF-17-MULTIEXPANDED-FILTER-14 was developed and described. One of the difficulties to solve was selection of training samples. It can be noticed that the recognition results depended on selected training samples. All samples is measured by one microphone. If the acoustic signal is measured by another type of microphone, then it can cause errors of recognition. The proposed acoustic-based approach should use one type of microphone for training as well as testing.

The second of the difficulties to solve was the testing (classification) of a new unknown test samples. It is difficult to recognize, for example, the acoustic signal of a car if we have training samples of an EID. To solve this problem the proposed acoustic-based approach used the NN classifier. The NN classifier found the nearest feature vector (analysed frequency bandwidths). If the acoustic signal of the car is measured, then it will be recognised as an unknown state of the EID. The training set consisted of acoustic signals of the EID and several unknown sounds of cars, ships, helicopters, animals, etc.

It can be noticed that the RMS was very good for recognition of acoustic signals of the EID with a shifted brush (motor off). This class of acoustic signal should be detected by the RMS. However, the RMS method was not good for similar sound intensity level values. The classes of acoustic signals "Healthy EID" and "EID with a rear ball bearing fault" had low values of TE_D. The classes of acoustic signals "CG-A with a heavily damaged rear sliding bearing" and "CG-A with a damaged shaft and heavily damaged rear sliding bearing" had lower values of $TE_{CG\text{-}A}$. The MSAF-17-MULTIEXPANDED-FILTER-14 method was good method of feature extraction for all analysed classes of acoustic signals.

5. Summary and Conclusions

This paper presented fault-detection techniques for an electric impact drill (EID), coffee grinder A (CG-A), and coffee grinder B (CG-B) using acoustic signals. Measurements of the acoustic signals of the EID, CG-A, and CG-B were carried out using a microphone. Five signals of the EID were analysed: healthy EID, EID with 15 broken rotor blades (faulty fan), EID with a bent spring, EID with a shifted brush (motor off), EID with a rear ball bearing fault. Four signals of the CG-A are analysed: healthy CG-A, CG-A with a heavily damaged rear sliding bearing, CG-A with a damaged shaft and heavily damaged rear sliding bearing, motor off. Three acoustic signals of the CG-B are analysed: healthy CG-B, CG-B with a light damaged rear sliding bearing, motor off.

Methods such as RMS, MSAF-17-MULTIEXPANDED-FILTER-14 were used for feature extraction. The MSAF-17-MULTIEXPANDED-FILTER-14 was also developed and described in the paper. The classification is carried out using the Nearest Neighbour (NN) classifier. An acoustic based analysis was carried out. The computed values of E_D and TE_D of the proposed approach were following: E_D = 88–100%, TE_D = 96% for the MSAF-17-MULTIEXPANDED-FILTER-14 and E_D = 56–100%, TE_D = 83.2% for the RMS. The computed values of $E_{CG\text{-}A}$ and $TE_{CG\text{-}A}$ of the proposed approach were following: $E_{CG\text{-}A}$ = 88–100%, $TE_{CG\text{-}A}$ = 97% for the MSAF-17-MULTIEXPANDED-FILTER-14 method and $E_{CG\text{-}A}$ = 92–100%, $TE_{CG\text{-}A}$ = 96% for the RMS. The computed values of $E_{CG\text{-}B}$ and $TE_{CG\text{-}B}$ of the proposed approach were following: $E_{CG\text{-}B}$ = 100%, $TE_{CG\text{-}B}$ = 100%.

The acoustic-based analysis was inexpensive. The experimental setup consisted of a microphone and computer. It cost about $500. Pros of this solution are instant measurement and online monitoring of the motor. Cons of this solution are the higher cost and size of the computer. The developed acoustic-based approach has many applications, for example in home and industrial appliances for fault detection. It can be used for electrical motors, engines, machinery and electric power tools [36–42]. It can also find applications in mining, oil, car, energy, and the steel industry. It can analyse acoustic signals in places with limited or no access. However, the proposed acoustic-based approach has one limitation. It cannot work for a machine that does not generate acoustic signals. Background noises can be also problem, if we analyse several motors in one place and at the same time.

In the future, the proposed acoustic-based approach can be further developed. Other faults of commutator motors can be added to an acoustic signal database. Measurements can be carried out using acoustic cameras and microphone arrays. Vibration-based methods can be added to the fault detection system of commutator motors. New feature extraction methods can also be developed in the future.

Funding: This research was funded by the AGH University of Science and Technology, grant No. 11.11.120.714.

Acknowledgments: This work has been supported by AGH University of Science and Technology, grant no. 11.11.120.714. The author thanks unknown reviewers for the valuable suggestions.

Conflicts of Interest: The author declares no conflict of interest.

References

1. Singh, M.; Shaik, A.G. Faulty bearing detection, classification and location in a three-phase induction motor based on Stockwell transform and support vector machine. *Measurement* **2019**, *131*, 524–533. [CrossRef]
2. Gutten, M.; Korenciak, D.; Kucera, M.; Sebok, M.; Opielak, M.; Zukowski, P.; Koltunowicz, T.N. Maintenance diagnostics of transformers considering the influence of short-circuit currents during operation. *Eksploat. I Niezawodn. Maint. Reliab.* **2017**, *19*, 459–466. [CrossRef]
3. Moujahed, M.; Benazza, H.; Jemli, M.; Boussak, M. Sensorless Speed Control and High-Performance Fault Diagnosis in VSI-Fed PMSM Motor Drive Under Open-Phase Fault: Analysis and Experiments. *Iran J. Sci. Technol. Trans. Electr. Eng.* **2018**, *42*, 419–428. [CrossRef]
4. Touti, W.; Salah, M.; Bacha, K.; Amirat, Y.; Chaari, A.; Benbouzid, M. An improved electromechanical spectral signature for monitoring gear-based systems driven by an induction machine. *Appl. Acoust.* **2018**, *141*, 198–207. [CrossRef]
5. Wang, P.P.; Chen, X.X.; Zhang, Y.; Hu, Y.J.; Miao, C.X. IBPSO-Based MUSIC Algorithm for Broken Rotor Bars Fault Detection of Induction Motors. *Chin. J. Mech. Eng.* **2018**, *31*, UNSP 80. [CrossRef]
6. Yan, X.A.; Jia, M.P. A novel optimized SVM classification algorithm with multi-domain feature and its application to fault diagnosis of rolling bearing. *Neurocomputing* **2018**, *313*, 47–64. [CrossRef]
7. Zhang, C.; Peng, Z.X.; Chen, S.; Li, Z.X.; Wang, J.G. A gearbox fault diagnosis method based on frequency-modulated empirical mode decomposition and support vector machine. *Proc. Inst. Mech. Eng. Part C J. Mech. Eng. Sci.* **2018**, *232*, 369–380. [CrossRef]
8. Taghizadeh-Alisaraei, A.; Mandavian, A. Fault detection of injectors in diesel engines using vibration time-frequency analysis. *Appl. Acoust.* **2019**, *143*, 48–58. [CrossRef]
9. Kong, Y.; Wang, T.Y.; Chu, F.L. Meshing frequency modulation assisted empirical wavelet transform for fault diagnosis of wind turbine planetary ring gear. *Renew. Energy* **2019**, *132*, 1373–1388. [CrossRef]
10. Caesarendra, W.; Tjahjowidodo, T. A Review of Feature Extraction Methods in Vibration-Based Condition Monitoring and Its Application for Degradation Trend Estimation of Low-Speed Slew Bearing. *Machines* **2017**, *5*, 21. [CrossRef]
11. Zhang, Y.L.; Duan, L.X.; Duan, M.H. A new feature extraction approach using improved symbolic aggregate approximation for machinery intelligent diagnosis. *Measurement* **2019**, *133*, 468–478. [CrossRef]
12. Li, Y.B.; Feng, K.; Liang, X.H.; Zuo, M.J. A fault diagnosis method for planetary gearboxes under non-stationary working conditions using improved Vold-Kalman filter and multi-scale sample entropy. *J. Sound Vib.* **2019**, *439*, 271–286. [CrossRef]
13. Sun, H.C.; Fang, L.; Guo, J.Z. A fault feature extraction method for rotating shaft with multiple weak faults based on underdetermined blind source signal. *Meas. Sci. Technol.* **2018**, *29*, 125901. [CrossRef]
14. Shanbhag, V.V.; Rolfe, B.F.; Arunachalam, N.; Pereira, M.P. Investigating galling wear behaviour in sheet metal stamping using acoustic emissions. *Wear* **2018**, *414*, 31–42. [CrossRef]
15. Appana, D.K.; Prosvirin, A.; Kim, J.M. Reliable fault diagnosis of bearings with varying rotational speeds using envelope spectrum and convolution neural networks. *Soft Comput.* **2018**, *22*, 6719–6729. [CrossRef]
16. Liu, J.; Hu, Y.M.; Wu, B.; Wang, Y. An improved fault diagnosis approach for FDM process with acoustic emission. *J. Manuf. Process.* **2018**, *35*, 570–579. [CrossRef]
17. Tang, J.; Qiao, J.F.; Liu, Z.; Zhou, X.J.; Yu, G.; Zhao, J.J. Mechanism characteristic analysis and soft measuring method review for ball mill load based on mechanical vibration and acoustic signals in the grinding process. *Miner. Eng.* **2018**, *128*, 294–311. [CrossRef]
18. Hao, Q.S.; Zhang, X.; Wang, K.W.; Shen, Y.; Wang, Y. A signal-adapted wavelet design method for acoustic emission signals of rail cracks. *Appl. Acoust.* **2018**, *139*, 251–258. [CrossRef]
19. Yao, Y.; Wang, H.L.; Li, S.B.; Liu, Z.H.; Gui, G.; Dan, Y.B.; Hu, J.J. End-To-End Convolutional Neural Network Model for Gear Fault Diagnosis Based on Sound Signals. *Appl. Sci.* **2018**, *8*, 1584. [CrossRef]
20. Vununu, C.; Moon, K.S.; Lee, S.H.; Kwon, K.R. A Deep Feature Learning Method for Drill Bits Monitoring Using the Spectral Analysis of the Acoustic Signals. *Sensors* **2018**, *18*, 2634. [CrossRef]
21. Vaimann, T.; Sobra, J.; Belahcen, A.; Rassolkin, A.; Rolak, M.; Kallaste, A. Induction machine fault detection using smartphone recorded audible noise. *Iet Sci. Meas. Technol.* **2018**, *12*, 554–560. [CrossRef]
22. Wang, Z.H.; Wu, X.; Liu, X.Q.; Cao, Y.L.; Xie, J.K. Research on feature extraction algorithm of rolling bearing fatigue evolution stage based on acoustic emission. *Mech. Syst. Signal Process.* **2018**, *113*, 271–284. [CrossRef]

23. Singh, G.; Naikan, V.N.A. Infrared thermography based diagnosis of inter-turn fault and cooling system failure in three phase induction motor. *Infrared Phys. Technol.* **2017**, *87*, 134–138. [CrossRef]
24. Mariprasath, T.; Kirubakaran, V. A real time study on condition monitoring of distribution transformer using thermal imager. *Infrared Phys. Technol.* **2018**, *90*, 78–86. [CrossRef]
25. Kim, D.; Youn, J.; Kim, C. Automatic Fault Recognition of Photovoltaic Modules Based on Statistical Analysis of Uav Thermography. In Proceedings of the International Archives of the Photogrammetry, Remote Sensing and Spatial Information Sciences, Volume XLII-2/W6, 2017 International Conference on Unmanned Aerial Vehicles in Geomatics, Bonn, Germany, 4–7 September 2017; pp. 179–182. [CrossRef]
26. Zhao, Y. *Oil Analysis Handbook*, 3rd ed.; Spectro Scientific: Chelmsford, MA, USA, 2017.
27. Jamadar, I.M.; Vakharia, D. Correlation of base oil viscosity in grease with vibration severity of damaged rolling bearings. *Ind. Lubr. Tribol.* **2018**, *70*, 264–272. [CrossRef]
28. Li, X.C.; Duan, F.; Mba, D.; Bennett, I. Multidimensional prognostics for rotating machinery: A review. *Adv. Mech. Eng.* **2017**, *9*, 1687814016685004. [CrossRef]
29. Gou, J.P.; Ma, H.X.; Ou, W.H.; Zeng, S.N.; Rao, Y.B.; Yang, H.B. A generalized mean distance-based k-nearest neighbor classifier. *Expert Syst. Appl.* **2019**, *115*, 356–372. [CrossRef]
30. Zhang, Y.Q.; Cao, G.; Wang, B.S.; Li, X.S. A novel ensemble method for k-nearest neighbor. *Pattern Recognit.* **2019**, *85*, 13–25. [CrossRef]
31. Bandaragoda, T.R.; Ting, K.M.; Albrecht, D.; Liu, F.T.; Zhu, Y.; Wells, J.R. Isolation-based anomaly detection using nearest-neighbor ensembles. *Comput. Intell.* **2018**, *34*, 968–998. [CrossRef]
32. Valis, D.; Zak, L.; Pokora, O.; Lansky, P. Perspective analysis outcomes of selected tribodiagnostic data used as input for condition based maintenance. *Reliab. Eng. Syst. Saf.* **2016**, *145*, 231–242. [CrossRef]
33. Glowacz, A. Acoustic-Based Fault Diagnosis of Commutator Motor. *Electronics* **2018**, *7*, 299. [CrossRef]
34. Gajewski, J.; Valis, D. The determination of combustion engine condition and reliability using oil analysis by MLP and RBF neural networks. *Tribol. Int.* **2017**, *115*, 557–572. [CrossRef]
35. Caesarendra, W.; Wijaya, T.; Tjahjowidodo, T.; Pappachan, B.K.; Wee, A.; Roslan, M.I. Adaptive neuro-fuzzy inference system for deburring stage classification and prediction for indirect quality monitoring. *Appl. Soft Comput.* **2018**, *72*, 565–578. [CrossRef]
36. Lu, S.L.; He, Q.B.; Wang, J. A review of stochastic resonance in rotating machine fault detection. *Mech. Syst. Signal Process.* **2019**, *116*, 230–260. [CrossRef]
37. Lu, S.L.; He, Q.B.; Zhang, H.B.; Kong, F.R. Rotating machine fault diagnosis through enhanced stochastic resonance by full-wave signal construction. *Mech. Syst. Signal Process.* **2017**, *85*, 82–97. [CrossRef]
38. Martin-Diaz, I.; Morinigo-Sotelo, D.; Duque-Perez, O.; Osornio-Rios, R.A.; Romero-Troncoso, R.J. Hybrid algorithmic approach oriented to incipient rotor fault diagnosis on induction motors. *Isa Trans.* **2018**, *80*, 427–438. [CrossRef]
39. Delgado-Arredondo, P.A.; Morinigo-Sotelo, D.; Osornio-Rios, R.A.; Avina-Cervantes, J.G.; Rostro-Gonzalez, H.; Romero-Troncoso, R.D. Methodology for fault detection in induction motors via sound and vibration signals. *Mech. Syst. Signal Process.* **2017**, *83*, 568–589. [CrossRef]
40. Pandiyan, V.; Caesarendra, W.; Tjahjowidodo, T.; Tan, H.H. In-process tool condition monitoring in compliant abrasive belt grinding process using support vector machine and genetic algorithm. *J. Manuf. Process.* **2018**, *31*, 199–213. [CrossRef]
41. Glowacz, A. Recognition of acoustic signals of commutator motors. *Appl. Sci.* **2018**, *8*, 2630. [CrossRef]
42. Glowacz, A. Fault diagnosis of single-phase induction motor based on acoustic signals. *Mech. Syst. Signal Process.* **2019**, *117*, 65–80. [CrossRef]

MDPI
St. Alban-Anlage 66
4052 Basel
Switzerland
Tel. +41 61 683 77 34
Fax +41 61 302 89 18
www.mdpi.com

Sensors Editorial Office
E-mail: sensors@mdpi.com
www.mdpi.com/journal/sensors

www.ingramcontent.com/pod-product-compliance
Lightning Source LLC
LaVergne TN
LVHW070101170726
843515LV00007B/1586

* 9 7 8 3 0 3 9 4 3 2 8 0 6 *